THERMODYNAMICS-I

AF656446

B.S. UBHI

SYLLABUS

THERMODYNAMICS-I

1. **Fundamental Concepts:** Thermodynamic state and system, boundary, surrounding, universe, thermodynamic systems - closed, open, isolated, adiabatic, homogeneous and heterogeneous, macroscopic and microscopic, properties of system - intensive and extensive, thermodynamic equilibrium, quasi - static process, reversible and irreversible processes, Zeroth law of thermodynamics, definition of properties like pressure, volume, temperature, enthalpy, internal energy.
2. **Laws of Perfect Gases :** Definition of gases, explanation of perfect gas laws - Boyle's law, Charle's law, Avogadro's law, Regnault's law, Universal gas constant, Characteristic gas constants, derivation specific heat at constant pressure, specific heat at constant volume of gas, derivation of an expression for specific heats with characteristics, simple problems on gas equation.
3. **Thermodynamic Processes on Gases:** Types of thermodynamic processes isochoric, isobaric, isothermal, hyperbolic, isentropic, polytropic and throttling processes, equations representing the processes. Derivation of work done, change in internal energy, change in entropy, rate of heat transfer for the above processes.
4. **Laws of Thermodynamics:** Laws of conservation of energy, first law of thermodynamics (Joule's experiment), Application of first law of thermodynamics to non-flow systems - Constant volume, constant pressure, Adiabatic and polytropic processes, steady flow energy equation, Application of steady flow energy to equation, turbines, pump, boilers, compressors, nozzles, evaporators, limitations. Heat source and heat sinks, statement of second laws of thermodynamics: Kelvin Planck's statement, Classius statement, equivalence of statements, Perpetual motion Machine of first kind, second kind, Carnot engine, Introduction of third law of thermodynamics, concept of irreversibility, entropy.
5. **Ideal and Real Gases :** Concept of ideal gas, enthalpy and specific heat capacities of an ideal gas, P—V—T surface of an ideal gas, triple point, real gases, Vander Wall's equation
6. **Properties of Steam:** Formation of steam and related terms, thermodynamics properties of steam, steam tables, internal latent heat, internal energy of stream, entropy of water, entropy of steam, T-S diagrams, Mollier diagram (H-S Chart), Expansion of steam. Hyperbolic, reversible adiabatic and throttling processes. Quality of steam.
7. **Steam Generators (dryness fraction):** Uses of steam, classification of boilers, comparison of fire tube and water tube boilers. Construction features of Lancashire boiler, nestler boiler, Babcock & Wilcox Boiler. Introduction to modern boilers.
8. **Air Standard Cycles :** Meaning of air standard cycle - its use, condition of reversibility of a cycle. Description of Carnot cycle, Otto cycle, Diesel cycle, simple problems on efficiency, calculation for different cycles. Comparison of Otto, Diesel cycles for same compression ratio or same peak pressure developed. Reasons for highest efficiency of Carnot cycle and all other cycles working between same temperature limits.
9. **Air Compressors :** Functions of air compressor - uses of compressed air, type of air compressors, Single stage reciprocating air compressor, its construction and working, representation of processes involved on P - V diagram, calculation of work done, air HP and BP prime mover, Multistage compressors - advantages over single stage compressors, use of air cooler–condition of minimum work in two stage compressor (without proof), simple problems, Rotary compressors—types, descriptive treatment of centrifugal compressor, axial flow compressor, vane type compressor.
10. **Introduction to Heat Transfer:** Modes of heat transfer, Fourier's law, steady state conduction, composite structures, Natural and forced convection, thermal radiation.

Preface to the Seventh Edition

This edition has been brought, after incorporating the changes in curriculum of the subject in various State Boards of Technical Education in the country w.e.f. 2007.

Every effort has been made to update and simplify the text. The contents have been added, wherever necessary. I feel this edition will cater to the needs of the students and teaching community, in a better way.

I express my heart-felt gratitude to M/s S.K. Kataria & Sons for giving a nice get up to this edition of the book and pricing it reasonably.

Constructive criticism and suggestions from all the esteemed readers will be highly acknowledged and incorporated.

January, 2009 **B.S. UBHI**

Dedicated to

Ramandeep

and

Taranjot

CONTENTS

1
CHAPTER
Fundamental Concepts

INSIDE THIS CHAPTER

1.1. INTRODUCTION

'Thermodynamics' is the branch of science which deals with energy transformations and the relationships between the properties of the working substance. These transformations are governed by the various 'Laws of Thermodynamics'. The working substance or the working fluid, which may be in gaseous or liquid form, convey energy from one point to another. If the working fluid remains in gaseous state and obeys certain 'Laws of Gases' throughout the energy transformation then we refer it as a *'Perfect Gas'*.

As the use of power in modern living is very widely involved, so every student of engineering, irrespective of their major area of interest, should have a fundamental knowledge of the subject to *'Thermodynamics'*.

In any working area, it is always necessary to define the basic terms used before a healthy discussion can be made. Therefore, it is being attempted in this chapter to present, in brief, the clear concept of basic terms used and the gas laws, which form the fundamentals of this subject.

1.1.1. Thermodynamic State

The instantaneous condition of a thermodynamic system is called its 'state'.

1.1.2. Thermodynamic System

A prescribed region in space or a finite quantity of matter under consideration for its thermodynamic study is known as a *'Thermodynamic System'*.

1.1.3. Boundary

An envelope enclosing the system is known as *'Boundary'* of the system.

1.1.4. Surrounding

Everything external to the system is known as *'Surrounding'*.

1.1.5. Universe

The space comprising of all the systems and boundaries together is called *'Universe'*.

1.1.6. Types of Thermodynamic System

There are three types of thermodynamic system.

(i) Closed System

It is the thermodynamic system in which the mass comprising the system is not allowed to cross the boundary of the system, so that no mass enters or leaves the system, although the energy (heat and work) may cross the boundary as shown in Fig. 1.1. For example, Piston cylinder.

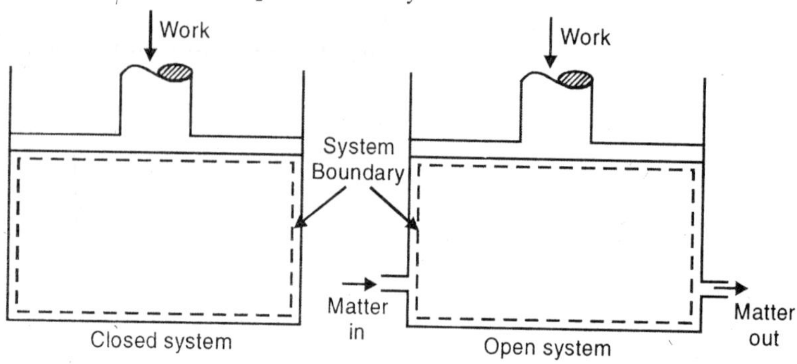

Fig. 1.1.

(ii) Open System

It is the thermodynamic system in which mass and energy of the system are allowed to cross the boundary of the system, as shown in the Fig. 1.1, e.g., Turbine.

(iii) Isolated System

It is the thermodynamic system in which the mass and energy, both of the system are not allowed to cross the boundary of the system.

(iv) Adiabatic System

It is the thermodynamic system, which is thermally insulated from its surroundings, although it can exchange work with its surroundings.

(v) Homogeneous System

It is the thermodynamic system which consists of a single phase throughout a thermodynamic process. For example, a mixture of air and water vapour.

(vi) Heterogeneous System

It is a thermodynamic system, which consists of two or more than two phases. For example, a mixture of water and steam or water and ice.

1.1.7. Macroscopic and Microscopic Thermodynamic Study

(i) Macroscopic Thermodynamic Study

The **'Macroscopic Thermodynamic Study'** is concerned with the overall behaviour of the matter under consideration it is also called *'Classical Thermodynamics'*.

In this study the analysis requires only simple mathematical formulae, taking the average values of the properties of the system.

(ii) Microscopic Thermodynamic Study

The '**Microscopic Thermodynamic Study**' is concerned with the structure of the matter under consideration. It is also called *'Statistical Thermodynamics'*.

In this study the analysis requires advanced statistical and mathematical methods/formulae. A large number of variables are needed to describe this complicated system approach.

1.1.8. Property

The *'property'* of a medium is defined as the fixed characteristic, when the working substance is in a particular state.

Pressure, Volume, Temperature, Internal Energy and Enthalpy are some of the examples of the properties of gases and vapours.

The property of a medium is fixed by its instantaneous state and is independent of the process by which it arrived at that state.

The state of a medium/system becomes fixed by fixing any two of its properties e.g. air at some definite temperature and pressure will have a definite volume and the amount of internal energy.

The properties of a system may be classified into two groups as;

(i) Intensive or Intrinsic Properties

The properties of a system, which are independent of its mass e.g., temperature, pressure and density etc. are known as *'Intensive* or *Intrinsic Properties'*.

(ii) Extensive or Extrinsic Properties

The properties of a system, which are dependent on its mass e.g., volume, mass, length and energy etc. are known as *'Extensive or Extrinsic Properties'*.

1.1.9. Thermodynamic Equilibrium

A thermodynamic system which is in equilibrium, thermally, mechanically as well as chemically is called in *'Thermodynamic Equilibrium'*.

1.1.10. Quasi-Static Process

The word *'quasi'* means *'almost'*. So, a *'quasi-static process'* may be defined as the process during which the equilibrium of the system is almost maintained, from its initial state to the final state.

1.1.11. Reversible and Irreversible Processes

(i) Reversible Process

A thermodynamic process, which can be operated in the reverse direction is called *'Reversible Process'*.

In this process, heat cannot flow from a body at lower temperature to a body at higher temperature *i.e.*, the process must obey the second law of thermodynamics.

(ii) Irreversible Process

A thermodynamic process which cannot be operated in the reverse direction is called *'Irreversible Process'*. In this process second law of thermodynamics does not holds good.

1.1.12. Zeroth Law of Thermodynamics

This law states that : *"If two bodies are in thermal equilibrium with a third body, separately, then they are also in thermal equilibrium with each other".*

This law not only provides the basis for the measurement of temperature, but it also enables us to compare the temperatures of two bodies, with the help of a third body, like a thermometer.

1.1.12.1. Explanation

Let us consider two bodies A (Hot) and B (Cold) are brought in contact with each other. The heat energy will be transferred from body A to the body B. After some time when there is no further transfer of heat between them, then these bodies are said to be in thermal equilibrium with each other.

Now, take a third body C (thermometer). When the body C is brought in contact with body A and B separately which are not in physical contact with each other, there is no change in the mercury level of the thermometer. So, we can say that the bodies A and B which are in thermal equilibrium with each other are also in thermal equilibrium with the body C, separately and hence the Zeroth law of thermodynamics proved.

1.1.13. Pressure

The 'pressure' is defined as the force per unit area exerted by body on its surface, in a direction normal to the surface. Where the pressure is continuously varying it must be defined as the force applied by the system on an infinitesimal piece of the boundary divided by the area of the piece. When a force is exerted on a fluid, this force is transmitted throughout the fluid and is exerted in all directions on its containing walls. If the containing walls are stationary, they exert the equal and opposite force on the fluid.

Thermodynamic relations are for the most part concerned with 'Absolute Pressure'. This is the pressure exerted by a system on its boundries. This word 'Absolute is used to differentiate it form the 'Gauge Pressure', because, in practice most pressure and vacuum gauges read only the difference between the absolute pressure and the atmospheric pressure. To obtain the absolute pressure from gauge pressure, the relation used is (Fig. 1.2).

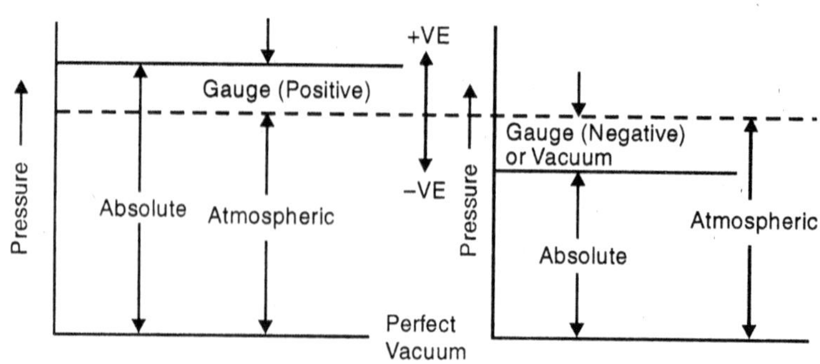

Fig. 1.2. Pressure measurement.

$$p_{abs} = p_{atm} + p_{vacuum}$$

Above relation is to be used for pressures above atmospheric. For pressures below atmospheric, the gauge pressure would be negative and is also termed as vacuum. Thus,

$$p_{abs} = p_{atm} - p_{vacuum}$$

In the engineering system of units, pressure is usually expressed in kilogramme (force) per square centimetre (kgf/cm^2). *The pressure is frequently expressed in terms of the height of a column of mercury which, it will support at a temperature of 20°C or 68°F.*

The standard atmospheric pressure is taken as 1.0332 kgf/cm^2. At this pressure the height of mercury column is 760 mm with the density of mercury as 13.5951 g/cm^3. Some other units being used now a days are;

1 microbar = 1 dyne/cm^2

1 milli meter = 1 mm = 1333.22 microbars

1 micron = 1 μ = 10^{-6} m Hg = 10^{-3} mm Hg

1 torr = 1 mm Hg.

Sometimes, the pressure is given in 'Atmospheres'. In FPS system an absolute atmosphere is that which has pressure equal to 14.6963 lb/in^2 or approximately 14.7 lb/in^2. Its equivalent in MKS system is 1.0332 kg/cm^2 and in S.I system 1.01326 bar.

In MKS system the atmospheric pressure is expressed as technical atmosphere commonly written as 'ata' in short form. One technical atmosphere has pressure equal to 1 kgf/cm^2 or 0.980665 bar.

In S.I units the pressure is expressed in Newton per square metre (N/m^2) or bar. The relations between various units of pressure measurement are given below :

1 kgf/cm^2 = 0.980665 bar
= 14.7 lb/in^2 = 0.977 ata = 1 atm.
= 735.5 Torr.

1 bar = 10^5 N/m^2
= 14.5038 lb/in^2
= 1.0197 kgf/cm^2
= 0.98692 atm.
= 750.06 Torr = 401.85 m H_2O

1 Technical atmosphere = 1 ata
= 735.5 mm Hg
= 735.5 Torr = 1 kgf/cm^2
= 14.225 lb/in^2

The relation between pressure 'p' and the height of the liquid column 'h' may be expressed as;

$$p = \rho . g . h$$

where p = Pressure in N/m^2 or Pa

ρ = Density of the liquid in kg/m^3

g = Acceleration due to gravity (9.81 m/s^2)

and h = Height of the liquid column in 'm'.

The density 'ρ' of water is 1000 kg/m^3.

1.1.14. Volume

It is the space occupied by any working substance in a thermodynamic process.

In S.I system of units it is expressed in 'm^3'.

1.1.15. Temperature

It is an intensive thermodynamic property which determines the hotness or the level of heat intensity of a body. A high temperature indicates high level of heat intensity and the body is said to be *'Hot'*. Similarly, a low temperature indicates low level of heat intensity and the body is said to be *'Cold'*. The temperature of a body is measured with the help of an instrument called *'Thermometer'*.

For the measurement of temperature there are two well known scales.

(i) Centrigrade Scale

This scale was first used by Celcius in 1742. This scale is used by scientists and engineers for all International work. The point at which water freezes under atmospheric pressure is taken as the *'Arbitrary Zero Point'* on the scale and the point at which water boils at this pressure is designated as 100. The distance between these two points is divided into one hundred equal units called 'Degree Centrigrade' or °C. It is now a days called as 'Degree Celcius'.

(ii) Fahrenheit Scale

This scale was first used in 1665. The official scale of temperature, used in English speaking countries is the 'Farenheit Scale'. On this scale the point at which water freezes under atmospheric pressure is marked as 32 and at which water boils at this pressure is represented by 212. Thus the distance between these two points is divided into 180 equal units called degrees **Fahrenheit** or **°F.**

The values on Fahrenheit scale can be readily converted into those on the Centigrade or Celcius Scale and vice versa by noting that 180 degrees of the Fahrenheit scale are equal to 100 degrees of the Centigrade or Celcius Scale.

Hence, $$°F = 1.8\,°C + 32$$

and $$°C = (°F - 32) \div 1.8$$

1.1.15.1. Absolute Temperature

The zero readings on the ordinary Centrigrade (Celcius) and Fahrenheit scale are chosen arbitrarily for simplicity.

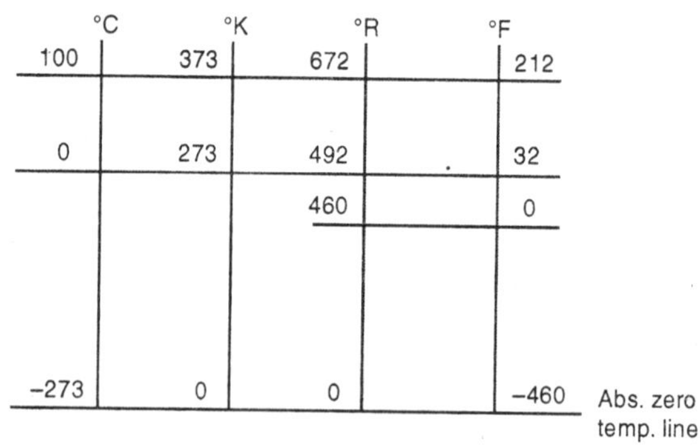

Fig. 1.3. Temperature relationship.

If it is desired to know only the changes of temperature that occur during a process or the temperature of a substance in relation to some known reference point, such readings will entirely suffice. However, when the readings are to be used in equations relating to fundamental laws, it is very necessary to use temperature readings whose reference point is the True or Absolute Zero of the temperature.

It will be seen later on that the *'Absolute Zero' of temperature is the zero below which the temperature of any substance cannot fall.* It is taken as –273 °C or –460 °F as shown in the Fig. 1.1 and *the temperature measured from this zero are called 'Absolute Temperature'.*

The absolute Fahrenheit scale is called degrees *'Rankine'* represented by °R where;

Degree Rankine, °R = °F + 460

The Absolute Centigrade or Celcius Scale is called degrees Kelvin, represented by K, where K = °C + 273.

Also, the relation between the Fahrenheit Absolute Scale and Centrigrade or Celcius Absolute Scale is

°R = 1.8 K.

1.1.15.2. Methods of Temperature Measurement

In 'Thermodynamics' the range of temperatures encountered is wide ranging from low temperatures in the vicinity of absolute zero in cryogenic engineering to high temperatures in furnaces or even higher temperatures like temperature of Sun in the studies of 'Solar Radiation' phenomenon. *In any thermometer, the basic principle is the measurement of a physical property.* Naturally for the measurement of such a wide range of temperatures we cannot bank upon one property. Some of the properties used for temperature measurement are given below :

(*i*) **Change in Volume :** A common 'Mercury-in-glass Thermometer' uses this principle. By confining the mercury in a glass tube of fixed diameter, the volumetric expansion is reduced to measurement of length of mercury column. Expansion of gases at constant pressure can also be used for temperature measurements.

Various substances, using volumetric expansion for temperature measurement are given below :

Substance	Range of Temp. Measurement
Mercury	— 39 °C to about 315 °C, upper limit may be extended to about 480 °C with Nitrogen above the Mercury.
Alcohol	— 73.3 °C to about 149 °C
Pentane	— 184 °C to about 20 °C

Glass begins to soften at about 480 °C. Upper working limit for gas thermometers is of the order of 1480 °C. Since volume changes are unlikely to be exactly proportional to temperature changes, so thermometers are calibrated for accuracy.

(*ii*) **Change in Pressure :** When the volume of a gas remains constant, its pressure changes directly with its temperature. This pressure change can be utilised for the measurement of temperature. A constant volume gas thermometer is an example of this type.

(*iii*) **Change in Electrical Resistivity :** The electrical resistivity of some metals increases almost in direct proportion to temperature increase. Thus, change in electrical resistance of a wire can be used to measure the temperature change.

Metals like Copper, Nickel, Platinum are used for this purpose. Also, semiconductors, which have a high sensitivity and a rapid thermal response can be used for low temperatures nearing absolute zero.

(*iv*) **Change in Electrical Potential :** *When two wires of different materials are joined together at their ends, an e.m.f. (electromotive-force) is generated, when the two ends are maintained at different temperatures.* The e.m.f. generated is a function of the temperature difference between the junction. A 'Thermocouple' uses this principle, in which one end is maintained at the reference temperature (usually 0° C with ice and water mixture) and the other end is maintained at the temperature to be measured. Some of the thermocouple combinations used are given below :

Combination of Metals	Temperature Range
Copper and Constantan	–185 °C to 340 °C
Iron and Constantan	–185 °C to 815 °C
Chromel and Alumel	–185 °C to 1200 °C
Platinum and Platinum Rohedium	Above 1200° (upto 1600 °C)

(*v*) **Optical Changes :** *A body radiates heat proportional to the fourth power of its absolute temperature.* An 'Optical Pyrometer' uses this principal in which the pyrometer is sighted at the hot body whose brightness is compared with the brightness of an adjustable and calibrated source of light within the instrument. When the brightness of the instrument has been adjusted to be the same as the body whose temperature is desired, the instrument 'read' the body temperature.

1.1.16. Enthalpy

The sum of internal energy and the external work done (pressure volume product) of the given quantity of a gas both expressed in the same units is called '*Enthalpy*' and is generally denoted by '*H*'.

$$\therefore \qquad H = E + \frac{pV}{J} \text{ in heat units} \qquad ...(1.1)$$

where '*E*' is the internal energy and $\frac{pV}{J}$ the external work done of the gas in heat units, in which p and V are the pressure and volume respectively of the given mass of the gas and '*J*' the mechanical equivalent of heat.

The product of pressure p and volume V has been divided by J (the mechanical equivalent of heat) to convert the mechanical work done into heat units.

Enthalpy should never be confused with heat energy. It is a function of the state of the gas under consideration. Since E, p and V are separately functions of the state of gas, so *enthalpy is property of the system.*

1.1.17. Internal Energy

The '*Internal Energy*' of a substance is the energy possessed by it, due to its molecular arrangement and activity. It is generally denoted by E or U.

When a certain amount of heat energy is supplied to a gas, some part of this heat energy is converted into mechanical work and the rest of it, is stored in the gas in the form of internal energy.

1.1.18. Units and Dimensions

In the present text, the S.I. (System International) system of units has been used. The basic units in this system are given in Table 1.1.

Table 1.1. Basic Units.

Quantity	Unit	Symbol
Length (L)	Metre	m
Mass (M)	Kilogramme	kg
Time (t)	Second	s
Amount of substance	Mole	mol
Temperature (T)	Kelvin	k
Electric current	Ampere	A
Luminous intensity	Candela	cd
Plane angle	Radian	rad
Solid angle	Steradian	sr

The dimensions of all other quantities are derived from these basic units which are given in Table 1.2.

It is often convenient and desirable to use multiple of various units, the standard list of which is given in Table 1.3.

Table 1.2. Derived Units.

Quantity	Unit	Symbol	Alternative unit	In Basic unit
Force (F)	Newton	N	—	$kg\ m/s^2$
Energy (E)	Joule	J	Nm	$kg\ m^2/s^2$
Power	Watt	W	J/s	$kg\ m^2/s^3$
Pressure	Pascal	Pa	N/m^2	$kg/(ms)^2$
Frequency	Hertz	Hz	—	S^{-1}
Electric Charge	Coulomb	C	—	As.
Electric Potential	Volt	V	W/A = J/C	$kg\ m^2/(s^3.A)$
Capacitance	Farad	F	C/V	$S^4A^2/(kg\ m^2)$
Electrical resistance	Ohm	W	V/A	$kg\ m^2\ (S^3A^2)$
Magnetic flux	Weber	Wb	Vs	$kg\ m^2/(S^2A)$
Magnetic flux density	Tesla		Wb/m^2	$kg/(S^2.A)$
Inductance	Henry	h	Wb/A	$kg\ m^2/(S^2A^2)$

Table 1.3. Standard Multipliers.

Factor	*Prefix*	*Factor*	*Prefix*
10^{12}	tera, T	10^{-3}	milli, m
10^{9}	giga, G	10^{-6}	micro, μ
10^{6}	mega, M	10^{-9}	nano, n
10^{3}	kilo, K	10^{-12}	pico, p

1.2. BASIC TERMS

1.2.1. Thermodynamics

The branch of science which deals with the various types of energy transformations and relationships between the various properties of the working substance, is known as 'Thermodynamics'.

The word thermodynamics means *'Heat Power'* or *'Power obtained from heat'*. Thermodynamics, when applied to the various fields, such as power generation and refrigeration etc. is referred as *'Applied Thermodynamics'*.

1.2.2. Path

The succession of states passed through, during a change of state of a working substance is called the *'path'*.

1.2.3. Process

When the path is completely specified, the change of state is called a *'process'* e.g. a constant volume process.

1.2.4. Cycle

A thermodynamic *'cycle'* is defined as a series of state changes such that the final state is identical with the initial state.

NUMERICAL EXAMPLES

Example 1.1. A barometer reads 760 mm of Hg. What would be the absolute pressure ? If ;

(*a*) A pressure gauge connected to a boiler reads 10 bar.

(*b*) A vacuum gauge connected to a condenser reads 650 mm of Hg.

Solution. We know that ;

$$\text{Absolute pressure} = \text{Atm. pressure} + \text{Gauge pressure} \quad ...(i)$$

(if Abs. Pr. > Atm. Pr.)

$$\text{and Absolute Pressure} = \text{Atm. pressure} - \text{Gauge pressure} \quad ...(ii)$$

(if Abs. Pr. < Atm. Pr.)

(*a*) Given; $P_{gauge} = 10$ bar

Also, $P_{atm.} = 1.01325$ bar

So, from Eqn. (*i*) $P_{abs.} = 1.01325 + 10 = 11.01325$ bar **Ans.**

(*b*) Given; $P_{gauge} = 650$ mm of Hg.

But, we know that;

$$\text{Atm. Pr.} = 760 \text{ mm of Hg} = 1.01325 \text{ bar}$$

$$\therefore \quad 650 \text{ mm of Hg} = \frac{1.01325 \times 650}{760}$$

$$= 0.8666 \text{ bar}$$

Now, from eqn. (*ii*)

$$P_{abs} = P_{atm.} - P_{gauge}$$

$$\therefore \quad P_{abs} = 1.01325 - 0.8666 = 0.1466 \text{ bar}$$ **Ans.**

Example 1.2. The gauge pressure of a steam generator is 10 kg/cm². The barometer pressure of the atmosphere is 760 mm of Hg. Find the absolute pressure in kg/cm², in bar and in N/m².

Solution. Given data;

$$\text{Gauge pressure} = 10 \text{ kg/cm}^2$$

Atm. Pressure $= 760$ mm of Hg $= 76$ cm of Hg

$$= \frac{76 \times 13.6}{1000} = 1.0336 \text{ kgf/cm}^2$$

We know that ; $P_{abs} = P_{atm.} + P_{gauge}$

$\therefore$ Abs. pressure, $P_{abs.} = 10 + 1.0336$

$= 11.0336 \text{ kgf/cm}^2$ **Ans.**

Also, $= \frac{11.0336}{1.0197} = 10.820$ bar **Ans.**

($\because$ 1 bar $= 1.0197 \text{ kgf/cm}^2$)

Also, $= 10.820 \times 10^5$

$= 1082000 \text{ N/m}^2$ **Ans.** ($\because$ 1 bar $= 10^5$ Pa or N/m^2)

Example 1.3. A steam generator has gauge pressure of 100 bar. The atmospheric pressure is 760 mm of Hg. Find the abs. pressure of the steam in :

(*a*) Bar (*b*) N/m² or Pa (*c*) MPa (*d*) N/m²

Solution. Given data :

Gauge pressure $= 100$ bar

Atm. pressure $= 760$ mm of Hg

$= 1.01325$ bar

But, we know that;

$$P_{abs.} = P_{atm.} + P_{gauge}$$

$\therefore$ (*a*) Abs. pressure,

$P_{abs} = 1.01325 + 100$

$= 101.01325$ bar **Ans.**

(*b*) $P_{abs.} = 101.01325 \times 10^5 \text{ N/m}^2$ or Pa

$= 10101325$ Pa **Ans.** ($\because$ $1 \text{ N/m}^2 = 1$ Pa)

(*c*) $P_{abs.} = \frac{10101325}{10^6} = 10.1013$ MPa **Ans.**

(*d*) $P_{abs.} = \frac{10101325}{1000 \times 1000} = 10.1013 \text{ N/mm}^2$ **Ans.**

Example 1.4. The condenser pressure as shown by the vacuum gauge is 720 mm of Hg. If the atmospheric pressure is 760 mm of Hg. Find the abs. pressure in the condenser in

(*a*) mm of Hg

(*b*) kN/m².

Solution. Given data :

Vacuum gauge pressure $= 720$ mm of Hg

Atm. pressure $= 760$ mm of Hg

We know that;

$P_{abs.} = P_{atm.} - P_{gauge}$ (For vacuum pressure)

$$\therefore \quad P_{abs.} = 760 - 720$$
$$= 40 \text{ mm of Hg}$$

(*a*) Abs. pressure in the condenser,

$$P_{abs} = 40 \text{ mm of Hg} \quad \textbf{Ans.}$$

(*b*) $\because$ 760 mm of Hg = 1.01325 bar

$$\therefore \quad 40 \text{ mm of Hg} = \frac{1.01325 \times 40}{760}$$
$$= 0.053329 \text{ bar}$$

Also $= 0.053329 \times 100$ $\quad (\because 1 \text{ bar} = 10^5 \text{ N/m}^2)$

$$= 5.3329 \text{ kN/m}^2 \quad \textbf{Ans.}$$

Example 1.5. The gauge pressure in a water tank is 100 mm of Hg. The atm. pressure is 110 KPa. What would be the abs. pressure in the tank ?

Solution. Given data :

Gauge pressure, P_{gauge} = 100 mm of Hg

Atm. pressure, $P_{atm.}$ = 110 KPa

We know that

$$P_{abs} = P_{atm.} + P_{gauge} \qquad ...(i)$$

Also, we know that;

$$760 \text{ mm of Hg} = 101.325 \text{ KPa}$$
$$\therefore \quad 100 \text{ mm of Hg} = \frac{101.325}{760} \times 100$$
$$= 13.3322 \text{ KPa.}$$

Now, substituting the values in Eqn. (*i*), we get

$$P_{abs.} = 110 + 13.3322$$
$$= 123.3322 \text{ KPa} \quad \textbf{Ans.}$$

Example 1.6. The gauge pressure of steam in a steam generator is 2.0 N/mm². The atmospheric pressure as read by the barometer is 760 mm of Hg. Find the abs. pressure in;

(*a*) N/mm² (*b*) KPa (*c*) Bar (*d*) N/m²

Solution. Given data :

Gauge pressure,

$$P_{gauge} = 2.0 \text{ N/mm}^2$$

Atm. pressure, P_{atm} = 760 mm of Hg

$$= 1.01325 \text{ bar}$$
$$= 1.01325 \times 10^5 \text{ N/m}^2 \text{ or Pa}$$
$$= 101325 \text{ N/m}^2 \text{ or Pa}$$
$$= 0.0101325 \text{ N/mm}^2$$

Also, we know that;

$$P_{abs.} = P_{atm.} + P_{gauge}$$

Substituting the values in above equation, we get

(*a*) $P_{abs} = 0.101325 + 2$
$= 2.101325\ N/mm^2$ **Ans.**

(*b*) $\frac{2.101325 \times 10^6}{1000} = 2101.325$ **KPa Ans.** $(\because\ 1\ Pa = 1\ N/m^2)$

(*c*) $\frac{2.101325 \times 10^6}{10^5} = 21.01325$ bar **Ans.** $(\because\ 1\ bar = 10^5\ N/m^2)$

(*d*) $2.101325 \times 10^6 = 2101325\ N/m^2$ **Ans.**

Example 1.7. The density of Hg is 13.6 gm/cm^3. Calculate the pressure in

(*a*) N/m^2

(*b*) Bar, when height of Hg column in the manometer is 260 mm.

Solution. Given data :

Density of Hg $= 13.6\ gm/cm^3$

Height of Hg column $h = 260$ mm

We know that;

$$\text{Pressure, } P = \rho \cdot g \cdot h \qquad ...(i)$$

where ρ = Density of the liquid in kg/m^3

g = Acceleration due to gravity, *i.e.* 9.81 m/s^2)

and h = Height of the liquid column

Now, substituting the values in Eqn. (*i*), we get

(*a*) $$\text{Pressure, } P = \frac{13.6 \times 100 \times 100 \times 100 \times 9.81 \times 260}{1000 \times 1000}$$

$$= 34688.16\ N/m^2 \text{ Ans.}$$

(*b*) $$\text{Pressure, } P = \frac{34688.16}{10^5} = 0.34688 \text{ bar Ans. } (\because\ 1\ bar = 10^5\ N/m^2)$$

Example 1.8. An air tank has three manometers connected to it with fluids :

(*i*) Oil (specific gravity = 0.8)

(*ii*) Water, and

(*iii*) Mercury (specific gravity = 13.6).

What will be height of liquid column in each of the manometer, if the abs. pressure in the tank is 1.5 bar.

Solution. Given data,

Abs. pressure in the tank,

$P_{abs} = 1.5$ bar

We know that

Atm. pressure = 760 mm of Hg
= 1.01325 bar

Also, $P_{abs} = P_{atm} + P_{gauge}$

∴ Gauge pressure,

$$P_{gauge} = P_{abs.} - P_{atm.}$$
$$= 1.5 - 1.01325$$
$$= 0.48675 \text{ bar}$$
$$= 48675 \text{ N/m}^2 \qquad (\because 1 \text{ bar} = 10^5 \text{ N/m}^2)$$

Also, we know that;

$$\text{Pressure, } P = \rho.g.h$$

(*i*) Substituting the corresponding values, we get;

$$48675 = (0.8 \times 1000) \times 9.81 \times h_{oil}$$

($\because$ Density of oil, ρ_0 = Sp. gr. of oil × density of water)

$$\therefore \quad h_{oil} = \frac{48675}{0.8 \times 1000 \times 9.81}$$
$$= 6.2022 \text{ m } \textbf{Ans.}$$

(*ii*) Substituting the corresponding values, we get;

$$48675 = 1000 \times 9.81 \times h_{water}$$
$$\therefore \quad h_{water} = 4.9618 \text{ m } \textbf{Ans.}$$

(*iii*) Again, substituting the corresponding values, we get;

$$48675 = (13.6 \times 1000) \times 9.81 \times h_{Hg}$$
$$\therefore \quad h_{Hg} = \frac{48675}{13.6 \times 1000 \times 9.81}$$
$$= 0.3648 \text{ m } \textbf{Ans.}$$

Example 1.9. Convert the following temperature into degree Fahrenheit.
(*i*) 60°C (*ii*) –15°C

Solution. We know that;

$$\frac{C}{100} = \frac{F - 32}{180}$$

or
$$°F = 1.8 \text{ °C} + 32$$

(*i*) ∴
$$°F = (1.8 \times 60) + 32$$
$$= 108 + 32 = 140 \text{ °F } \textbf{Ans.}$$

(*ii*)
$$°F = 1.8 \times (-15) + 32$$
$$= -27 + 32$$
$$= 5°F \textbf{ Ans.}$$

Example 1.10.

(*a*) Determine, the temp. which is having the same value on centigrade and Fahrenheit scales, both.

(*b*) Find the temperature on which the centigrade scale has twice the value on Fahrenheit scale.

(*c*) Find the temp. at which the Fahrenheit scale has twice the value on centigrade scale.

Solution. We know that :

$$\frac{C}{100} = \frac{F-32}{180} \text{ or } \frac{C}{5} = \frac{F-32}{9}$$

(*a*) For some value, put C = F in above equation

$$\therefore \quad \frac{C}{5} = \frac{C-32}{9}$$

$$\text{or} \quad 9C = 5C - 160$$

$$\text{or} \quad 9C - 5C = -160 \text{ or } C = -40$$

Hence, at temp. –40 °F = –40°C on both the scales. **Ans.**

(*b*) Now, put $C = 2F$, we get;

$$\frac{2F}{100} = \frac{F-32}{180} \text{ or } \frac{2F}{5} = \frac{F-32}{9}$$

$$\text{or} \quad 18F = 5F - 160$$

$$\text{or} \quad 13F = -160$$

$$\text{or} \quad F = -12.3°$$

∴ Fahrenheit value –12.3,

and centigrade value = $2 \times -(12.3) = -23.6$ **Ans.**

(*c*) Now, put $F = 2C$, we get,

$$\frac{C}{5} = \frac{2C-32}{9}$$

$$\text{or} \quad 9C = 10C - 160$$

$$\text{or} \quad 9C - 10C = -160 \text{ or } C = 160$$

∴ Centigrade value = 160°

and Fahrenheit value = $2C = 2 \times 160 = 320°$ **Ans.**

ASSIGNMENT

Objective Type Questions :

(A) Fill in the Blanks :

1. The instantaneous condition of a thermodynamic system is called ________.
2. A prescribed region in space or a finite quantity of matter under consideration for its thermodynamic study is known as a ________.
3. An envelope enclosing the system is known as ________ of the system.
4. Everything external to the system is known as ________.
5. The space comprising of all the systems and boundries is called ________.
6. The thermodynamic system in which the mass comprising the system is not allowed to cross the boundary of the system, although the energy may cross the boundary of the system is called ________ system.
7. The thermodynamic system in which the mass and energy of the system are allowed to cross the boundary of the system is called ________ system.

8. The thermodynamic system in which the mass and energy of the system, both are not allowed to cross the boundary of the system is called ________ system.
9. The thermodynamic system, which is thermally insulated from its surroundings, although it can exchange work with its surroundings is called ________ system.
10. The thermodynamic system which consists of a single phase throughout a thermodynamic process is called ________ system.
11. The thermodynamic system which consists of two or more than two phases is called ________ system.
12. The fixed characteristic, when the working substance is in a particular state is called ________ of the system.
13. The properties which are independent of its mass e.g. temperature pressure etc. are known as ________ properties.
14. The properties which are dependent on its mass e.g. volume, mass, energy etc. are known as ________ properties.
15. A thermodynamic system which is in equilibrium thermally, mechanically as well as chemically is called in ________.
16. The process in which the equilibrium of the system is almost maintained, from its initial state to the final state is called ________ process.
17. A process which can be stopped and reversed at any stage so that the system and surroundings are exactly restored to their initial states is called ________ process.
18. A process in which heat is transferred through a finite temperature is called ________ process.
19. "If two bodies are in thermal equilibrium with a third body separately, then they are also in thermal equilibrium with each other", it is true as per ________ .
20. The ________ is defined as the force per unit area exerted by body on its surface, in a direction normal to the surface.

(B) State True or False :

1. To obtain the absolute pressure above atmospheric the relation used is;
$$p_{abs} = p_{atm} + p_{vacuum}$$
2. To obtain absolute pressure, below atmospheric the relation used is;
$$p_{abs} = p_{atm} - p_{vacuum}$$
3. In S.I. system of units the pressure is expressed in Newton per square metre (N/m^2) or Pascal (pa).
4. Temperature is the thermodynamic property, which determines the hotness or the level of heat intensity of a body.
5. The point at which water freezes, under atmospheric pressure is not taken as the 'Arbitrary zero point'.
6. The 'Absolute zero' of temperature is the zero below which the temperature of any substance can fall.
7. The absolute contigrade or celcius scale is not called degrees 'Kelvin'.
8. The sum of internal energy and the external work done of the given quantity of a gas both expressed in the same units is not called 'Enthalpy'.
9. 'Enthalpy' is the property of the system.

10. The 'internal energy' of a substance is the energy possessed by it, due to its molecular arrangement and activity.
11. The branch of science which deals with the various types of energy transformations and relationships between the various properties of the working substance is known as 'Thermodynamics'.
12. Thermodynamics, when applied to the various fields, such as power generation and refrigeration etc. is reffered as 'Applied Thermodynamics'.
13. The succession of states passed through, during a change of state of a working substance is not called the 'path'.
14. When the path of a thermodynamic process is completely specified, the change of state is not called a 'process'.
15. If the working fluid remains in gaseous state and obeys certain 'Laws of Gases' throughout the energy transformation, then we do not refer it as a 'Perfect Gas'.
16. The instantaneous condition of a thermodynamic system is not called its state.
17. An envelope enclosing the system is known as 'Boundary' of the system.
18. Everything external to the system is known as 'Surroundings'.
19. The space comprising of all the system and boundaries is called 'Universe'.
20. The study which is concerned with the overall behaviour of the matter under consideration is called 'Macroscopic Thermodynamic Study'.

(C) Multiple Choice Questions :

1. A thermodynamic system refers to
   (*a*) any defined region in space
   (*b*) a specified mass in fluid flow
   (*c*) a specific region of constant vol.
   (*d*) a prescribed and identifiable quantity of matter.
2. A closed thermodynamic system manifests when
   (*a*) matter is not allowed to cross the boundary but energy does occur across the boundary.
   (*b*) there is transfer of both mass and energy across the system boundaries.
   (*c*) there is only transfer of mass, but no heat and work energy are transferred.
   (*d*) there is absolutely no interaction of the system with surroundings across the boundaries.
3. In an open thermodynamic system
   (*a*) mass content of the system under consideration remains same
   (*b*) transfer of mass and/or energy takes place
   (*c*) there is only mass transfer even though there may not be any exchange of energy with the system environment
   (*d*) the system exchanges energy with the surroundings in the form of heat energy only.
4. A thermodynamic system is referred to be an isolated system when there is transfer of ____ across the system boundaries.
   (*a*) only mass (*b*) only energy
   (*c*) both mass and energy (*d*) neither mass nor energy.

5. Choose the open thermodynamic system
   (*a*) manual ice cream freezer
   (*b*) centrifugal pump
   (*c*) steam generator
   (*d*) automobile storage battery.
6. Which one of the following represents a closed system ?
   (*a*) bomb calorimeter (*b*) steam generator
   (*c*) universe (*d*) exhaust stroke of an I.C. engine.
7. Which aspect is not true regarding thermodynamic property of a system ?
   (*a*) a characteristic that describes the condition of the system.
   (*b*) a variable whose value depends upon the particular path in going from one state to another.
   (*c*) an intensive property is independent of the extent or mass of the system.
   (*d*) an extensive property can be made intensive, if it is divided by mass and is called specific property.
8. The value of an extensive property is essentially dependent on
   (*a*) mass or extent of the system
   (*b*) interaction of the system with its surroundings
   (*c*) path followed by the system in going from one state to another.
   (*d*) nature of boundaries, rigid or flexible.
9. Which one of the properties given below is an intensive property of the system ?
   (*a*) composition (*b*) volume
   (*c*) kinetic energy (*d*) entropy.
10. All of the following are intensive properties of a system, except.
   (*a*) viscosity (*b*) temperature
   (*c*) pressure (*d*) potential energy.
11. Choose the specific property of a thermodynamic system
   (*a*) viscosity (*b*) density
   (*c*) pressure (*d*) temperature.
12. Which amongst the following is not a property of the system ?
   (*a*) pressure (*b*) composition
   (*c*) thermal conductivity (*d*) heat.
13. Choose the correct statement
   (*a*) temperature is an extensive property
   (*b*) mass remains same in an open system
   (*c*) the system boundaries may expand or collapse
   (*d*) an isolated system allows exchange of energy in the form of heat only.
14. For a system to be in thermodynamic equilibrium, the system and its surroundings are to be in;
   (*a*) thermal equilibrium
   (*b*) chemical equilibrium
   (*c*) mechanical equilibrium
   (*d*) mechanical thermal and chemical equilibrium.

15. Absolute zero temperature is taken as
    (*a*) –273 °C (*b*) 273 °C
    (*c*) 237 °C (*d*) –373 °C
16. The value of one bar in S.I units is equal to;
    (*a*) 100 N/m^2 (*b*) 1000 N/m^2
    (*c*) 1×10^4 N/m^2 (*d*) 1×10^5 N/m^2.
17. Which of the following is the extensive property of the thermodynamic system;
    (*a*) pressure (*b*) volume
    (*c*) temperature (*d*) density.
18. The unit of energy in S.I. units is
    (*a*) joule (J) (*b*) joule metre (Jm)
    (*c*) watt (W) (*d*) None of the above.
19. One 'Watt' is equal to;
    (*a*) 1 Nm/s (*b*) 1 N/min
    (*c*) 10 N/s (*d*) 100 Nm/s.
20. One 'Joule' is equal to;
    (*a*) 1 Nm (*b*) 1 KNm
    (*c*) 10 Nm/s (*d*) 10 KNm/s.

ANSWERS

(A) Fill in the Blanks :

1. thermodynamic state **2.** thermodynamic system
3. boundary **4.** surrounding
5. universe **6.** closed
7. open **8.** isolated
9. adiabatic **10.** homogeneous
11. heterogeneous **12.** property
13. intensive **14.** extensive
15. thermodynamic equilibrium
16. quasi-static **17.** reversible
18. irreversible **19.** zeroth law of thermodynamics
20. pressure

(B) State True or False :

1. True	**2.** True	**3.** True	**4.** True	**5.** False
6. False	**7.** False	**8.** False	**9.** True	**10.** True
11. True	**12.** True	**13.** False	**14.** False	**15.** False
16. False	**17.** True	**18.** True	**19.** True	**20.** True.

(C) Multiple Choice Questions :

1. (*d*)	**2.** (*a*)	**3.** (*b*)	**4.** (*d*)	**5.** (*b*)
6. (*a*)	**7.** (*b*)	**8.** (*a*)	**9.** (*a*)	**10.** (*d*)
11. (*b*)	**12.** (*d*)	**13.** (*c*)	**14.** (*d*)	**15.** (*a*)
16. (*d*)	**17.** (*b*)	**18.** (*a*)	**19.** (*a*)	**20.** (*a*).

(D) Short Answer Type Questions

Q 1. What is 'Thermodynamics' ?

Ans. 'Thermodynamics' is the branch of science which deals with energy transformations and the relations between the properties of the working substance.

Q 2. What do you understand about temperature ? How it is measured ?

Ans. It is an intensive thermodynamic property, which determines the hotness or the level of heat intensity of a body. The temperature of a body is measured generally with the help of an instrument called 'Thermometer'.

Q 3. What is the 'Absolute temperature' ?

Ans. The temperature, below which the temperature of any substance cannot fall is called 'Absolute Zero' and the temperature measured from this absolute zero is called 'Absolute Temperature. The absolute zero is generally taken as –273 °C or –460 °F.

Q 4. What do you know about 'Pressure' ? What is the relation between the Absolute pressure, Atmospheric pressure and Gauge Pressure ?

Ans. The 'Pressure' is defined as the force per unit area exerted by the body on its surface, in a direction normal to the surface. The relation between the 'absolute pressure', atmospheric pressure and the gauge pressure is;

$$p_{abs} = p_{atm} + p_{gauge}$$

In the above relation '+' is used when the gauge pressure is above atmospheric, whereas '–' is used when the gauge pressure is below atmospheric *i.e.*; vacuum.

Q 5. Define path and cycle as applied to thermodynamics.

Ans. **Path :** The succession of states passed through, during a change of state of a working substance is called the 'path'.

Cycle : A thermodynamic 'cycle' is defined as a series of state changes such that the final state is same or identical with the initial state.

Q 6. Define 'process' as applied to thermodynamics. What is a 'Quasi-Static Process' ?

Ans. **Process :** When the path is completely specified, the change of state is called a 'process' e.g. a constant volume process.

Quasi static process : It may be defined as the process during which the equilibrium of the system is almost maintained, from its initial state to the final state.

Q 7. What is a 'property' of a medium ? Write down its types, also.

Ans. The 'property' of a medium is defined as the fixed characteristic, when the working substance is in a particular state e.g., pressure, temperature and volume of a gas.

The properties of a system may be classified into two groups as;

(*i*) **Intensive or Intrinsic Properties.** The properties of a system, which are independent of its mass e.g., temperature, pressure, and density etc., are known as 'Intensive or Intrinsic Properties'.

(*ii*) **Extensive or Extrinsic Properties.** The properties of a system, which are dependent on its mass e.g., volume, mass, length and energy etc. are known as 'Extensive or Extrinsic Properties'.

Q 8. Define, surrounding and boundary as applied to thermodynamics.

Ans. **Surrounding :** Everything external to the system is known as surrounding.

Boundary : An envelope enclosing the system is known as the 'boundary' of the system.

Q 9. Define the 'Thermodynamic System'. What are its important types ?

Ans. A prescribed region in space or a finite quantity of matter under consideration for its thermodynamic study is known as a 'Thermodynamic System'.

There are three types of thermodynamic systems :

(*i*) **Closed System :** It is the thermodynamic system is which the mass comprising the system is not allowed to cross the boundary of the system, so that no mass enters or leaves the system, although, the energy (heat and work) may cross the boundary e.g., piston and cylinder of an I.C. Engine.

(*ii*) **Open System :** It is the thermodynamic system in which mass and energy of the system are allowed to cross the boundary of the system e.g., turbine.

(*iii*) **Isolated System :** It is the thermodynamic system in which the mass and energy both, of the system are not allowed to cross the boundary of the system.

Q 10. Define the 'Heat' and 'Work' as applied to thermodynamics.

Ans. 'Heat' is defined as the energy transfered without the transfer of mass, across the boundary of the system, due to a difference of temperatures between the system and the surroundings.

'Work' is defined as the energy transfered without the transfer of mass, across the boundary of a system due to an intensive property difference, other than temperature, existing between the system and surroundings.

Q 11. What is 'Universe' as applied to 'Thermodynamics' ?

Ans. The space comprising of all the systems and boundaries togather is called 'Universe'.

Q 12. Define Adiabatic System.

Ans. It is the thermodynamic system which is thermally insulated from its surroundings, although it can exchange work with its surroundings.

Q 13. What is a hemogeneous system ?

Ans. A thermodynamic system consisting of a single phase throughout a thermodynamic process is called 'Homogeneous System'. For example, a mixture of air and water vapour.

Q 14. What is a Hetrogeneous system ?

Ans. A thermodynamic system consisting of two or more than two phases is called 'Hetrogeneous system' e.g., a mixture of water and steam or water and ice.

Q 15. What is 'Macroscopic Thermodynamic Study' ?

Ans. The study concerned with overall behaviour of the matter under consideration is called 'Macroscopic Thermodynamic Study or Classical Thermodynamics'.

Q 16. What is 'Microscopic Thermodynamic Study' ?

Ans. The study concerned with the structure of matter under consideration is called 'Microscopic Thermodynamic Study' or 'Statistical Thermodynamics'.

Q 17. What is 'Thermodynamic Equilibrium' ?

Ans. A thermodynamic system which is in equilibrium thermally, mechanically as well as chemically is called in 'Thermodynamic Equilibrium'.

Q 18. Define 'Zeroth Law of Thermodynamics'.

Ans. This law states that, "If two bodies are in thermal equilibrium with a third body separately, then they are also in thermal equilibrium with each other".

Q 19. Define the term 'Enthalpy'.

Ans. The sum of internal energy and the external work done *i.e.*, pressure volume product of the given quantity of a gas both expressed in the same units is called 'Enthalpy' and it is generally denoted by H. Enthalpy can be expressed as :

$$H = E + \frac{pV}{J} \text{ in heat units.}$$

Q 20. Define the term 'Internal Energy'.

Ans. The 'Internal energy' of a substance is the energy possessed by it due to its molecular arrangement and activity. It is generally denoted by 'E' or 'U'.

REVIEW QUESTIONS

1. Define the following :
   (*i*) Thermodynamics (*ii*) System
   (*iii*) Boundary (*iv*) Surroundings.
   (SBTE Haryana, May 1978)
2. Discuss the various properties of a thermodynamic system.
3. Define a thermodynamic system. Differentiate between open system, closed system and an isolated system.
4. Define Homogeneous and Heterogeneous system.
5. Explain the following terms :
   (*i*) State (*ii*) Process (*iii*) Cycle.
6. Explain briefly the 'Zeroth Law of Thermodynamics'.
7. What is a 'Quasi-static' process ?
8. Define 'Reversible and Irreversible Process' as applied to thermodynamics.

❑❑❑

2

CHAPTER

Laws of Perfect Gases

INSIDE THIS CHAPTER

2.1. INTRODUCTION

The various laws of perfect gases are based upon the experimental results. These laws govern the behaviour of perfect gases undergoing changes of the following three variables :

(*i*) The pressure 'p' exerted by the gas.

(*ii*) The volume 'V' occupied by it.

(*iii*) Its absolute temperature 'T'.

If any two of the above variables of a given mass of a perfect gas are known, the third can be calculated by applying the various gas laws, which are discussed below.

It should be carefully noted that, these laws are not applicable to the 'vapours'.

Perfect gas : A perfect gas is one, which strictly obeys all the five gas, laws, *i.e.*, Charle's law, Boyle's law, Joule's law, Avogadro's law and Gay-Lussac law, under all conditions of pressure and temperature.

2.1.1. Boyle's Law

Robert Boyle in 1662 A.D. discovered this law. It states that, *"The volume of a given mass of a perfect gas, varies inversely as its absolute pressure, provided the temperature is kept constant."*

Let p = Absolute pressure of the given gas

and V = Volume of this gas at pressure 'p'.

Then, as per this law :

$$V \propto \frac{1}{p} \qquad \text{(Temperature constant)}$$

or

$$pV = \text{Constant} \qquad ...(2.1)$$

From the above eqn. (2.1), it is clear that *'At constant temperature, the product of absolute pressure and volume of a given mass of a perfect gas is constant'*.

It is represented by 'Hyperbola' (isothermal curve) as shown in Fig. 2.1.

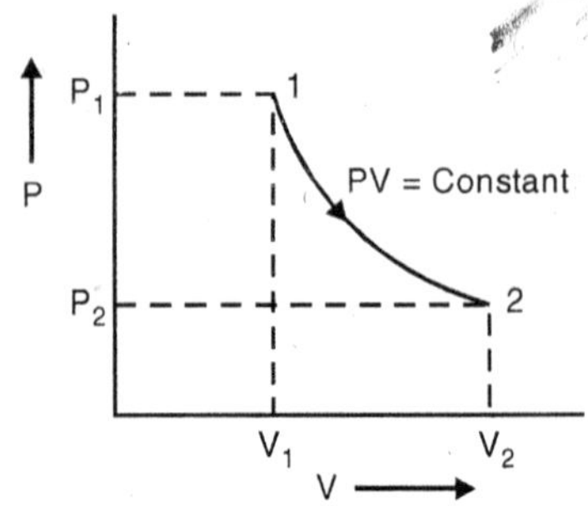

Fig. 2.1.

At condition (1)

$$p_1 V_1 = \text{constant}$$

At condition (2)

$$p_2 V_2 = \text{constant}$$

$$\therefore \quad p_1 V_1 = p_2 V_2 \quad \text{...(2.2)}$$

No known gas obeys 'Boyle's Law' perfectly. There is always a deviation (up to 3 percent), but it can be overlooked in case of air, Hydrogen and Nitrogen etc. except at very low temperatures or at very high pressures.

2.1.2. Charle's Law

Charle's in 1787 A.D. discovered this law. It states that, "*The volume of a given mass of a perfect gas varies directly as its absolute temperature, provided the pressure is kept constant*".

The variation in volume is the same for all gases.

Let V = Volume of the given gas in m^3

T = Absolute temperature in k.

Then as per this law;

$$V \propto T \quad \text{(Pressure constant)}$$

or

$$\frac{V}{T} = A \text{ constant} \quad \text{...(2.3)}$$

From the above eqn. (2.3), it is clear that '*At constant pressure, the ratio of the volume of a given gas to its absolute temperature remains constant*'.

It is represented by a horizontal line as shown in Fig. 2.2.

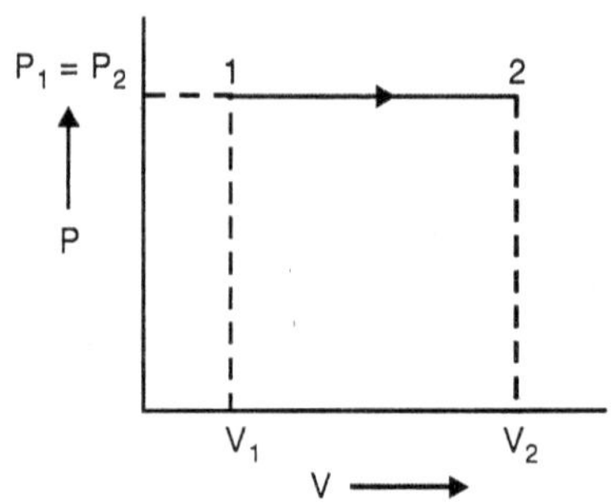

Fig. 2.2.

Let T_1 and T_2 be the absolute temperatures at conditions (1) and (2). Then,

At condition (1)

$$\frac{V_1}{T_1} = \text{A constant}$$

At condition (2)

$$\frac{V_2}{T_2} = \text{A constant}$$

$$\therefore \quad \frac{V_1}{T_1} = \frac{V_2}{T_2} \quad \text{...(2.4)}$$

In eighteenth century, it was found by Regnault and Gay Lussac, that at constant pressure, the change in volume of any perfect gas, corresponding to every unit degree change in its temperature is given by 1/273 of its volume at 0 °C.

Let, V_0 = Volume of the given gas at 0 °C

V_t = Volume of this gas at 't' °C

then $$V_t = V_0 + V_0 \times \frac{t}{273}$$

or $$V_t = V_0\left(1+\frac{t}{273}\right) \quad ...(2.5)$$

Now, if $t = -273$ °C

then $V_t = 0$.

Which shows that at –273 °C, the volume of any gas is zero. This is taken as the point of 'Absolute Zero Temperature'.

From eqn. (2.5)

$$V_t = V_0\left(\frac{273+t}{273}\right)$$

or $$V_t = V_0 \times \frac{T}{T_0} \quad (\because T = t + 273)$$

or $$\frac{V_t}{V_0} = \frac{T}{T_0} \quad ...(2.6)$$

where T = Absolute temperature of the given gas at t °C in k.

T_0 = Absolute temperature of this gas 0 °C in k.

The eqn. (2.6) proves the Charle's Law.

Hence, the Charle's Law may also be expressed as follows. "*At constant pressure, any perfect gas changes in volume by 1/273th of its original volume at 0 °C for every 1 °C change in its temperature*".

2.2. CHARACTERISTIC GAS EQUATION

Let a given mass 'm' of a perfect gas expands from its initial condition (1) to the final condition (2).

Let the first part of expansion *i.e.*, from, 1 to 1′ occurs at constant temperature and the second part *i.e.*, from 1′ to 2 occurs at constant pressure, as shown in Fig. 2.3. Now, for the first part of expansion, at constant temperature, applying Boyle's Law;

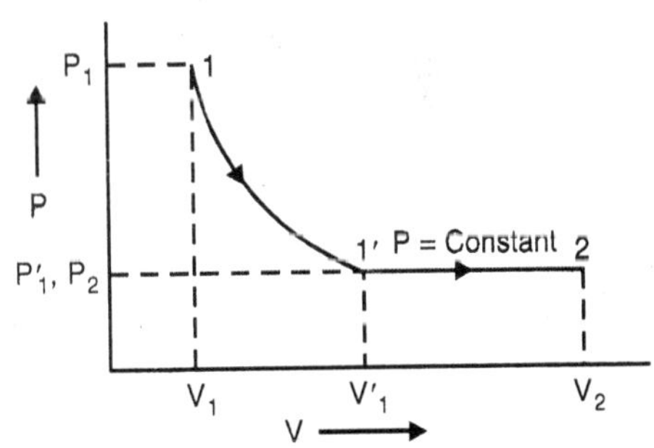

Fig. 2.3.

$$p_1V_1 = p_2V_1' \quad (\because p_1' = p_2) \quad ...(i)$$

Also, for the second part of expansion, at constant pressure, *i.e.*, p'_1, or p_2 applying Charle's Law.

$$\frac{V_1'}{T_1} = \frac{V_2}{T_2} \quad (\because T_1' = T_1) \quad ...(ii)$$

or $$V_1' = \frac{V_2}{T_2} \times T_1$$

Now substituting this value of V_1' in eqn. (*i*), we get

$$p_1V_1 = p_2 \times \frac{V_2}{T_2} \times T_1$$

or $$\frac{p_1V_1}{T_1} = \frac{p_2V_2}{T_2} \quad \text{or} \quad \frac{pV}{T} = \text{Constant}$$

or $$\frac{pV}{T} = R \text{ ('}R\text{' is a constant)}$$

or $$PV = RT \qquad ...(2.7)$$

Equation (2.7) is called the *'Equation of a Perfect Gas' or the 'Characteristic gas equation'.*

If m = Mass of the given gas

Then, the eqn. (2.7) becomes;

$$pV = m\,RT \qquad ...(2.8)$$

where p = Absolute pressure of the given gas in N/m^2.

V = Volume of this gas in m^3.

m = Mass of gas in kg.

R = Characteristic gas constant in Nm/kg k or J/kg k

T = Absolute temperature of gas in k.

2.2.1. Characteristic Gas Constant

The constant 'R' used in the characteristic gas equation $PV = RT$, has a constant value for a particular gas and is called *'Characteristic gas constant'* or *'specific gas constant'* or *Simply 'gas constant'*. Its value depends upon the temperature scale used and the properties of the gas, under consideration.

From eqn. (2.8), the units of 'R' will be

$$R = \frac{pV}{mT} = \frac{N}{m^2} \times \frac{m^3}{1} \times \frac{1}{kg} \times \frac{1}{k} = \text{Nm/kg k}$$

$$= \text{J/kg k} \qquad (\because 1 \text{ Nm} = 1 \text{ Joule})$$

If the units of pressure 'p' are bar and that of the volume 'V' m^3, then eqn. (2.8) becomes,

$$10^5\, pV = m\,RT \qquad ...(2.9)$$

For atmospheric air,

$$R = 287 \text{ J/kg/k} \qquad \text{(In S.I. system of units)}$$

2.2.2. Universal Gas Constant

If the molecular weight of any gas is multiplied by its specific gas constant 'R', it is found that this product is 'constant' for all gases.

This 'constant' is known as *'Universal Gas Constant', Ru.*

The value of Ru is as follows,

$$Ru = 8314 \text{ J/kg mol/k}$$

So, $$pV = n\,Ru\,.\,T \qquad (n = \text{No. of kg moles}) \quad ...(2.10)$$

This is known as 'Universal Gas Equation'.

2.2.3. Joule's Law

Joule's law states that *"The internal energy of a given quantity of a gas depends only on the temperature"*.

This is an inference from the fact, established by experiments of Joule, that when a gas expands without doing external work and without taking in or giving out heat (and therefore without changing its stock of internal energy), its temperature does not change.

2.2.4. Regnault's Law

Regnault's law states that, *"The two specific heats. C_p, (specific heat at constant pressure) and C_v (specific heat at constant volume) of a gas do not change with the change of temperature and pressure"*.

In other words, *C_p and C_v of a gas always remain constant.*

The above law holds good only approximately because it has been observed that the specific heats of a gas vary with the temperature. This variation in two specific heats due to the rise in temperature may be as high as 20% in engineering practice of the day.

The Tables 2.1 and 2.2 below gives the value of the density, volume and the characteristic gas constant (R) for various gases;

Table 2.1.

	Density kg per Cubic metre	Volume		R	
		Cu. metre per kg.	Cu. ft. per lb.	Kgfm per kg °k	J/kg °k
Dry air	1.293	0.773	12.39	29.27	287
Hydrogen	0.099	11.126	178.2	421.1	4124.4
Oxygen	1.429	0.700	11.21	26.49	259.8
Nitrogen	1.251	0.799	12.80	30.25	296.8

Values of the Molecular weight, specific heats, R and γ.

Table 2.2.

	Molecular Weight	C_p (KJ/kgk)	C_v (KJ/kgk)	R (KJ/kgk)	$\gamma = \frac{C_p}{C_v}$
Dry air	29	1.005	0.718	0.287	1.4
Hydrogen	2	14.32	10.163	4.157	1.41
Oxygen	32	0.92	0.66	0.260	1.4
Nitrogen	28	1.033	0.736	0.297	1.4

2.2.5. Avogadro's Law

According to this law 'Equal volume of different perfect gases at the same temperature and pressure contain equal number of molecules'.

Thus it two ideal gases are contained in two vessels of equal volume and at the same temperature and pressure each gas will have the same number of molecules say, n. If M_1 and M_2 are the moleculer weights and m_1, m_2 the masses of the two gases, then

$$m_1 = k \cdot M_1 \cdot n \quad ...(i)$$

and $$m_2 = k \cdot M_2 \cdot n \quad ...(ii)$$

where k is a proportionality factor

From eqns. (*i*) and (*ii*) we get, $\frac{m_1}{m_2} = \frac{M_1}{M_2}$

or $$\frac{\rho_1}{\rho_2} = \frac{M_1}{M_2}$$

($\because m = \rho.v$ and v *i.e.*; sp. volume for two gases)

If v_1 and v_2 are the specific volumes of the gases, then

$$\frac{v_2}{v_1} = \frac{M_1}{M_2} \qquad \left(\because \ \rho = \frac{1}{v}\right)$$

or $$M_1 v_1 = M_2 v_2 = M.v = \text{Constant.}$$

From above we can see that the product of molecular weight and the specific volume is constant for all ideal gases at the same temperature and pressure. The product of mole and specific volume is known as kg molecular volume or molar volume which obviously is same for all gases, when the pressure and temperature are same. The molar volume at N.T.P. may be expressed as,

$$V = \frac{MRT}{P} = \frac{848 \times 273}{1.0332 \times 10^4} = 22.41 \text{ m}^3/\text{kg mole at N.T.P.}$$

This value of molecular volume will be same for all ideal gases.

Thus, 22.41 m^3 of H_2 at NTP has a mass of 2 kg as its molecular weight is 2.

Similarly, 22.41 m^3 of O_2 at NTP, has a mass of 32 kg as its molecular weight is 32.

And 22.41 m^3 of N_2 at NTP, has a mass of 28 kg as its molecular weight is 28.

2.3. MOLAR HEAT

'Molar heat' may be defined as *'The quantity of heat required to raise the temperature of one mole of a substance through one degree.* Since, the mass of one mole is M kg so, Molar heat C_{mol} = M.C.

2.3.1. Heat

'Heat' is defined as the energy transferred without transfer of mass, across the boundary of a system due to a difference of temperatures between the system and the surroundings.

From the above definition the following points should be kept in mind.

(*i*) Heat is a transient quantity, which can be identified only as it crosses the boundary of a system.

(*ii*) Heat is not a property of a system.

It will, therefore, be incorrect to say heat contained in a system but the correct term to be used in this connection is internal energy (Kinetic or potential).

The units of heat in M.K.S. system are Kcal and in S.I. system 'Joule' or 'KJ'. The other units of heat are Centrigate Heat Unit (CHU) and British Thermal Unit (BTU) in C.G.S. and F.P.S. system respectively.

2.3.2. Work

'*Work*' is defined as the energy transferred without the transfer of mass, across the boundary of a system due to an intensive property difference, other than temperature, existing between the system and the surroundings.

In general the intrinsic or intensive property difference encountered in engineering problems is the pressure difference. The pressure difference at the surface of the system gives rise to a force and the action of this force, over a distance is termed as 'Mechanical Work'.

The units of 'Work' are kgm in M.K.S. system and N-m or Joule in S.I. system. *It should be noted here that in S.I. system of units the units of heat and work are the same i.e. Joule.*

2.3.3. Specific Heat

The '*specific heat*' of a substance is defined as the amount of heat, required to raise the temperature of its unit mass through a unit degree.

Solids and liquids have only one specific heat but gases have two specific heats as stated below. Units of sp. heat are KJ/kg k or KJ/kg °C.

(*i*) **Specific Heat at Constant Pressure (C_p) :** It is the amount of heat required to raise the temperature of a unit mass of a gas through unit degree, when its '*pressure is kept constant*'.

If '*m*' kg of a gas heated at constant pressure from an initial temperature T_1 to final temperature T_2, then the amount of heat, supplied to '*m*' kg of this gas is = Mass × Sp. heat at constant pressure × Rise in temperature.

or $$Q = m \times C_p \times (T_2 - T_1)$$

The value of 'C_p' for air in S.I. system is 1.0 KJ/kg k and in M.K.S. system, 0.24 Kcal/kg k.

(*ii*) **Specific Heat at Constant Volume (C_v) :** It is the amount of heat required to raise the temperature of a unit mass of a gas though unit degree when its volume is kept constant.

If '*m*' kg of a gas be heated at constant volume from an initial temperature T_1 to the final temperature T_2 then,

Amount of heat supplied, Q is

= Mass × Sp. heat at constant volume × Rise in temperature

or $Q = m \times C_v \times (T_2 - T_1)$.

The value of 'C_v' for air in S.I. system is 0.7128, KJ/kgk and in MKS system 0.17 kcal/kgk.

The specific heat 'C_p' and 'C_v' of a perfect gas are assumed to remain constant at all temperatures, but for real gases they increase considerably with rise in their temperature.

For any particular gas, the specific heat at constant pressure is always greater than the specific heat at constant volume and their ratio remains constant, which is generally denoted by 'γ' or 'K'.

It may be expressed as;

$$\gamma = \frac{C_p}{C_v} \qquad ...(2.11)$$

The value of 'γ' for air is 1.4.

Relation between the specific heats of a gas and the gas constant

Consider one kg of a gas to be heated at constant pressure.

Let, p = Absolute pressure of the gas

T_1 = Initial absolute temperature

T_2 = Final absolute temperature

C_p = Specific heat of gas at constant pressure

C_v = Specific heat of gas at constant volume.

Initially, for the gas,

$$pV_1 = RT_1$$

and finally after heating

$$pV_2 = RT_2$$

Subtracting,

$$pV_2 - pV_1 = RT_2 - RT_1$$

or $$p(V_2 - V_1) = R(T_2 - T_1) \quad ...(i)$$

Work done by the gas,

$$W = \frac{p(V_2 - V_1)}{J} \text{ in heat units.}$$

Substituting the value of $P(V_2 - V_1)$ from eqn. (*i*)

$$W = \frac{R(T_2 - T_1)}{J} \text{ in heat units} \quad ...(ii)$$

Now change in internal energy

$$\Delta U = C_v (T_2 - T_1) \text{ in heat units} \quad ...(iii)$$

Total heat supplied to the gas,

$$Q = C_p (T_2 - T_1) \text{ in heat units} \quad ...(iv)$$

According to the law of conservation of energy.

Total heat supplied to the gas = work done by gas (in heat units) + change in internal energy.

or $$Q = W + \Delta U \quad ...(v)$$

Now, substituting the values of W, ΔU and Q from (*ii*), (*iii*) and (*iv*) respectively in eqn. (*v*) we get

$$C_p (T_2 - T_1) = \frac{R(T_2 - T_1)}{J} + C_v (T_2 - T_1)$$

Dividing throughout by $(T_2 - T_1)$

$$\therefore \quad C_p = \frac{R}{J} + C_v$$

or $$C_p - C_v = \frac{R}{J} \text{ in (M.K.S. units)} \quad ...(2.12)$$

and $$C_p - C_v = R \text{ in (S.I. units)}$$

or $$C_p = R + C_v \qquad ...(vi)$$

or $$\frac{C_p}{C_v} = \frac{R}{C_v} + \frac{C_v}{C_v} \qquad \text{(Dividing both sides by } C_v\text{)}$$

or $$\gamma = \frac{R}{C_v} + 1$$

or $$\gamma = 1 + \frac{R}{C_v}.$$

It should be noted that in S.I. system the units of heat and work are the same *i.e.* Joule, so, the mechanical equivalent of heat '*J*' from eqn. (2.12) should be droped. So, the relation of specific heats of a gas in S.I. system of units becomes,

$$C_p - C_v = R \qquad ...(2.13)$$

Also, if M = Molecular weight of the gas

and Ru = Universal gas constant

Then, $$Ru = M.R. \text{ or } R = \frac{Ru}{M}$$

The above equation will become;

$$C_p - C_v = \frac{Ru}{M} \text{ (in S.I. units)} \qquad ...(2.14)$$

NUMERICAL EXAMPLES

Example 2.1. 2.5 m^3 of a gas at 8 bar absolute pressure expand at constant temperature to a volume of 10 m^3. Find the final pressure of the gas.

Solution. Given data :

p_1 = 8 bar absolute

V_1 = 2.5 m^3

V_2 = 10 m^3

p_2 = ?

Since, the temperature remains constant, so applying Boyle's law.

$$\therefore \quad p_1 V_1 = p_2 V_2$$

or $$p_2 = \frac{p_1 V_1}{V_2} = \frac{8 \times 2.5}{10} = 2 \text{ bar absolute.}$$ **Ans.**

Example 2.2. 0.14 m^3 of a gas at 40 °C receive heat at constant pressure so that the final temperature is 200 °C. Find the final volume and write down the increase in volume during the supply of heat.

Solution. Given data,

T_1 = 273 + 40 = 313 k

T_2 = 273 + 200 = 473 k

V_1 = 0.14 m^3

V_2 = ?

Now, since heat is supplied at constant pressure.

So, applying Charle's law :

$$\therefore \quad \frac{V_1}{T_1} = \frac{V_2}{T_2} \quad \text{or,} \quad V_2 = \frac{V_1}{T_1} \times T_2$$

or
$$V_2 = \frac{0.14}{313} \times 473 = 0.212 \text{ m}^3 \textbf{ Ans.}$$

$\therefore$ Increase in Volume$= V_2 - V_1 = 0.212 - 0.14 = 0.072 \text{ m}^3$ **Ans.**

Example 2.3. Find the volume occupied when a given quantity of a gas exists at a temperature of 400 °C, if the volume of this quantity of gas at 0 °C is 0.57 m^3, the pressure being same at both temperatures.

Solution. Charle's law is applicable here, since the pressure is same in both the cases.

If the given quantity of the gas is heated from 0 °C to 400 °C keeping the pressure constant.

$$\therefore \quad \text{Increase in Volume} = \left(\frac{1}{273} \times \text{vol. at } 0\ ^\circ\text{C}\right) \times \text{Rise in temp.}$$

or
$$= \frac{1}{273} \times 0.57 \times (400 - 0) = \frac{0.57 \times 400}{273} = 0.835 \text{ m}^3.$$

Hence, Volume at 400 °C = 0.57 + 0.835 = 1.405 m^3. **Ans.**

Example 2.4. A vessel of capacity 5 m^3 contains 20 kg of an ideal gas having a molecular weight of 25. The temperature is 15 °C. Find out the pressure.

Solution. $\because$
$$R = \frac{Ru}{M}$$

$$\therefore \quad R = \frac{8314}{25} = 332.56 \text{ J/kg k}$$

Now, applying characteristic gas equation,

$$pV = mRT$$

$$\therefore p = \frac{mRT}{V} = \frac{20 \times 332.56 \times (15 + 273)}{5} \text{ N/m}^2\text{(Pa)}$$

$$= \frac{20 \times 332.56 \times 288}{10^5 \times 5} = 3.831 \text{ bar} \qquad (\because 1 \text{ bar} = 10^5 \text{ N/m}^2 \text{ or Pa}) \textbf{ Ans.}$$

Example 2.5. How much heat must be supplied to 1.5 kg of a gas to raise its temperature from 90 °C to 225 °C at constant pressure ? Find also the external work done during the supply of heat. Sp. heat at constant pressure is 1.0 KJ/ kg k and that at constant volume is 0.7128 KJ/kg k.

Solution. Given data :

$$m = 1.5 \text{ kg}$$
$$T_1 = 225 + 273 = 498 \text{ k}$$
$$C_p = 1.0 \text{ KJ/kg k}$$
$$C_v = 0.7128 \text{ KJ/kg k}$$

$\therefore$ The heat supplied, at constant pressure,

$$Q = m \,.\, C_p\,(T_2 - T_1) = 1.5 \times 1 \times (498 - 363)$$

$$C_p = 1.5 \times 135 = 202.5 \text{ KJ}$$

Change in internal energy,

$$\Delta U = m.C_v (T_2 - T_1) = 1.5 \times 0.7128 \times (498 - 363)$$
$$= 1.5 \times 0.7128 \times 135 = 144.34 \text{ KJ}$$

Now, $Q = W + \Delta U$

$\therefore$ External work done,

$$W = Q - \Delta U = 202.5 - 144.34 = 58.16 \text{ KJ}$$ **Ans.**

Example 2.6. The state of a quantity of gas enclosed in a gas engine cylinder, at the beginning of compression is, volume 0.05 m^3, temperature 80 °C, and pressure 0.98 bar. What will be the temperature after the volume is reduced to 0.014 m^3 and the pressure raised to 7.35 bar, at the end of compression ?

Solution. Given data :

$$V_1 = 0.05 \text{ m}^3$$
$$T_1 = 80 + 273 = 353 \text{ k}$$
$$p_1 = 0.98 \text{ bar}$$
$$V_2 = 0.014 \text{ m}^3$$
$$p_2 = 7.35 \text{ bar}$$

Now applying the gas equation,

$$\frac{p_1 V_1}{T_1} = \frac{p_2 V_2}{T_2}$$

$$\therefore \quad T_2 = \frac{T_1}{p_1 V_1} \times p_2 V_2 = \frac{353}{0.98 \times 0.05} \times 7.35 \times 0.014$$

or $$T_2 = 741.3 \, k = (741.3 - 273)^\circ\text{C} = 468.3 \text{ °C}$$

Hence, temperature of gas at the end of compression = 468.3 °C. **Ans.**

Example 2.7. The values of C_p and C_v for a gas are 1.0 and 0.7128 KJ/kg k respectively. Find the density of this gas at 15 °C and 101.325 k Pa absolute pressure.

Solution. Given data :

$$C_p = 1.0 \text{ KJ/kg k}$$
$$C_v = 0.7128 \text{ KJ/kg k}$$
$$T = 15 + 273 = 288 \text{ k}$$
$$p = 101.325 \text{ k Pa}$$

In S.I. system of units,

$$\therefore \quad C_p - C_v = R$$

or $$1.0 - 0.7128 = R = 0.2872 \text{ KJ/kg K}$$

Also, for one kg mass of gas the gas = n is,

$$pV = RT$$

$$\therefore \quad V = \frac{RT}{p} = \frac{0.2872 \times 288}{101.325} = 0.816 \text{ m}^3/\text{kg}$$

and $$\text{Density of gas} = \frac{1}{\text{Volume of one kg of the gas}} = \frac{1}{0.816}.$$

Hence $$\text{Density} = 1.22 \text{ kg/m}^3$$ **Ans.**

Example 2.8. One kg of an ideal gas is heated from 18.3 °C to 93.4 °C. Assuming $R = 0.287$ KJ/kg k and $\gamma = 1.4$ for the gas. Find the specific heats, at constant pressure and at constant volume both.

Solution. We know that :

$$C_p - C_v = R \quad ...(i)$$

Also $$\frac{C_p}{C_v} = \gamma \quad ...(ii)$$

From (*ii*) $$C_p = C_v \times \gamma = 1.4\, C_v \quad ...(iii)$$

Put in value of C_p in (*i*)

$$\therefore \quad 1.4\, C_v - C_v = R$$

or $$0.4\, C_v = R$$

or $$0.4\, C_v = 0.287$$

$$\therefore \quad C_v = \frac{0.287}{0.4} = 0.7175 \text{ KJ/kg k}$$

Also from (*iii*)

$$C_p = 1.4 \times C_v = 1.4 \times 0.7175 = 1.0 \text{ KJ/kg k}$$

Hence, $$C_p = 1.0 \text{ and } C_v = 0.7175 \text{ KJ/kg k } \textbf{Ans.}$$

Example 2.9. 1.5 kg of a gas at 8.4 bar absolute and 87 °C is placed in a cylinder. What is the volume of the cylinder if the characteristic constant is 0.287 KJ/kg k.

Solution. Given data :

$$m = 1.5 \text{ kg}$$
$$p = 8.4 \text{ bar absolute}$$
$$= 8.4 \times 10^5 \text{ N/m}^2 \text{ absolute}$$
$$T = 87 + 273 = 360 \text{ k}$$
$$R = 0.287 \text{ KJ/kg k} = 287 \text{ J/kg k}$$

We know that characteristic gas = n is,

$$pV = mRT$$

or $$V = \frac{m.R.T}{p} = \frac{1.5 \times 287 \times 360}{8.4 \times 10^5} = 0.1845 \text{ m}^3. \textbf{ Ans.}$$

Example 2.10. The characteristic equation of a certain gas is $pV = 290\ T$, 'p' being in N/m² (Pa) absolute, 'V' in m³, while 'T' in degree kelvin. Find the weight of this gas which occupies 5.5 m³ at 7×10^5 N/m² (Pa) and 200 °C.

Solution. Given = n is ; $pV = 290\ T$

Substituting the values of p and T

$$\therefore \quad 7 \times 10^5 \times V = 290 \times (200 + 273)$$

or $$V = \frac{290 \times 473}{7 \times 10^5} = 0.196 \text{ m}^3/\text{kg}.$$

$\therefore$ Weight of the gas which occupied 5.5 m³

$$= \frac{5.5}{0.196} = 28.07 \text{ kg. } \textbf{Ans.}$$

Example 2.11. Determine the mass of a gas occupying 4.5 m^3 at 6 bar abs. pressure at 200°C temperature. Take, gas constant, R = 287 J/kg k.

Solution. Given data;

Volume of the gas, $V = 4.5 \text{ m}^3$

Pressure, $P = 6$ bar abs.

$= 6 \times 10^5 \text{ N/m}^2$ or Pa

Gas constant, $R = 287$ J/kg k

and Temp. $T = 200°C = 200 + 273 \text{ k} = 473 \text{ k}$

We know that; $PV = m \,.\, r \,.\, T$

$$\therefore \quad m = \frac{PV}{RT}$$

Substituting the values, we get

$$m = \frac{6 \times 10^5 \times 4.5}{287 \times 473}$$

or Mass of gas, $m = 19.889$ kg **Ans.**

Example 2.12. A reservoir of capacity 5 m^3 contains 10 kg of an ideal gas having molecular weight of 28.97. The temperature of the gas is 200 k. Determine its pressure.

Solution. Given data;

Volume of gas, $V = 5 \text{ m}^3$

Mass of gas, $m = 10$ kg

Molecular weight of gas,

$M = 28.97$

and Temp. = 200 k

We know that;

Universal gas constant, $R_u = R \times M$

$$\text{or} \quad R = \frac{R_u}{M} = \frac{8314}{28.97} = 286.98 \text{ J/kg k}$$

($\because$ R_u = 8314 J/kg mole/k)

Also, $PV = m.R.T$

$$\therefore \quad \text{Pressure } P = \frac{m.R.T}{V}$$

$$= \frac{10 \times 286.98 \times 200}{5}$$

$= 114792$ Pa $= 114.792$ KPa **Ans.**

Example 2.13. The initial pressure and temperature of the air in a truck tyre are 250 KPa and 35°C respectively. After, certain journey the pressure has increased to 280 KPa. Determine, the temperature of the air inside the tyre after the journey. Take, R = 287 J/kg k and C_V = 718 J/kg k for air.

Solution. Given data;

Initial pressure, P_1 = 250 KPa

Initial temperature, T_1 = 35°C

= 35 + 273 = 308 K

Final pressure, P_2 = 280 KPa

As, the volume before and after the journey remains the same, therefore,

$$V_1 = V_2$$

Now, applying gas equation;

$$\frac{P_1}{T_1} = \frac{P_2}{T_2}$$

or
$$T_2 = \frac{P_2 \times T_1}{P_1}$$

$$= \frac{280 \times 308}{250}$$

$$= 344.96 \simeq 345 \text{ K}$$

or
$$= 72\ °\text{C} \quad \textbf{Ans.}$$

Example 2.14. A gas occupying 0.15 m³, at a pressure of 1.8 bar and temperature of 35°C, is expanded to 1 bar pressure at 0 °C. Find :

(*i*) Mass of the gas

(*ii*) Volume, and

(*iii*) Density, if its molecular weight is 28.

Solution. Given data :

Initial volume, V_1 = 0.15 m³

Initial pressure, p_1 = 1.8 bar = 1.8×10^5 N/m²

Initial temp., t_1 = 35°C = 35 + 273 = 308 K

Final pressure, p_2 = 1 bar = 1×10^5 N/m²

Final temp., t_2 = 0°C = 273 K

We know that, Universal gas equation is

$$PV = n.R_u.T$$

∴
$$n = \frac{PV}{R_u.T} = \frac{1.8 \times 10^5 \times 0.15}{8314 \times 308} = 0.0105 \text{ moles}$$

(*i*) But, mass of gas, m = Mole × Molecular weight

∴ Mass = 0.0105 × 28 = 0.2952 kg **Ans.**

(*ii*) We know that :

$$\frac{P_1V_1}{T_1} = \frac{P_2V_2}{T_2}$$

or
$$V_2 = \frac{P_1}{P_2} \times \frac{T_2}{T_1} \times V_1$$

or
$$= \frac{1.8 \times 10^5}{1 \times 10^5} \times \frac{273}{308} \times 0.15$$

$$= 0.2393 \text{ m}^3 \quad \textbf{Ans.}$$

(*iii*) $$\text{Density} = \frac{\text{Mass}}{\text{Volume}}$$

$$= \frac{0.2952}{0.2393} = 1.234 \text{ kg/m}^3 \textbf{ Ans.}$$

Example 2.15. A gas at a pressure of 4 bar, occupies a volume of 1.2 m³ at 300 K. It expands to a pressure of 0.5 bar and temp. 150 K. Find;

(*i*) Mass of the gas

(*ii*) Final volume of gas

(*iii*) Molecular weight of gas per kg mole.

Assume R = 320 Nm/kg k.

Solution. Given data :

Initial pressure, $p_1 = 4 \text{ bar} = 4 \times 10^5 \text{ N/m}^2$

Initial volume, $V_1 = 1.2 \text{ m}^3$

Initial temp., $T_1 = 300 \text{ K}$

Final pressure, $P_2 = 0.5 \text{ bar} = 0.5 \times 10^5 \text{ N/m}^2$

Characteristic gas constant,

$$R = 320 \text{ Nm/kg k.}$$

(*i*) From gas equation,

$$PV = m.R.T$$

or $$\text{Mass, } m = \frac{PV}{RT}$$

$$= \frac{4 \times 10^5 \times 1.2}{320 \times 300} = 5 \text{ kg } \textbf{Ans.}$$

(*ii*) From gas equation

$$\frac{P_1V_1}{T_1} = \frac{P_2V_2}{T_2}$$

$$V_2 = \frac{P_1V_1}{T_1} \times \frac{T_2}{P_2}$$

$\therefore$ $$\text{Final volume, } V_2 = \frac{4 \times 10^5 \times 1.2}{300} \times \frac{150}{0.5 \times 10^5}$$

$$= 4.8 \text{ m}^3 \textbf{ Ans.}$$

(*iii*) We know that;

Characteristic gas constant,

$$R = \frac{\text{Universal gas constant}}{\text{Molecular weight}}$$

$\therefore$ $$R = \frac{R_u}{M}$$

or $$M = \frac{R_u}{R} = \frac{8314}{320} = 25.981 \text{ kg/kg mole } \textbf{Ans.}$$

Example 2.16. **Determine the specific volume and density of a gas at a pressure of 8 bar and a temp. of 300°C. The molecular weight of the gas be 42 kg/kg mole.**

Solution. Given data :

$$\text{Pressure, } P = 8 \text{ bar} = 8 \times 10^5 \text{ N/m}^2$$

$$\text{Temp., } T = 300°\text{C} = 300 + 273 \text{ K}$$

$$= 573 \text{ K}$$

$$\text{Molecular weight, } M = 42 \text{ kg/kg mole}$$

We know that;

$$R = \frac{R_u}{M}$$

$$= \frac{8314}{42} = 197.95 \text{ Nm/kg k}$$

Also,

$$PV = m \,.\, R \,.\, T$$

$$\therefore \quad V = \frac{mRT}{P} = \frac{1 \times 197.95 \times 573}{8 \times 10^5}$$

$$= 0.1418 \text{ m}^3/\text{kg} \quad \textbf{Ans.}$$

Hence, density,

$$\rho = \frac{m}{V} = \frac{1}{0.1418} = 7.052 \text{ kg/m}^3 \quad \textbf{Ans.}$$

Example 2.17. **An Oxygen cylinder of capacity 250 litres, contains oxygen, at a pressure of 3.5 MPa and temp. of 21°C. After certain use of gas from this cylinder, the pressure and temp. of the oxygen left in the cylinder falls to 1.5 MPa and 14°C respectively. Determine the mass of oxygen used. M = 32 kg/kg mole.**

Solution. Given data :

$$\text{Initial pressure, } P_1 = 3.5 \text{ MPa} = 3 \times 10^6 \text{ Pa or N/m}^2$$

$$\text{Initial temp., } T_1 = 21°\text{C} = 21 + 273 \text{ K} = 294 \text{ K}$$

$$\text{Initial volume, } V_1 = 250 \text{ litres}$$

$$= 350 \times 10^{-3} \text{ m}^3$$

$$= 0.25 \text{ m}^3$$

$$\text{Final pressure, } P_2 = 1.5 \text{ MPa} = 1.5 \times 10^6 \text{ Pa or N/m}^2$$

$$\text{Final Temp., } T_2 = 14°\text{C}$$

$$= 14 + 273 \text{ K}$$

$$= 287 \text{ K}$$

$$\text{Final Volume, } V_2 = V_1 = 0.25 \text{ m}^3 \quad (\because \text{ Volume of cylinder is constant})$$

Now, we know that :

$$R = \frac{R_u}{M}$$

$$= \frac{8314}{32} = 259.81 \text{ Nm/kg k}$$

From characteristic gas equation,

or $$m_1 = \frac{P_1V_1}{R.T_1} = \frac{3 \times 10^6 \times 0.25}{259.81 \times 294} = 9.818 \text{ kg}$$

Also, $$P_2V_2 = m_2RT_2$$

or $$m_2 = \frac{P_2V_2}{R.T_2} = \frac{1.5 \times 10^6 \times 0.25}{259.81 \times 287} = 5.029 \text{ kg}$$

Hence, mass of oxygen used

$$= m_1 - m_2$$
$$= 9.818 - 5.029 = 4.789 \text{ kg} \quad \textbf{Ans.}$$

Example 2.18. The density of Hydrogen (H_2) gas is 0.1 kg/m^3 at N.T.P. What will be the value of characteristic gas constant 'R' and both the sp. heats if the value of γ is 1.4.

Solution. Given data : Density of hydrogen gas = 0.1 kg/m^3

Adiabatic index, γ = 1.4

As : $$\text{Volume} = \frac{1}{\text{Density}}$$
$$= \frac{1}{0.1} = 10 \text{ m}^3/\text{kg}$$

At N.T.P.; we know

Temperature, T = 273 K

Pressure, P = 1.01325×10^5 N/m^2

Also, $$PV = RT$$

or $$R = \frac{PV}{T}$$
$$= \frac{1.01325 \times 10^5 \times 10}{273}$$
$$= 3711.54 \text{ J/kg k}$$
$$= 3.71154 \text{ KJ/kg k} \quad \textbf{Ans.}$$

Also, $$r = \frac{C_p}{C_V} = 1.4 \text{ (Given)}$$

and $$C_p - C_V = R = 3.7115$$

or $$\frac{C_p}{C_V} - 1 = \frac{R}{C_V}$$ (Dividing both sides by 'C_V')

or $$\gamma - 1 = \frac{R}{C_V}$$

or $$1.4 - 1 = \frac{3.7115}{C_V}$$ (Substituting the values)

or $$C_V = \frac{3.7115}{0.4} = 9.279 \quad \textbf{Ans.}$$

Now, $\quad C_P - 9.279 = R = 3.7115$

or $\quad C_P = 3.7115 + 9.279 = 12.990$ **Ans.**

Example 2.19. A certain quantity of air occupying 0.08 m³ has initial temp. of 200°C. Then, it is expanded at const. pressure to a final volume of 0.3 m³. Determine ;

(*i*) Final temp. (*ii*) Value of C_p if R = 287 J/kg k and C_V = 0.7128 KJ/kg k

Solution. Given data :

$$\text{Initial volume, } V_1 = 0.08 \text{ m}^3$$

$$\text{Initial temp., } T_1 = 200°\text{C} = 200 + 273 \text{ K} = 473 \text{ K}$$

Pressure is constant

$$\text{Final volume, } V_2 = 0.3 \text{ m}^3$$

$$R = 287 \text{ J/kg k} = 0.287 \text{ KJ/kg k}$$

$$C_V = 0.7128 \text{ KJ/kg k}$$

Since, the expansion is at constant pressure, therefore

$$\frac{V_1}{T_1} = \frac{V_2}{T_2}$$

or

$$T_2 = \frac{V_2}{V_1} \times T_1 = \frac{0.3}{0.08} \times 473 = 1773.75 \text{ K}$$

$$= 1500.75°\text{C} \quad \textbf{Ans.}$$

Also, we know that;

$$C_p - C_V = R$$

or

$$C_P = C_V + R = 0.7128 + 0.287 = 0.998 \text{ KJ/kg k} \quad \textbf{Ans.}$$

Example 2.20. The values of C_P and C_V for a gas are 1 KJ/kg k and 0.713 KJ/kg k respectively. Determine, the density of gas at 25°C and 1.2 bar abs. pressure.

Solution. Given data :

Sp. heat at const. pressure, C_P = 1.0 KJ/kg k

Sp. heat at constant volume, C_V = 0.713 KJ/kg k

$$\text{Temp. of gas, } T = 25°\text{C} = 25 + 273 \text{ K} = 298 \text{ K}$$

and $\quad$ Abs. pressure = 1.2 bar = 1.2×10^5 N/m²

As we know that;

$$PV = RT \qquad \text{(for unit mass of gas)}$$

and

$$C_P - C_V = R$$

$\therefore$

$$R = 1 - 0.713 = 0.287 \text{ KJ/kg k}$$

Also $$V = \frac{RT}{P} = \frac{0.287 \times 298 \times 1000}{1.2 \times 10^5} = 0.713 \text{ m}^3/\text{kg}$$

Hence, $$\text{density}, \rho = \frac{\text{Mass}}{\text{Volume}} = \frac{1}{0.713} = 1.403 \text{ kg/m}^3 \textbf{ Ans.}$$

ASSIGNMENT

Objective Type Questions :

(A) Fill in the Blanks :

1. The gas laws are applicable to all __________ gases.
2. Boyle's law is applicable, when __________ is kept constant.
3. Charle's law is applicable, when __________ is kept constant.
4. In the characteristic gas equation, $pV = mRT$, the units of 'p' are __________ in S.I. system of units.
5. The value of characteristic gas constant, R for air in S.I. system of units is __________ J/kg k.
6. The gas equation is applicable to all __________ gases.
7. A prescribed region in space for its thermodynamic study is known as __________.
8. Properties of a system are __________.
9. Intensive properties like __________ and __________ remain the same, whether one describes the whole system or part of it.
10. The properties of a thermodynamic system, which are dependent on its __________ are known as extensive properties.
11. The boundaries of a closed system are not necessarily be __________ but they may be __________.
12. The system and its surroundings taken together consititute an __________ system.
13. Everything external to the system is known as __________.
14. In a __________ system the mass does not pass through the boundaries.
15. A heat pump delivers heat from __________ temperature to a __________ temperature.
16. A heat engine converts __________ energy into __________ energy.
17. Unit of heat in S.I. system of units is __________.
18. In a constant volume process, the specific heat at __________ is to be used, for calculating heat transfer.
19. The value of 'C_p' is always __________ than 'C_v'.
20. A tank containing 5 kg of certain gas is supplied 240 KJ of heat and that causes increase in temperature, from 15° to 60 °C. The sp. heat of the gas would be __________.

(B) State True or False :

1. The vapours do not obey gas laws.
2. Gas laws cannot be applied to perfect gases.
3. Boyle's and Charle's laws are applied, when mass is considered constant.
4. Boyle's law is applicable when temperature is kept constant.
5. Charle's law is applicable when pressure is kept constant.

6. Charle's law may also be expressed as, At constant pressure, any perfect gas changes in volume by $\frac{1}{273}$th of its original volume at 0 °C for every one degree centigrade change in its temperature.
7. The value of gas constant, 'R' is same for different gases.
8. The value of universal gas constant 'R_u' is same for all the gases.
9. The value of gas constant 'R' for the atmospheric air in S.I. system of units is 287 J/kg k.
10. The value of universal gas constant, 'R_u' in S.I. system of units is 8314 J/kg mol k.
11. The properties of a system which depend on mass of the system are known as intensive properties.
12. Mass transfer may take place in an open system.
13. In a closed system neither the mass nor the energy transfer takes place across the system boundaries.
14. There is a temperature change in the surroundings in a thermodynamic system.
15. Heat and work are not interchangeable.
16. In a constant volume process, the specific heat at constant volume is to be used for calculating heat transfer.
17. The ratio of specific heats is constant for all gases.
18. The heat units cannot be converted into work units.
19. The value of 'C_p' for a gas is always less than 'C_v'.
20. The specific heat of a gas changes with the change in its temperature and pressure.

(C) Multiple Choice Questions :

1. Choose the false statement in the context of 'macroscopic' approach followed in classical thermodynamics.
   (*a*) considers gross behaviour of matter
   (*b*) requires no hypothesis on the detailed structure of matter
   (*c*) predicts the behaviour of individual fluid particles by statistical averaging
   (*d*) deals with characteristics which can be perceived by senses and measured directly.
2. A closed thermodynamic system manifests when;
   (*a*) matter is not allowed to cross the boundary, but energy transfer does occur across the boundary
   (*b*) there is transfer of both mass and energy across the system boundaries
   (*c*) there is only transfer of mass, but no heat and work energy are transferred.
   (*d*) there is absolutely no interaction of the system with surroundings across its boundaries.
3. In an open thermodynamic system;
   (*a*) mass content of the system under consideration remains same
   (*b*) transfer of mass and/or energy takes place
   (*c*) there is only mass transfer even though there may not be any exchange of energy with the system environment.
   (*d*) the system exchanges energy with the surroundings in the form of heat energy only.

4. A thermodynamic system is referred to be an isolated system when there is transfer of ________ across the system boundaries
   (*a*) only mass (*b*) only energy
   (*c*) both mass and energy (*d*) neither mass nor energy.
5. A system and its environment put together constitute;
   (*a*) an adiabatic system (*b*) an isolated system
   (*c*) a segregated system (*d*) a homogeneous system.
6. Which one of the following represents a closed system ?
   (*a*) bomb clorimeter (*b*) steam generator
   (*c*) universe (*d*) exhaust smoke of an I.C. engine.
7. Identify the hetrogeneous system;
   (*a*) atmospheric air
   (*b*) mixture of hydrogen and oxygen
   (*c*) cooling fluid in a radiator
   (*d*) mixture of ice, water and steam.
8. Which aspects is not true regarding thermodynamic property of a system ?
   (*a*) a characteristic that describes the condition of the system
   (*b*) a variable whose value depends upon the particular path in going from one state to another.
   (*c*) an intensive property is independent of the extent or mass of the system
   (*d*) an extensive property can be made intensive, if it is divided by mass and is called specific property.
9. Choose the specific property of a thermodynamic system
   (*a*) viscosity (*b*) density
   (*c*) pressure (*d*) temperature.
10. Identify the false statement in the context of heat and work;
    (*a*) they exist only during interaction between the system and the boundary
    (*b*) are path functions and exact differential
    (*c*) may or may not occur simultaneously
    (*d*) occur due to some unbalanced potential which tends to promote a change of state.
11. For a system to be in thermodynamic equilibrium the system and its surroundings are to be in;
    (*a*) thermal equilibrium
    (*b*) chemical equilibrium
    (*c*) mechanical equilibrium
    (*d*) thermal, chemical and mechanical equilibrium.
12. The 'Absolute Zero' temperature corresponds to the condition when;
    (*a*) all the substances exist only as solids
    (*b*) volume of gas reduces to zero
    (*c*) kinetic energy of gas molecules becomes zero
    (*d*) no pressure is exerted by the gas.
13. A 'Quasi-static' process is one in which all the states through which a system passes are very close to
    (*a*) equilibrium state (*b*) original state
    (*c*) same temperature (*d*) each other.

14. Which of the following is a non-quasi-static process ?
    (*a*) gradual compression of gas inside a piston-cylinder arrangement
    (*b*) expansion of gas in a cylinder under constant pressure
    (*c*) free or unrestricted expansion of gas
    (*d*) rapid leakage of air from a bicycle tyre.
15. Boyle's law states that
    (*a*) at constant volume, the pressure of a gas varies directly with temperature
    (*b*) at constant pressure, the volume of a gas varies directly with temperature
    (*c*) under constant temperature conditions the product of pressure and volume of a given mass of gas is always constant
    (*d*) the volume of a given mass of gas held at constant temperature is directly proportional to its pressure.
16. It is decided to verify Boyle's law over a wide range of pressure and temperature. The most suitable gas to be selected for this purpose is
    (*a*) oxygen (*b*) hydrogen
    (*c*) helium (*d*) air.
17. For all perfect gases, the change in volume is $\frac{1}{273}$ of original volume at 0 °C for every 1 °C change in temperature while pressure remains constant. This statement is referred to as
    (*a*) Boyle's law (*b*) Charle's law
    (*c*) Gay-Lussac's law (*d*) Joule's law.
18. In the characteristic gas equation $pV = mRT$ the gas constant R for air has the value
    (*a*) 29.27 mkgf/kg/k (*b*) 287 mkgf/kg/k
    (*c*) 427 mkgf/kg/k (*d*) 848 mkgf/kg/k.
19. Which one of the following identity is correct ?
    (*a*) $G = MR$ (*b*) $G = R/M$
    (*c*) $G = M/R$ (*d*) $G = M^2R$.
    where 'G' is universal gas constant 'R' is characteristic gas constant and 'M' is the molecular weight.
20. The value of universal gas constant is
    (*a*) 427 mkgf/kg mole/k (*b*) 735 mkgf/kg mole/k
    (*c*) 848 mkgf/kg mol/k (*d*) 8314 mkgf/kg mol/k.

ANSWERS

(A) Fill in the Blanks :

1. perfect **2.** temperature **3.** pressure **4.** N/m^2 or Pascal
5. (287) **6.** perfect **7.** thermodynamic system
8. intensive and extensive **9.** temperature, pressure
10. mass **11.** fixed, movable **12.** open
13. surrounding **14.** closed **15.** low, high
16. heat, mechanical **17.** joule **18.** constant volume
19. greater **20.** (1.07).

(B) State True or False :

1. True **2.** False **3.** True **4.** True **5.** True
6. True **7.** False **8.** True **9.** True **10.** True

11. False	**12.** True	**13.** False	**14.** False	**15.** False
16. True	**17.** False	**18.** False	**19.** False	**20.** True.

(C) Multiple Choice Questions :

1. (*c*)	**2.** (*a*)	**3.** (*c*)	**4.** (*d*)	**5.** (*b*)
6. (*a*)	**7.** (*d*)	**8.** (*b*)	**9.** (*b*)	**10.** (*b*)
11. (*d*)	**12.** (*c*)	**13.** (*d*)	**14.** (*c*)	**15.** (*c*)
16. (*c*)	**17.** (*b*)	**18.** (*a*)	**19.** (*a*)	**20.** (*c*)

(D) Short Answer Type Questions :

Q 1. What is a 'Perfect Gas' ?

Ans. A 'perfect gas' is one, which strictly obeys all the gas laws *i.e.*, Boyle's law, Charle's law, Joule's law, Avogadro's law and the Gay-Lussac's law.

Q 2. Define Boyle's law.

Ans. The Boyle's law states that, "The volume of a given mass of a perfect gas, varies inversely as its absolute pressure, provided the temperature is kept constant".

So, pV = Constant (When T = Constant)

Q 3. Define Charle's law.

Ans. The Charle's law states that, "The volume of a given mass of a perfect gas varies directly as the absolute temperature, provided the pressure is kept constant".

So, as per this law, $\frac{V}{T}$ = Constant (When, p = Constant)

Q. 4. What is the 'Equation of a perfect gas' or the 'characteristic gas equation' ? Also, give brief idea about the 'characteristic gas constant'.

Ans. The equation which gives the relation between, absolute pressure volume mass and absolute temperature of a perfect gas under the stated conditions, is called the 'Equation of a perfect gas' or the 'Characteristic Gas Equation'.

It may be expressed as :

$$pV = m.R.T.$$

where p = Absolute pressure of the given perfect gas in N/m^2

V = Volume of the gas in m^3

m = Mass of the gas in kg

R = Characteristic gas constant in N-m/kg K

T = Absolute temperature of the gas in k.

The constant 'R' used in the characteristic gas equation $pV = mRT$ has a constant value for a particular gas and is called 'Characteristic gas constant' or simply 'Gas constant'. Its value depends upon the temperature scale used and the properties of the gas, under consideration.

Q 5. What is the 'Universal Gas Constant' ?

Ans. If the molecular weight of any gas is multiplied by its characteristic gas constant 'R', it is found that this product is always 'constant' for all the gases. This 'constant' is known as 'Universal Gas Constant' generally denoted by 'R_u'.

The value of 'R_u' = 8314 J/kg mol/k (in S.I. system of units)

and 'Ru' = 848 mkgf/kg mole/k (in M.K.S. system of units)

The equation using 'Ru', *i.e.*, $pV = m.Ru.\ T$ is known as the 'Universal Gas Equation'.

Q 6. Define the Joule's Law.

Ans. Joule's Law states that, "The internal energy of a given quantity of a perfect gas depends only on the temperature".

Q 7. Define the Regnault's Law.

Ans. Regnault's law states that, "The two specific heats, *i.e.*, sp. heat at constant pressure, 'C_p' and sp. heat at constant volume, 'C_v' of a perfect gas do not change with the change of its temperature and pressure". In other words, 'C_p' and 'C_v' of a perfect gas always remain constant.

Q 8. Define the Avogadro's law.

Ans. According to the Avogadro's law, "Equal volume of different perfect gases at the same temperature and pressure contain equal number of molecules".

Q 9. What is the 'Molar Heat' ?

Ans. The 'Molar Heat' may be defined as, "The quantity of heat required to raise the temperature of one mole of a substance through one degree".

Q 10. What is 'specific heat of a gas' ? What are its types as applied to gases ?

Ans. The 'specific heat of a gas' may be defined as the amount of heat, required to raise the temperature of its unit mass through a unit degree. Gases have two specific heats as stated below :

(*i*) **Specific heat at constant pressure (C_p) :** It is the amount of heat required to raise the temperature of a unit mass of a gas through unit degree, when its pressure is kept constant. The value of C_p for air in S.I. system of units is 1.0 KJ/kg k and in M.K.S. system 0.24 Kcal/kg °C.

(*ii*) **Specific heat at constant volume (C_v) :** It is the amount of heat required to raise the temperature of a unit mass of a gas through unit degree when its volume is kept constant. The value of 'C_v' for air in S.I. system of units is 0.7128 KJ/kg K and in M.K.S. System 0.17 Kcal/kg °C.

Q 11. What is 'Joule's Mechanical equivalent of heat' J ?

Ans. If 'W' is the amount of mechanical work converted from heat energy 'H' then; $W \propto H$

or $$\frac{W}{H} = J, \text{ the mechanical equivalent of heat.}$$

In the above equation the ratio of mechanical work 'W' to the heat energy 'H' is a constant 'J' called the 'Mechanical equivalent of heat.

The value of 'J' = 427 kgf m/kcal (In M.K.S. system)
= 4.1868 KJ/kcal (In S.I. system)

Q 12. Define 'Heat' as applied to thermodynamics.

Ans. 'Heat' is defined as the energy transfered without the transfer of mass, across the boundary of a system due to a difference of temperature between the system and its surroundings.

Q 13. What are the two main characteristics of 'Heat' ?

Ans. Main characteristics of 'Heat' are as below :

(*i*) Heat is a transient quantity, which can be identified only as it crosses the boundary of the system.

(*ii*) Heat is not a property of a system.

Q 14. What are the units of heat in S.I. system of units ?

Ans. In S.I. system of units, the units of the heat are 'Joule (J) or 'Kilo Joule (KJ).

Q 15. Define the term 'work' as applied to thermodynamics.

Ans. 'Work' is defined as the energy transferred without the transfer of mass, across the boundary of the system due to an intensive property difference, other than temperature, existing between the system and the surroundings.

Q 16. What are units of work in S.I. system of units ? Are they differ from the units of heat in the same system ?

Ans. In S.I. system of units the units of work are Newton-metre (N-m) or Joule (J). No, the units of heat and work in S.I. system of units are the same, *i.e.*, Joule (J) or Kilo Joule (KJ).

Q 17. In a 'Constant Pressure Process', write the formula to find out the 'Amount of Heat' transferred.

Ans. In a constant pressure process;

Heat transfered, $Q = m \times C_p \times (T_2 - T_1)$

where m = Mass

C_p = Sp. heat at constant pressure

$(T_2 - T_1)$ = Change in temperature.

Q 18. Write the formula for finding the heat transferred in a 'Constant Volume Process'.

Ans. In a 'Constant Volume Process' ;

The heat transferred, $Q = m \times C_v \times (T_2 - T_1)$

where m = Mass

C_v = Sp. heat at constant volume

$(T_2 - T_1)$ = Change in temperature.

Q 19. What is the ratio of two sp. heats called ? What is the value for air ?

Ans. The ratio of two specific heats *i.e.*, sp. heat at constant pressure 'C_p' and sp. heat at constant volume (C_v) is called Adiabatic index ('γ').

So,
$$\gamma = \frac{C_p}{C_v}$$

The value of 'γ' for air is 1.4.

Q 20. What are the units of 'γ' in S.I. system ?

Ans. As 'γ' is the ratio of two sp. heats *i.e.*, sp. heat at constant pressure and sp. heat at constant volume. So it has no units in any system of units.

REVIEW QUESTIONS

1. (*a*) Differentiate between;
   (*i*) Vapour and a perfect gas.
   (*ii*) Open system and closed system. **(SBTE Pb. Dec. 1999)**
   (*b*) Explain Boyle's law and Charle's law for gases.
2. Derive the characteristic gas equation for gases.
3. Differentiate between :
   (*i*) Heat and Work.
   (*ii*) Specific gas constant and Universal Gas Constant.
   (SBTE Haryana, Nov. 1995)
4. Define the following;
   (*i*) Thermodynamics (*ii*) System
   (*iii*) Boundary (*iv*) Surroundings.
   (SBTE Haryana, May 1978)
5. Discuss the various properties of a thermodynamic system.
6. There are two specific heats for gases, why ? Explain.
7. What are the different types of systems ?
8. Derive an expression for the relationship of two specific heats of a perfect gas and the characteristic gas constant.

PROBLEMS FOR PRACTICE

1. Air in a vessel having a volume of 5 m^3 at 3.0 bar is compressed to 3 m^3 without any change of temperature. Calculate the pressure of air after compression. **[Ans.** 5 bar]

2. Determine the volume occupied by a given mass of air at a temperature of 245 °C, if the same mass of air occupies 2 m^3 at 23 °C. The pressure of the air remains unchanged. **[Ans.** 3.5 m^3]

3. Find the mass of a gas occupying 5.5 m^3 at 7 bar absolute and at 200 °C. Take gas constant, R = 287 J/kg k. **[Ans.** 28.36 kg]

4. A certain gas occupies 2 m^3, at a temperature of 127 °C. The pressure of the gas is 7 bar. The gas expands in such a manner that the volume becomes 8 m^3 and the temperature 15 °C. What will be the pressure of gas after expansion ? **[Ans.** 1.26 bar]

5. A steel cylinder of 2 m^3 capacity stores inert gas, N_2 at a pressure of 50 bar and temperature of 30 °C. Calculate the
   (*i*) mass of the gas
   (*ii*) density of the gas. **[Ans.** 111.15 kg, 55.57 kg/m^3]

6. A quantity of air at an initial temperature of 175 °C occupies 0.08 m^3. Then, it is expanded at constant pressure to a final volume of 0.25 m^3. Find;
   (*i*) temperature at the end of expansion
   (*ii*) the value of C_p,
   Take, R = 287 J/kg k, C_v = 0.7128 KJ/kg k.
   [Ans. 1127 °C, 0.9998 KJ/kg k]

7. One kg of air at N.T.P. occupies a volumes 768 litres. What will be the volume and density of this air at 20 °C and at an absolute pressure of 740 mm of Hg ? **[Ans.** 0.8465 m^3, 1.18 kg/m^3]

8. Determine the gas constant for hydrogen if the sp. volume of H_2 at N.T.P. is 11 m^3/kg. Also, find the sp. heats of the above gas if the value of 'γ' is 1.407.
   [Ans. 4.08 KJ/kg k, 10.03 KJ/kg k, 14.1 KJ/kg k]

9. Density of H_2 gas is 0.09 kg/m^3 at N.T.P. and the value of 'γ' is 1.4. Find the characteristic gas constant for H_2 gas and its both sp. heats.
   [Ans. 4.12 KJ/kg k, 10.31 KJ/kg k, 14.4 KJ/kg k]

10. 0.15 m^3 of a gas at a pressure of 600 KPa expands to a pressure of 200 KPa, the final volume being 0.2 m^3. Calculate the change of internal energy of the gas. Take C_p = 1 KJ/kg k, C_v = 0.714 KJ/kg k. **[Ans.** –124.8 KJ]

11. A gas contained in a closed vessel, is having initial pressure and temperature as 20 bar and 50 °C respectively. If it is heated to 120 °C, then find the final pressure of this gas ? **[Ans.** 24.33 bar]

12. One kg of air at normal temperature and pressure occupies a volume of 800 litres. If the temperature and pressure of this air are increased to 32 °C and 730 mm of Hg respectively then what will be the final volume of the air ?
    [Ans. 0.93 m^3]

❑❑❑

3
CHAPTER

Thermodynamic Processes on Gases

INSIDE THIS CHAPTER

3.1. INTRODUCTION

The subject of thermodynamics is based upon the experimental observations and laws like all other branches of science. The dealing of problems in thermodynamics is done wholly on the basis of these laws. The results of this study are generally highly accurate. Therefore, the science of thermodynamics has got an important place in 'Engineering'.

The performance of the various thermodynamic processes is measured with the help of various thermodynamic laws. In this chapter workdone, change in internal energy, change in entropy, rate of heat-transfer for various thermodynamic processes is being dealt with.

3.2. VARIOUS THERMODYNAMIC PROCESSES

3.2.1. Constant Volume or Isochoric Process

The process in which the volume of the given mass of the gas is not allowed to change, is known as 'Constant Volume Process'.

Let us consider a given mass 'm' of a gas be heated in a container of constant volume. As, there is no change in volume, so the process of heating can be represented by a vertical line on the pressure-volume (P-V) diagram, as shown in the Fig. 3.1.

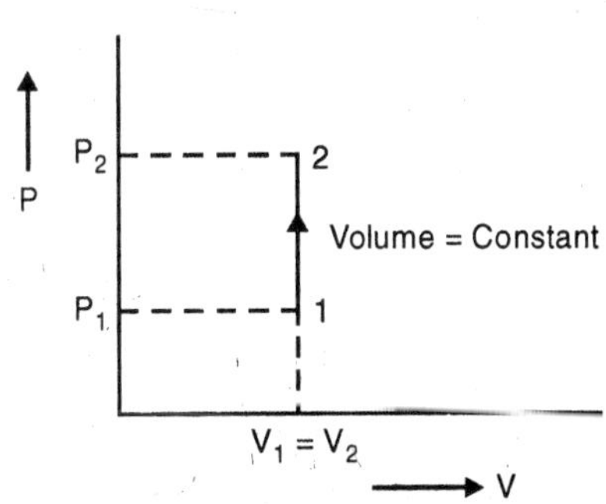

Fig. 3.1.

As, the work done, during the process 1–2,

$$W_{12} = \int_1^2 p\,dv$$

or $$W_{12} = p(V_2 - V_1)$$

As, the change in volume,

$$dv = 0 \text{ or } V_2 - V_1 = 0$$

$\therefore$ Work done, $$W_{12} = 0 \quad \text{...(3.1)}$$

Now, applying the first law of thermodynamics,

$$Q_{12} - W_{12} = \Delta U_{12}$$

or $$Q_{12} = \Delta U_{12} \qquad (\because W_{12} = 0) \quad \text{...(3.2)}$$

Which means that in constant volume process;

Heat transferred = Change in internal energy

Also, for a perfect gas, we know that

Heat transferred, Q_{12} = Mass × Sp. Heat × Rise in temperature

$\therefore$ $$Q_{12} = m.C_v.(T_2 - T_1)$$

$\therefore$ Change in internal energy,

$$\Delta U_{12} = m.C_v.(T_2 - T_1) \quad \text{...(3.3)}$$

3.2.2. Constant-Pressure or Isobaric Process

The process in which the pressure of the given mass of a gas is not allowed to change is known as 'Constant pressure process'.

Consider a given mass 'm' of a gas be heated at constant pressure 'p', as shown in Fig. 3.2.

The work done during the process 1-2,

$$W_{12} = \int_1^2 p.dv$$

or $$W_{12} = p(V_2 - V_1)$$

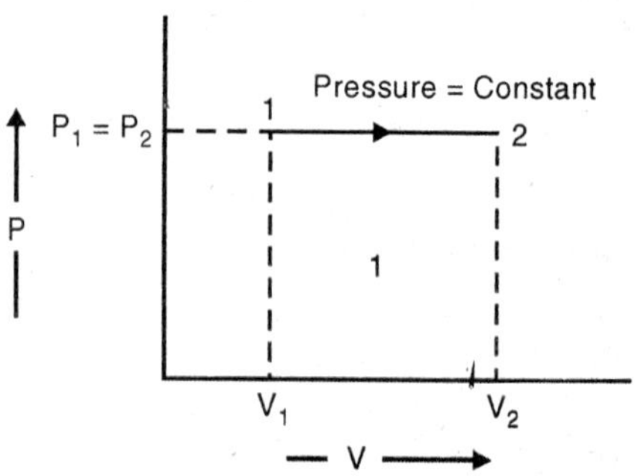

Fig. 3.2.

But, from Gas Equation :

$$pV_1 = mRT_1$$

and $$pV_2 = mRT_2$$

or $$p(V_2 - V_1) = mR(T_2 - T_1)$$

$\therefore$ Work done, $$W_{12} = m.R(T_2 - T_1) \quad \text{...(3.4)}$$

Now, applying first law of thermodynamics to this process,

$$Q_{12} - W_{12} = \Delta U_{12}$$

or Heat transferred, $$Q_{12} = \Delta U_{12} + W_{12}$$

or $$Q_{12} = (U_2 - U_1) + p(V_2 - V_1)$$

or $$Q_{12} = (U_2 + pV_2) - (U_1 + pV_1)$$

But, $$U + pV = \text{Enthalpy}, h$$

$\therefore$ $$Q_{12} = h_2 - h_1$$

or $$Q_{12} = \Delta h_{12} \quad \text{...(3.5)}$$

Which means, that in Constant-Pressure process :

Heat Transferred = Change in Enthalpy

Also, for a perfect gas, we know that :

Heat transferred, Q_{12} = Mass × Specific Heat at Constant Pressure × Rise in temp.

or $$Q_{12} = m \cdot C_p \cdot (T_2 - T_1)$$

∴ Change in internal energy,

$$U_{12} = m.C_V.(T_2 - T_1) \quad ...(3.6)$$

3.2.3. Constant-Temperature or Isothermal Process

The process in which, the temperature of the given mass of a gas is not allowed to change, is known as 'constant-temperature process' or 'isothermal process'.

In this process, because of constant temperature, the change in thermal energy, which is the function of temperature only remains zero.

∴ Change in internal energy, $\Delta U = 0$ ...(3.7)

Now, consider a given mass 'm' of a gas be heated from an initial state '1' to state '2' at constant temperature as shown in Fig. 3.3.

The work done during the process,

$$W_{12} = \int_1^2 p_1\, dv.$$

Because, in this process the temperature is constant, so applying Boyle's law,

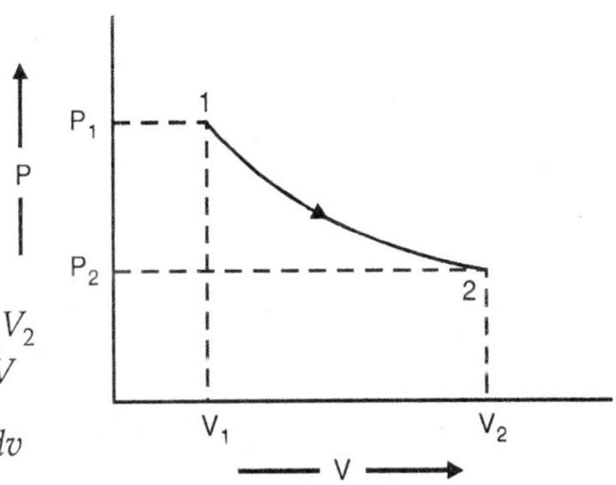

Fig. 3.3.

∴ $$p_1V_1 = p_2V_2 = pV = \text{Constant}$$

∴ $$p = \frac{p_1V_1}{V} = \frac{p_2V_2}{V}$$

and Work done, $$W_{12} = \int_1^2 \frac{p_1V_1}{V}.dv$$

or $$W_{12} = p_1V_1 \int_1^2 \frac{dv}{V}$$

or $$W_{12} = p_1V_1 \log_e \frac{V_2}{V_1}$$

or $$W_{12} = p_1V_1 \log_e r \quad \left(\because \frac{V_2}{V_1} = r\right) \quad ...(3.8)$$

or Work done, $$W_{12} = mRT \log_e r$$

Also, as per the first law of thermodynamics,

$$Q_{12} - W_{12} = \Delta U_{12}$$

But, change in internal energy, $\Delta U_{12} = 0$

∴ $$Q_{12} = W_{12}$$

or Heat transferred, $$Q_{12} = mRT \log_e r \quad ...(3.9)$$

3.2.3.1. Hyperbolic Process

The process in which the product of the pressure of the gas and its volume remains constant is known as Hyperbolic Process. In the above case of isothermal process, it is clear that pV = Constant.

So, the work done and heat transferred during hyperbolic process can be calculated by using the formulae for the above isothermal process.

3.2.4. Adiabatic Process

The process in which, no heat is added or withdrawn from the system under consideration is known as an 'Adiabatic Process'. The process is shown in the Fig. 3.4 by the line 1-2.

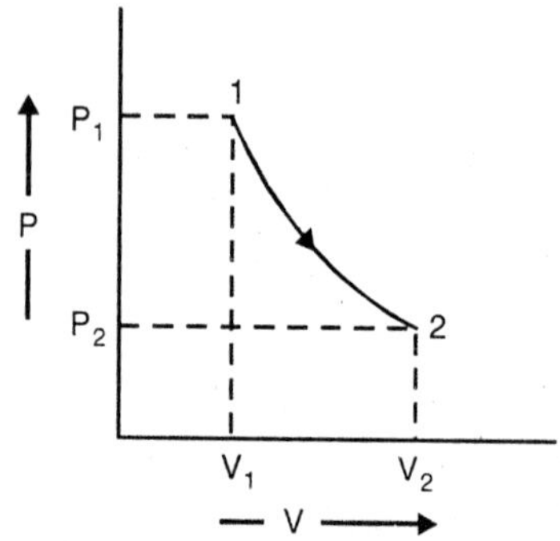

Fig. 3.4.

During this process there is no heat transfer across the boundary of the system.

Therefore, Heat transferred, during the process 1-2

$$Q_{12} = 0 \quad ...(3.10)$$

Also, as per first law of thermodynamics,

$$Q_{12} - W_{12} = \Delta U_{12}$$

But, as $Q_{12} = 0$

$$\therefore \quad -W_{12} = \Delta U_{12} \quad ...(3.11)$$

which means, work done, W_{12} = Change in internal energy.

Also, for any perfect gas, we know that;

$$\Delta U_{12} = \text{Mass} \times \text{Sp. heat} \times \text{Change in temperature.}$$

or

$$\Delta U_{12} = m \cdot C_V \cdot (T_2 - T_1) \quad ...(3.12)$$

The work done during the process, $W_{12} = \int_1^2 pdv$

$$\therefore \quad \int_1^2 pdv = m.C_V.(T_2 - T_1)$$

or

$$p\,dv = -m.C_V.(dT). \quad (\because dT = T_2 - T_1)$$

But,

$$pV = mRT \quad \text{(Gas Equation)}$$

$$\therefore \quad p = \frac{m.R.T}{V}$$

$$\therefore \quad \frac{m.R.T}{V}.dv = -m.C_V.dT$$

or

$$\frac{dT}{T} = -\frac{R}{C_v} \cdot \frac{dv}{V}$$

Also, as

$$\frac{R}{C_v} = \frac{C_P - C_v}{C_v} \quad (\because C_P - C_v = R)$$

or

$$= \frac{C_P}{C_v} - 1 \quad \text{or} \quad \gamma - 1 \quad \left(\because \frac{C_P}{C_v} = \gamma\right)$$

$$\therefore \quad \frac{dT}{T} = -(\gamma - 1) \cdot \frac{dv}{V}$$

Integrating both sides between the limits, we get

$$\log_e \frac{T_2}{T_1} = -(\gamma - 1) \log_e \frac{V_2}{V_1}$$

or

$$\frac{T_2}{T_1} = \left(\frac{V_2}{V_1}\right)^{-\gamma - 1}$$

or $$\frac{T_2}{T_1} = \left(\frac{V_1}{V_2}\right)^{\gamma - 1}$$

or $$TV^{\gamma - 1} = \text{Constant}$$

But, $$\frac{V_1}{V_2} = \frac{p_2 . T_1}{p_1 . T_2} \qquad \left(\because \frac{p_1 V_1}{T_1} = \frac{p_2 V_2}{T_2}, \text{Gas} = n\right)$$

$$\therefore \quad \left(\frac{T_2}{T_1}\right)^{\frac{1}{\gamma - 1}} = \frac{p_2 . T_1}{p_1 . T_2}$$

or $$\frac{T_2}{T_1} = \left(\frac{p_2}{p_1}\right)^{\frac{\gamma - 1}{\gamma}}$$

or $$Tp^{\frac{1-\gamma}{\gamma}} = \text{Constant}$$

From above equations of $\frac{T_2}{T_1}$

$$\left(\frac{p_2}{p_1}\right)^{\frac{\gamma - 1}{\gamma}} = \left(\frac{V_1}{V_2}\right)^{\gamma - 1}$$

or $$\left(\frac{p_2}{p_1}\right) = \left(\frac{V_1}{V_2}\right)^{\gamma}$$

$$pV^{\gamma} = \text{Constant}$$

Now, work done during the process 1-2,

$$W_{12} = \int_1^2 p.dv$$

Also, $$pV^{\gamma} = p_1 V_1^{\gamma} = p_2 V_2^{\gamma}$$

$$\therefore \quad p = \left(\frac{p_1 V_1^{\gamma}}{V^{\gamma}}\right)$$

$$\therefore \quad W_{12} = \int_1^2 p_1 \left(\frac{V_1}{V}\right)^{\gamma} . dv = p_1 V_1^{\gamma} \int_1^2 V^{-\gamma} . dv$$

or $$W_{12} = . p_1 V_1^{\gamma} \left[\frac{V^{-\gamma + 1}}{-\gamma + 1}\right]^2 = \frac{p_1 V_1^{\gamma}}{1 - \gamma} \left[V_2^{1-\gamma} - V_1^{1-\gamma}\right]$$

$$= \frac{p_1 V_1^{\gamma} . V_2^{1-\gamma} - p_1 . V_1^{\gamma} . V_1^{1-\gamma}}{1 - \gamma}$$

$$= \frac{p_2 V_2 - p_1 V_1}{1 - \gamma} \qquad (\because p_1 V_1^{\gamma} = p_2 V_2^{\gamma})$$

or $$W_{12} = \frac{p_1 V_1 - p_2 V_2}{\gamma - 1} \qquad \text{...(3.13)}$$

or $$W_{12} = \frac{m . R(T_1 - T_2)}{\gamma - 1}$$

3.2.4.1. Isentropic Process

The process, such as reversible adiabatic process, in which the entropy of the system remains constant is known as an 'isentropic process'.

Here are some of the conditions which must be fulfilled to make the adiabatic process reversible *i.e.* isentropic.

(*i*) The process should be frictionless.

(*ii*) No heat should be transferred to or from the system.

(*iii*) Work should be done by the gas or on the gas.

Here, it is important to note that, *an isentropic process is not possible in practice, because the friction cannot be eliminated completely from the system. Moreover, some heat is transferred between the gas and the cylinder walls, so the heat transfer from the system or to the system is also unavoidable.*

3.2.5. Polytropic Process

The process in which the relation pV^n = Constant holds good, is known as 'Polytropic Process', where 'p' is the pressure, 'V' the volume of the gas and 'n' is the polytropic exponent, the value of which varies from zero to infinity for various gases, as shown in the Fig. 3.5.

For constant volume process, $n = \infty$

For constant pressure process, $n = 0$

For an isothermal or constant temperature process

$$n = 1$$

Also, from Ist law of thermodynamics, we have

$$Q_{12} - W_{12} = \Delta U_{12}$$

But work done during process 1-2

$$W_{12} = \int_1^2 p\,dv$$

For polytropic process,

$$pV^n = p_1 V_1^n$$

or
$$p = \frac{p_1 V_1^n}{V^n}$$

$\therefore$ Work done during the process 1-2,

$$W_{12} = p_1 V_1^n \int_1^2 \frac{dv}{V^n}$$

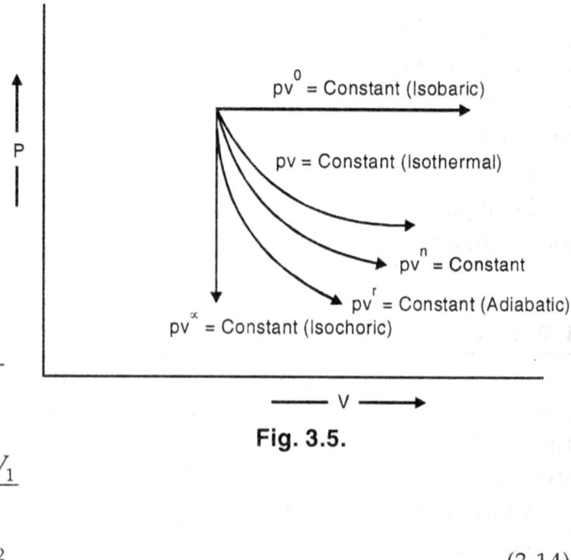

Fig. 3.5.

Which gives,

$$W_{12} = \frac{p_2 V_2 - p_1 V_1}{1-n}$$

or
$$W_{12} = \frac{p_1 V_1 - p_2 V_2}{n-1} \qquad ...(3.14)$$

Also, $$W_{12} = \frac{m.R(T_1 - T_2)}{n-1}$$

Change in internal energy, during process 1-2.

$$\Delta U_{12} = \text{Mass} \times \text{Sp. heat} \times \text{Change in temperature}$$

or $$\Delta U_{12} = m.C_v.(T_2 - T_1) \quad ...(3.15)$$

$$\therefore \quad Q_{12} - \frac{mR(T_2 - T_1)}{(1-n)} = m.C_v.(T_2 - T_1)$$

or Heat transferred,

$$\therefore \quad Q_{12} = m.C_v.(T_2 - T_1) + \frac{mR(T_2 - T_1)}{(1-n)}$$

or $$Q_{12} = m.\frac{R}{(\gamma - 1)}.(T_2 - T_1) + \frac{mR(T_2 - T_1)}{(1-n)} \qquad \left(\because \frac{R}{C_v} = \gamma - 1\right)$$

or $$= \frac{m.R(T_2 - T_1)}{(\gamma - 1)}\left(1 + \frac{\gamma - 1}{1 - n}\right) = \frac{m.R(T_2 - T_1)}{(\gamma - 1)}\left(\frac{1 - n + \gamma - 1}{1 - n}\right)$$

$$= \frac{(\gamma - n)}{(1-n)(\gamma - 1)}(m.R(T_2 - T_1) = \frac{(\gamma - n)}{\gamma - 1}.\frac{m.R(T_2 - T_1)}{(1-n)}$$

or $$Q_{12} = \frac{\gamma - n}{\gamma - 1}.W_{12} \quad ...(3.16)$$

3.2.6. Throttling Process

When a gas or vapour is expanded through an aperture of minute (narrow) dimensions like a valve, which is very slightly opened, the process is known as 'throttling process'.

The fluid is forced through the narrow aperture under pressure, but the velocity of the fluid flowing out is reduced up to a negligible amount, by the high resistance offered by the friction, between the fluid and the walls of the aperture. Hence, the kinetic energy of the flowing fluid is very small.

In fact the kinetic energy of the fluid is converted into heat energy, which warms up the fluid to its initial temperature. Thus, no heat is supplied from or rejected to an external source and there is no external work done. Also, there is no change in temperature in the case of a perfect gas. So, in the throttling process;

Work done, $W = 0$ ...(3.17)

Heat supplied, $Q = 0$ ...(3.18)

and change in internal energy,

$$\Delta U = 0 \quad ...(3.19)$$

3.2.7. Free Expansion

In this process the fluid expands suddenly into a vacuum chamber, through an orifice of large dimension. No heat is supplied or rejected, there is no work done and during the process the temperature remains constant. So, in free expansion process, also;

Work done, $W = 0$ ...(3.20)

Heat supplied, $Q = 0$ ...(3.21)

and Change in internal energy,

$$\Delta U = 0 \quad ...(3.22)$$

3.2.8. Summary of Various Methods of Heating and Expanding Gases and Vapours

Table 3.1.

S. No.	*Process*	*Workdone, W (N-m) or J*	*Change in Internal Energy, ΔU(J)*	*Heat transferred, Q(J)*	*Relation between p, V and T*
1.	Constant Volume *i.e.* ; (V = Constant)	Zero	$m.C_v.(T_2 - T_1)$	$m.C_v.(T_2 - T_1)$	$\frac{p_1}{T_1} = \frac{p_2}{T_2}$
2.	Constant Pressure *i.e.* ; p = Constant	$p(V_2 - V_1)$	$m.C_v.(T_2 - T_1)$	$m.C_p.(T_2 - T_1)$	$\frac{V_1}{T_1} = \frac{V_2}{T_2}$
3.	Constant Temp. or isothermal *i.e.*, PV = Constant	$p_1 V_1 \log_e r$	Zero	$p_1 V_1 \log_e r$	$p_1V_1 = p_2V_2$
4.	Adiabatic pV^γ = Constant	$\frac{p_1V_1 - p_2V_2}{\gamma - 1}$	$m.C_v.(T_2 - T_1)$	Zero	$p_1V_1^\gamma = p_2V_2^\gamma$ $T_1V_1^{\gamma-1} = T_2V_2^{\gamma-1}$ $T_1p_1^{\frac{1-\gamma}{\gamma}} = T_2p_2^{\frac{1-\gamma}{\gamma}}$
5.	Polytropic pV^n = Constant	$\frac{p_1V_1 - p_2V_2}{n - 1}$	$m.C_v.(T_2 - T_1)$	$\frac{\gamma - n}{\gamma - 1} \cdot \frac{mR(T_2 - T_1)}{(1 - n)}$	$p_1V_1^n = p_2V_2^n$ $T_1V_1^{n-1} = T_2V_2^{n-1}$ $T_1p_1^{\frac{1-n}{n}} = T_2p_2^{\frac{1-n}{n}}$
6.	Throttling	Zero	Zero	Zero	—
7.	Free Expansion	Zero	Zero	Zero	—

3.3. GENERAL EXPRESSION FOR THE CHANGE OF ENTROPY OF A PERFECT GAS

Let, p_1, v_1, T_1 and S_1 apply to the initial conditions of a certain amount of a gas. And p_2, v_2, T_2 and S_2 apply to the final conditions of the gas after adding some heat to it. Now, as per the law of conservation of energy.

$$dQ = dW + dU$$

But $$dW = p \,.\, dv$$

Thus for a small change in state of one kg of a perfect gas.

$$dQ = p \,.\, dv + dU$$

Since $$dU = C_v \,.\, dT$$

$\therefore$ $$dQ = p \,.\, dv + C_v \,.\, dT$$

Dividing both sides by T.

$\therefore$ $$\frac{dQ}{T} = \frac{p.dv}{T} + C_v . \frac{dT}{T}$$

Also, $$ds = \frac{dQ}{T}$$

$\therefore$ $$ds = \frac{p}{T} \times dv + C_v . \frac{dT}{T}$$

From characteristic gas equation

$$pv = RT \text{ (for unit mass of gas)}$$

or $$\frac{p}{T} = \frac{R}{V}$$

$\therefore$ $$ds = R.\frac{dv}{V} + C_v . \frac{dT}{T}$$

Integrating both sides, assuming C_v to remain constant for temperature range of T_1 to T_2

$$\int_{s_1}^{s_2} ds = R \int_{v_1}^{v_2} \frac{dv}{V} + C_v \int_{T_1}^{T_2} \frac{dT}{T}$$

or $$S_2 - S_1 = R \log_e \frac{V_2}{V_1} + C_v \log_e \frac{T_2}{T_1}$$

The above equation can be represented in terms of pressure and volumes, Applying characteristic gas equation

$$\frac{p_1 V_1}{T_1} = \frac{p_2 V_2}{T_2}$$

or $$\frac{T_2}{T_1} = \frac{V_2}{V_1} \cdot \frac{p_2}{p_1}$$

Putting the value of $\frac{T_2}{T_1}$ in above equation, we get

$$S_2 - S_1 = R \log_e \frac{V_2}{V_1} + C_v \log_e \left(\frac{V_2}{V_1} \cdot \frac{p_2}{p_1} \right)$$

or $$= R \log_e \frac{V_2}{V_1} + C_v \log_e \frac{V_2}{V_1} + C_v \log_e \frac{p_2}{p_1}$$

or $$= (C_v + R) \log_e \frac{V_2}{V_1} + C_v \log_e \frac{p_2}{p_1}$$

$$S_2 - S_1 = C_p \log_e \frac{V_2}{V_1} + C_v \log_e \frac{p_2}{p_1} \qquad (\because C_p = C_v + R)$$

Now again this equation can also expressed in terms of pressure and temperature.

So, from characteristic gas equation

$$\frac{p_1 V_1}{T_1} = \frac{p_2 V_2}{T_2}$$

$$\therefore \quad \frac{V_2}{V_1} = \frac{p_1}{p_2} \cdot \frac{T_2}{T_1}$$

Putting the value of $\frac{V_2}{V_1}$ in above equation, we get

$$S_2 - S_1 = R \log_e \left(\frac{p_1}{p_2} \times \frac{T_2}{T_1} \right) + C_v \log_e \frac{T_2}{T_1}$$

$$= R \log_e \frac{p_1}{p_2} + R \log_e \frac{T_2}{T_1} + C_v \log_e \frac{T_2}{T_1}$$

$$= (R + C_v) \log_e \frac{T_2}{T_1} - R \log_e \left(\frac{p_2}{p_1} \right)$$

$$= C_p \log_e \frac{T_2}{T_1} - R \log_e \frac{p_2}{p_1}.$$

3.3.1. Isothermal and Reversible Adiabatic Process

During an isothermal process the temperature remains constant so the amount of heat supplied is equal to the work done.

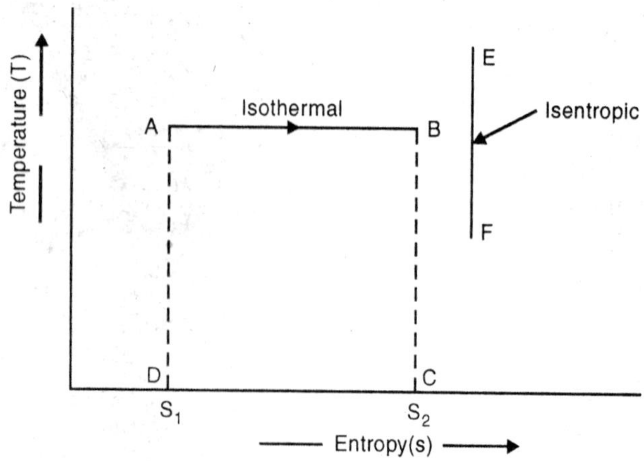

Fig. 3.6.

Referring to the Fig. 3.6, *AB* represents an isothermal expansion of a gas at a constant temperature *T*, entropy increasing from S_1 to S_2 heat is absorbed by the gas during expansion. Heat absorbed will be given by the area *ABCD*.

Thus, $$Q = W\int_{s_1}^{s_2} T.ds = (S_2 - S_1)T$$

or $$S_2 - S_1 = \frac{Q}{T}$$

But, work done during an isothermal process is equal to

$$= p_1V_1 \log_e \frac{V_2}{V_1}$$

$$\therefore \quad Q = W = p_1V_1 \log_e \frac{V_2}{V_1}$$

or $$S_2 - S_1 = \frac{Q}{T_1} = \frac{p_1V_1}{T_1} \log_e \frac{V_2}{V_1} = \frac{mRT_1}{T_1} \log_e \frac{V_2}{V_1}$$

$$= mR \log_e \frac{V_2}{V_1}.$$

This equation can also be obtained from equation, $S_2 - S_1 = \frac{Q}{T}$ by putting $T_1 = T_2$.

During reversible adiabatic process there is no heat supplied or rejected and hence

$$dQ = 0$$

Thus, $$ds = \frac{dQ}{T} = 0.$$

This means that during reversible adiabatic process entropy remain constant and thus the process is known as 'Isentropic'. From this it follows that the graph on *T-S* chart representing reversible adiabatic process will be a straight vertical line, *EF* (Refer Fig. 3.6). Since no heat is supplied or rejected during this process therefore the diagram gives no area under the expansion line *EF*.

3.3.2. Heating at Constant Pressure

When a gas is heated at constant pressure, a part of heat supplied is utilised in doing external work and rest of it is stored in the gas in the form of internal energy.

Let *m* kg of a gas be heated at constant pressure '*p*' such that V_1, T_1, and S_1, change to V_2, T_2, S_2 respectively.

Then $Q = m \,.\, C_p\, (T_2 - T_1)$.

Differentiating, to find the small quantity of heat added *dQ* during a small rise of temperature *dT*.

Then $dQ = m \,.\, C_p \,.\, dT$.

Dividing both sides by *T*

$$\frac{dQ}{T} = m.C_p.\frac{dT}{T}$$

Also, $$ds = \frac{dQ}{T}$$

$$\therefore \quad ds = m.C_p.\frac{dT}{T}$$

Integrating both sides,

$$\int_{S_1}^{S_2} ds = m.C_p \int_{T_1}^{T_2} \frac{dT}{T}$$

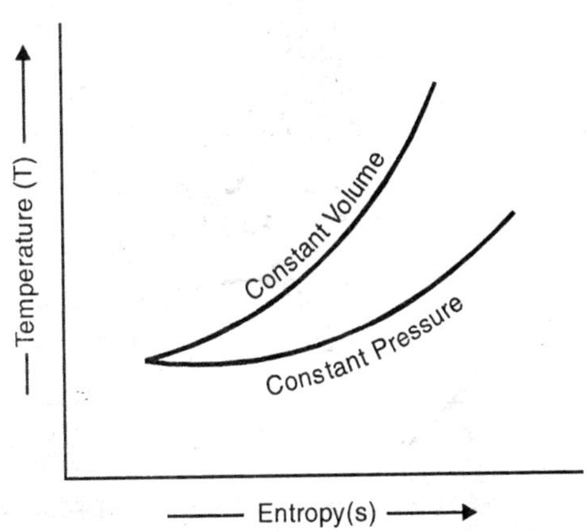

Fig. 3.7.

$\therefore$ Change in entropy, $S_2 - S_1 = m.C_p.\log_e \frac{T_2}{T_1}$.

It will be seen that the curve when plotted with abs. temperature as ordinate and entropy as abscissa will give a logarithmic curve as shown in Fig. 3.7.

3.3.3. Heating at Constant Volume

Let 'm' kg of a gas be heated at a constant volume 'V' so that entropy increases from S_1 to S_2 and absolute temperature from T_1 to T_2.

Then, $Q = m.C_v.(T_2 - T_1)$ ($\because$ No work is done)

Differentiating to find the small amount of heat added dQ corresponding to a small rise in temperature dT.

$$dQ = m \,.\, C_v \,.\, dT$$

Dividing both sides by T

$$\frac{dQ}{T} = m.C_v.\frac{dT}{T}$$

Also
$$ds = \frac{dQ}{T}$$

$\therefore$
$$ds = m.C_v.\frac{dT}{T}$$

Integrating for the total change in entropy

$$\int_{S_1}^{S_2} ds = m.C_v.\int_{T_1}^{T_2} \frac{dT}{T}$$

$\therefore$
$$S_2 - S_1 = m.C_v.\log_e \frac{T_2}{T_1}$$

It will be noticed that this curve when plotted with abs. temperature as ordinate and entropy as abscissa will also be a logarithmic curve. Also, since C_p is greater than C_v the constant volume lines will be steeper that the constant pressure lines, as shown in Fig. 3.7.

3.3.4. Change of Entropy during a Polytropic Process

$$S_2 - S_1 = C_v \log_e \frac{T_2}{T_1} + R \log_e \frac{V_2}{V_1} \quad ...(i)$$

Since $pV^n = \text{Constant}$ (in polytropic process)

$\therefore$
$$pV^n = p_1V_1^n = p_2 V_2^n$$

or
$$\frac{p_1}{p_2} = \left(\frac{V_2}{V_1}\right)^n \quad ...(ii)$$

From characteristic gas equation

$$\frac{p_1 V_1}{T_1} = \frac{p_2 V_2}{T_2}$$

or
$$\frac{p_1}{p_2} = \frac{V_2}{V_1} \cdot \frac{T_1}{T_2}$$

∴ From eqn. (*ii*)

$$\left(\frac{V_2}{V_1}\right)^n = \frac{V_2}{V_1}\cdot\frac{T_1}{T_2}$$

or
$$\frac{V_2}{V_1} = \left(\frac{T_1}{T_2}\right)^{\frac{1}{n-1}}$$

Putting the value of $\frac{V_2}{V_1}$ in above eqn. (*i*)

$$\therefore \quad S_2 - S_1 = C_v \log_e \frac{T_2}{T_1} + R\log_e\left(\frac{T_1}{T_2}\right)^{\frac{1}{n-1}}$$

$$= C_v \log_e \frac{T_2}{T_1} + R\left(\frac{1}{n-1}\right)\log_e \frac{T_1}{T_2}$$

$$= \left[C_v - R.\left(\frac{1}{n-1}\right)\right].\log_e.\frac{T_2}{T_1}$$

$$= \left(C_v - \frac{C_p - C_v}{n-1}\right).\log_e \frac{T_2}{T_1} \qquad (\because C_p - C_v = R)$$

$$= \left(C_v - \frac{C_v.\gamma - C_v}{n-1}\right).\log_e \frac{T_2}{T_1} \qquad \left(\because \frac{C_p}{C_v} = \gamma\right)$$

$$= C_v\left(1 - \frac{\gamma-1}{n-1}\right)\log_e \frac{T_2}{T_1} = C_v\left(\frac{n-\gamma}{n-1}\right)\log_e \frac{T_2}{T_1}$$

If '*m*' is the mass of the gas considered, then,

$$S_2 - S_1 = m.C_v\left(\frac{n-\gamma}{n-1}\right)\log_e \frac{T_2}{T_1}$$

NUMERICAL EXAMPLES

Example 3.1. During an expansion of a gas, it is observed that 10,000 N-m of work is done on the piston and 7.0 KJ of heat is absorbed. Calculate the work required to compress the gas to its initial condition if 2.0 KJ of heat is rejected to the surroundings.

Solution. We know :

$$Q_{12} - W_{12} = \Delta U_{12} \text{ or } U_2 - U_1$$

During the expansion process (say 1-2),

Work done, W_{12} = +10000 N-m = 10000 Joule (∵ 1 Nm = 1 Joule)

or
$$W_{12} = \frac{10000}{1000} = 10 \text{ KJ}$$

Heat absorbed, Q_{12} = +7 KJ

∴ $U_2 - U_1$ *i.e.*;

Change in internal energy = 7 – 10 = –3 KJ.

–ve sign shows that there is loss of internal energy during the expansion process.

Also, during the compression process (2 – 1)

$$Q_{21} - W_{21} = \Delta U_{21} \text{ or } U_1 - U_2$$

or Work done, $W_{21} = -(U_1 - U_2) + Q_{21} = -(-3) + 2$

$$= 3 + 2 = 5 \text{ KJ}$$

or Work done, $W_{21} = 5 \times 1000 = 5000$ J or N-m. **Ans.**

Example 3.2. How much heat must be supplied to 1.5 kg of a gas to raise its temperature from 90 °C to 225 °C at constant pressure ? Find also the external work done during the supply of heat. Take $C_p = 1.0$ KJ/kg k, $C_v = 0.7128$ KJ/kg k.

Solution.

$$m = 1.5 \text{ kg}$$
$$T_1 = 90 + 273 = 363 \text{ K}$$
$$T_2 = 225 + 273 = 498 \text{ K}$$
$$C_p = 1 \text{ and } C_v = 0.7128 \text{ KJ/kg K}$$

Total heat supplied,

$$Q = m.C_p.(T_2 - T_1) = 1.5 \times 1 \times (498 - 363)$$
$$= 1.5 \times 135 = 202.5 \text{ KJ. } \textbf{Ans.}$$

Change in internal energy,

$$\Delta U = m.C_v.(T_2 - T_1) = 1.5 \times 0.7128 \times (498 - 363)$$
$$= 144.34 \text{ KJ}$$

Now, $Q = W + \Delta U$

$\therefore$ External Work done,

$$W = Q - \Delta U = 202.5 - 144.34$$
$$= 58.16 \text{ KJ. } \textbf{Ans.}$$

Example 3.3. A quantity of a gas occupying 0.08 m^3 at 8.4 bar abs. pressure receives heat at constant pressure until the volume increases to 0.16 m^3. The initial temperature is 200 °C and $C_p = 1$, $C_v = 0.7128$ KJ/kg k. Find the change in internal energy and external work done during the heat supply.

Solution.

$$V_1 = 0.08 \text{ m}^3$$
$$p_1 = 8.4 \text{ bar} = 8.4 \times 10^5 \text{ N/m}^2 \text{ or Pa.}$$
$$V_2 = 0.16 \text{ m}^3$$
$$C_p = 1, \quad C_v = 0.7128 \text{ KJ/kg k}$$

$\because$ $C_p - C_v = R$

$\therefore$ $1 - 0.7128 = R$

or $R = 0.287$ KJ/kg k

or $R = 287$ J/kg k

Now, $p_1 V_1 = m \,.\, R \,.\, T_1$

So, $$m = \frac{p_1 V_1}{RT_1} = \frac{8.4 \times 10^5 \times 0.08}{287 \times 473}$$

or Mass, $m = 0.495$ kg.

Now, applying Charle's law :

$$\frac{V_1}{T_1} = \frac{V_2}{T_2}$$

$$\therefore \qquad T_2 = \frac{V_2}{V_1} \times T_1 = \frac{0.16}{0.08} \times 473 = 946 \text{ K}$$

Change in internal energy,

$$\Delta U = m.C_v.(T_2 - T_1)$$

or $$\Delta U = 0.495 \times 0.7128\ (946 - 473)$$

$$= 0.495 \times 0.7128 \times 473$$

or $$\Delta U = 166.89 \text{ KJ. } \textbf{Ans.}$$

Total heat supplied, $$Q = m.C_p.(T_2 - T_1) = 0.495 \times 1 \times (946 - 473)$$

$$= 0.495 \times 473$$

or $$Q = 234.13 \text{ KJ}$$

Now, $$Q = W + \Delta U$$

$\therefore$ External work done,

$$W = Q - \Delta U = 234.13 - 166.89$$

$$= 67.24. \textbf{ Ans.}$$

Example 3.4. A closed vessel contains 2.0 kg of a gas at 20 °C and 0.7 bar abs. pressure. Heat is supplied to the vessel till the gas acquires a pressure of 1.4 bar abs. Find out :

(*i*) Final temperature of gas

(*ii*) Work done

(*iii*) Heat added, and

(*iv*) Change in internal energy.

Take $$C_v = \mathbf{0.7128 \text{ KJ/kg k.}}$$

Solution.

$$p_1 = 0.7 \text{ bar abs.}$$

$$T_1 = 20 + 273 = 293 \text{ K}$$

$$p_2 = 1.4 \text{ bar abs.}$$

$$T_2 = ?$$

As the gas is heated in a closed vessel.

$$\therefore \qquad V_1 = V_2$$

(*i*) During constant volume process

$$\frac{p_1}{T_1} = \frac{p_2}{T_2}$$

or $$T_2 = \frac{p_2}{p_1} \times T_1 = \frac{1.4}{0.7} \times 293$$

or $$T_2 = 586 \text{ K}$$

$$\therefore \qquad T_2 = 586 - 273 = 313 \text{ °C.}$$

(*ii*) Work done, $W = 0$ (as the change in volume, $dv = 0$), (*iii*) Heat added, $Q_{12} = m.C_v.(T_2 - T_1)$

$= 2 \times 0.7128\,(586 - 293)$

or $Q_{12} = 417.7$ KJ. **Ans.**

(*iv*) Change in internal energy

$\Delta U_{12} = Q_{12}$

or $\Delta U_{12} = 417.7$ KJ. **Ans.**

Example 3.5. The value of C_p and C_v for an ideal gas are 1.0 and 0.7128 KJ/kg k respectively. Find the values of R and ratio of specific heats, γ for the gas. If one kg of this gas is heated at constant pressure from 25 °C to 200 °C, estimate the heat added, ideal work done and change in internal energy. Also, calculate the pressure and final volume if initial volume is 2 m^3.

Solution. $C_p = 1$

and $C_v = 0.7128$ KJ/kg k

$V_1 = 2\text{ m}^3$

$T_1 = 25 + 273 = 298$ k

$T_2 = 200 + 273 = 473$ k

We know $C_p - C_v = R$

$\therefore$ $1 - 0.7128 = R$

or $R = 0.2872$ KJ/kg k

or $R = 0.287$ J/kg k. **Ans.**

Again $\gamma = \frac{C_p}{C_v} = \frac{1}{0.7128} = 1.4.$ **Ans.**

During the constant pressure process :

Heat transferred, $Q_{12} = C_p(T_2 - T_1) = 1(473 - 298)$

or $Q_{12} = 175$ KJ. **Ans.**

Change in internal energy,

$\Delta U_{12} = C_v.(T_2 - T_1) = 0.7128\,(473 - 298)$

or $\Delta U_{12} = 124.74$ KJ. **Ans.**

We know $Q_{12} - W_{12} = \Delta U_{12}$

$\therefore$ Work done, $W_{12} = Q_{12} - \Delta U_{12} = 175 - 124.74$

$= 50.26$ KJ. **Ans.**

Now, from characteristic gas equation,

$10^5.p_1V_1 = m.RT_1$

$\therefore$ $p_1 = \frac{mRT_1}{10^5 \times V_1} = \frac{1 \times 287 \times 298}{10^5 \times 2}$

or $p_1 = 0.4276$ bar. **Ans.**

Final pressure, $p_2 = p_1$ as it is constant pressure process,

Further during constant pressure process.

$$V_2 = \frac{T_2}{T_1} \times V_1 = \frac{473}{298} \times 2 = 3.1744 \text{ m}^3/\text{kg}. \textbf{ Ans.}$$

Example 3.6. A gas occupies 0.35 m^3 at a pressure of 1 bar absolute. Find the work done on the gas if it is compressed iso thermally to a pressure of 16 bar abs.

Solution.

$$p_1 = 1 \text{ bar}$$
$$p_2 = 16 \text{ bar}$$
$$V_1 = 0.35 \text{ m}^3$$

We know, during an isothermal process

Work done, $W = p_1V_1 \log_e (r)$

Also, $p_1V_1 = p_2V_2$

or $\dfrac{V_2}{V_1} = \dfrac{p_1}{p_2} = \gamma$ *i.e.* Compression ratio

$$= \frac{16}{1} = 16.$$

$\therefore$ Work done, $W = 10^5 \times 1 \times 0.35 \log_e 16$ $(\therefore 1 \text{ bar} = 10^5 \text{ N/m}^2)$

$$= 10^5 \times 0.35 \times 2.3 \log 16$$
$$= 96931.66 \text{ Joule or N-m.}$$
$$= 0.969 \text{ KJ. } \textbf{Ans.}$$

Example 3.7. One kg of a gas at 240 °C, expands adiabatically, so that its volume is doubled and the temperature falls to 115 °C. The work done during expansion is 90.0 KJ. Calculate the two specific heats.

Solution.

$$T_1 = 240 + 273 = 513 \text{ k}$$
$$T_2 = 115 + 273 = 388 \text{ k}$$

During an adiabatic process the work done is numerically equal to change in internal energy.

$\therefore$ Work done, $W = C_v.(T_1 - T_2)$ $(\because m = 1 \text{ kg})$

or $90 = C_v (513 - 388) = C_v \times 125$

$\therefore$ $C_v = \dfrac{90}{125} = 0.72 \text{ KJ/kg k.}$ **Ans.**

Now, $\dfrac{T_1}{T_2} = \left(\dfrac{V_2}{V_1}\right)^{\gamma - 1}$

or $\dfrac{513}{388} = (2)^{\gamma - 1}$ $(\because V_2 = 2V_1)$

$\therefore$ $1.32 = (2)^{\gamma - 1}$

Taking log of both sides

$$\log 1.32 = (\gamma - 1) \log 2$$

or $$0.1206 = (\gamma - 1) \times 0.301$$

or $$0.301\gamma - 0.301 = 0.1206$$

or $$\gamma = \frac{(0.1206 + 0.301)}{0.301} = \frac{0.4216}{0.301}$$

or $$\gamma = 1.4$$

But, $$\frac{C_p}{C_v} = \gamma$$

$$\therefore \quad C_p = \gamma . C_v = 1.4 \times 0.72 = 1.01 \text{ KJ/kg k. } \textbf{Ans.}$$

Example 3.8. Gas at 1.5 bar abs. and 20 °C in a closed vessel is compressed to 10 bar abs. Its temperature then becomes 180 °C. If the compression follows the law $pV^n = C$. Find the value of 'n'.

Solution.

$$p_1 = 1.5 \text{ bar abs.}$$
$$T_1 = 20 + 273 = 293 \text{ k}$$
$$p_2 = 10 \text{ bar abs.}$$
$$T_2 = 180 + 273 = 453 \text{ k}$$

We know, during polytropic process,

$$\frac{T_2}{T_1} = \left(\frac{p_2}{p_1}\right)^{\frac{n-1}{n}}$$

or $$\frac{453}{293} = \left(\frac{10}{1.5}\right)^{\frac{n-1}{n}}$$

or $$1.546 = (6.67)^{\frac{n-1}{n}}$$

Taking log of both sides.

$$\log 1.546 = \frac{n-1}{n} . \log 6.67$$

or $$\frac{n-1}{n} = \frac{\log 1.546}{\log 6.67} = \frac{0.1892}{0.8241}$$

or $$\frac{n-1}{n} = 0.2295$$

or $$n - 1 = 0.2295\, n$$

or $$n = \frac{1}{1 - 0.2295}$$

$$n = 1.297. \textbf{ Ans.}$$

Example 3.9. One kg of an ideal gas is heated from 18.3 °C to 93.4 °C. Assuming $R = 0.287$ KJ/kg k and $\gamma = 1.38$ for the gas, find out :

(*i*) Specific heats.

(*ii*) Change in internal energy.

(*iii*) Change in enthalpy.

Solution.

$$R = 0.287 \text{ KJ/kg k}$$
$$\gamma = 1.38$$

(*i*) **Specific heats :**

We know $C_p - C_v = R$

or $R = C_v\left(\frac{C_p}{C_v} - 1\right) = C_v(\gamma - 1)$

$\therefore$ $C_v = \frac{R}{(\gamma - 1)} = \frac{0.287}{(1.38 - 1)} = 0.755 \text{ KJ/kg k}$. **Ans.**

$\therefore$ $C_p = C_v \times \gamma$

$\therefore$ $C_p = 1.59 \times 1.18$

$= 1.04$ KJ/kg k. **Ans.**

(*ii*) **Change in internal energy :**

We know, $\Delta U_{12} = m.C_v.(T_2 - T_1)$

$T_1 = 18.3 + 273 = 291.3$ K

$T_2 = 93.4 + 273 = 366.4$ K

$\therefore$ For unit mass,

$\Delta U_{12} = 0.755\,(366.4 - 291.3)$

$= 56.70$ KJ/kg. **Ans.**

(*iii*) **Change in enthalpy :**

$\Delta h = C_p(T_2 - T_1) = 1.04\,(75.1)$

or $\Delta h = 78.1$ KJ/kg. **Ans.**

Example 3.10. **2 kg of air at a pressure of 7 bar abs. occupies a volume of 0.28 m^3, this air then expanded to a volume of 1.4 m^3 either isothermally or by** law **$pV^{1.2} = C$. Assume, $R = 287$ J/kg k $C_p = 1.0$, $C_v = 0.7128$ KJ/kg k, $\gamma = 1.4$. Find :**

(*a*) the final temperatures,

(*b*) the work done

(*c*) heat absorbed or rejected for both the cases mentioned above.

Solution. $m = 2$ kg

$p_1 = 7$ bar abs.

$= 7 \times 10^5$ N/m^2

$V_1 = 0.28$ m^3

$V_2 = 1.4$ m^3

(*i*) When the gas is expanded isothermally

(*a*) $\therefore$ Final temperature t_2 = Initial temperature t_1

$\therefore$ Applying characteristic gas equation;

$$p_1V_1 = m.R.T_1$$

or $T_1 = \frac{p_1V_1}{m.R} = \frac{7 \times 10^5 \times 0.28}{2 \times 287}$

or $= 341.5$ k

or $= 68.46$ °C

$\therefore \quad T_2 = T_1 = 68.46\ °C.$ **Ans.**

(*b*) Work done during isothermal process.

$$W = p_1 V_1 \log_e r = 7 \times 10^5 \times 0.28 \log_e^{(1.4/0.28)}$$
$$= 7 \times 10^5 \times 0.28 \times 1.61 = 315449.83 \text{ Joules}$$
$$= 315.45 \text{ KJ. } \textbf{Ans.}$$

(*c*) Heat absorbed during isothermal expansion is equal to the work done, as change in internal energy is zero.

i.e. $\quad Q = W = 315.45$ KJ. **Ans.**

(*ii*) When the gas is expanded according to $pV^{1.2} = C$

(*a*) We know $\quad \dfrac{T_2}{T_1} = \left(\dfrac{V_1}{V_2}\right)^{n-1} = \left(\dfrac{0.28}{1.4}\right)^{1.2-1} = 0.7246$

or $\quad T_2 = 341.5 \times 0.7246 = 247.45$ K

$= -25.55\ °C.$ **Ans.**

(*b*) Work done during polytropic process.

$$W = \frac{p_1V_1 - p_2V_2}{n-1} = \frac{mR(T_1 - T_2)}{(n-1)} \text{ N-m}$$
$$= \frac{2 \times 287\,(341.5 - 247.45)}{(1.2 - 1)}$$
$$= 269923.5 \text{ N-m or } J = 269.9 \text{ KJ. } \textbf{Ans.}$$

(*c*) Heat transferred during the polytropic process;

$$Q = W \times \frac{\gamma - n}{\gamma - 1} = 269.9 \times \frac{(1.4 - 1.2)}{(1.4 - 1)}$$
$$= 269.9 \times \frac{0.2}{0.4} = 134.95 \text{ KJ. } \textbf{Ans.}$$

Example 3.11. Calculate the workdone for a non-flow quasi-static process, represented by $P = (-3V + 20)$ bar where 'V' is the volume in m^3 and changes from 2 to 8 m^3.

Solution. Given data :

Non-flow quasi-static process, represented by :

Pressure, $P = (-3V + 20)$ bar

Since,

$$\text{Work done, } W_{(1-2)} = \int_1^2 P.DV$$

or

$$= \int_{V_1}^{V_2} (-3V + 20) \times 10^5 \, .dV$$
$$= \left[\frac{-3V^2}{2} + 20\,V\right]^2 \times 10^5 \text{ J}$$
$$= \left[\frac{-3}{2}(8^2 - 2^2) + 20(8 - 2)\right] \times 10^5 \text{ J}$$

$$= \left[\frac{-3}{2}(64-4)+(20\times 6)\right]\times 10^5$$

$$= 30 \times 10^5 \text{ J}$$

Hence, work done, $W = 3$ MJ **Ans.**

Example 3.12. One kg of an ideal gas is heated from 20°C to 70°C. Assuming, $R = 287$ J/kg k and $r = 1.4$, find out;

(i) Both the specific-heats i.e. C_P and C_V.

(ii) Change in internal energy.

(iii) Change in enthalpy.

Solution. Given data :

Initial temp., $T_1 = 20°C = 20 + 273 \text{ K} = 293 \text{ K}$

Final temp., $T_2 = 70°C = 70 + 273 \text{ K} = 343 \text{ K}$

Gas constant, $R = 287 \text{ J/kg K} = 0.287 \text{ KJ/kg k}$

$r = 1.4$

(i) We know that;

$$C_P - C_V = R$$

or $$R = C_V(r-1)$$

or $$C_V = \frac{R}{r-1}$$

or $$C_V = \frac{0.287}{1.4-1} = 0.7175 \text{ KJ/kg k}$$ **Ans.**

Also, $$\frac{C_P}{C_V} = r$$

$\therefore$ $$C_P = r \times C_V$$

$$= 1.4 \times 0.7175 = 1.0045 \text{ KJ/kg k}$$ **Ans.**

(ii) $\because$ Change in internal energy;

$$\Delta_U = m \,.\, C_V(T_2 - T_1)$$

$\therefore$ $$\Delta_U = 1 \times 0.7175\,(343 - 293)$$

or $$\Delta_U = 35.875 \text{ KJ/kg}$$ **Ans.**

(iii) $\because$ Change in enthalpy;

$$\Delta_h = m \,.\, C_P(T_2 - T_1)$$

or $$íu_h = 1 \times 0045\,(343 - 293)$$

or $$\Delta_h = 50.225 \text{ KJ/kg}$$ **Ans.**

Example 3.13. A quantity of air with 0.5 m^3 of volume has a pressure of 4 bar and temp. of 70°C. It is expanded at a constant temperature to the final pressure of 1 bar. Determine, the work done during the expansion.

Solution. Given data :

Initial volume, $V_1 = 0.5 \text{ m}^3$

Initial pressure, $P_1 = 4 \text{ bar} = 4 \times 10^5 \text{ N/m}^2$

Initial temp., $T_1 = 70°\text{C} = 70 + 273 \text{ K} = 343 \text{ K}$

Process is constant temperature

Final pressure, $P_2 = 1 \text{ bar} = 1 \times 10^5 \text{ N/m}^2$

We know that; in constant temp. process;

$$P_1V_1 = P_2V_2 \qquad \text{(Boyle's law)}$$

$$\therefore \quad V_2 = \frac{P_1V_1}{P_2}$$

$$= \frac{4 \times 10^5 \times 0.5}{1 \times 10^5} = 2 \text{ m}^3$$

Now, work done in an isothermal process;

$$W_{1-2} = 2.3\, P_1V_1 \log r$$

where, 'r' is the expansion ratio *i.e.,*

$$\frac{V_2}{V_1} = \frac{2}{0.5} = 4$$

So,

$$W_{1-2} = 2.3 \times 4 \times 10^5 \times 0.5 \log 4 \text{ J}$$

$$= 276947.59 \text{ J}$$

$$= 276.95 \text{ KJ} \textbf{ Ans.}$$

Example 3.14. 0.15 m^3 of air at a pressure of 1.5 bar is expanded, isothermally to 0.5 m^3. Calculate the final pressure of the gas and heat-supplied during the process.

Solution. Given data :

Initial volume, $V_1 = 0.15 \text{ m}^3$

Initial pressure, $P_1 = 1.5 \text{ bar} = 1.5 \times 10^5 \text{ N/m}^2$

Final volume, $V_2 = 0.5 \text{ m}^3$

As the expansion is isothermal, so,

$$P_1V_1 = P_2V_2 \qquad \text{(Boyle's law)}$$

or

$$P_2 = \frac{P_1V_1}{V_2} = \frac{1.5 \times 10^5 \times 0.15}{0.5}$$

$$= 0.45 \times 10^5 \text{ N/m}^2$$

or

$$P_2 = 0.45 \text{ bar} \textbf{ Ans.}$$

Also, expansion ratio, $r = \dfrac{V_2}{V_1}$

$$= \frac{0.5}{0.15} = 3.33$$

$\therefore$ Work done during the process,

$$W_{1-2} = 2.3\, P_1V_1 \log r$$

$$= 2.3 \times 1.5 \times 10^5 \times 0.15 \times \log 3.33$$

$$= 27036.49 \text{ J}$$

or

$$W_{1-2} = 27.036 \text{ KJ} \textbf{ Ans.}$$

Example 3.15. 1.5 kg of air is heated at constant volume from a pressure of 2 bar abs. and temp. of 60°C to a pressure of 6 bar abs. Calculate the change in entropy. Take C_V = 0.71 KJ/kg k and R = 0.287 KJ/kg k.

Solution. Given data :

Mass of air = 1.5 kg

Process is a 'constant pressure process'.

Initial pressure of air,

$$P_1 = 2 \text{ bar} = 2 \times 10^5 \text{ N/m}^2$$

Final pressure of air, $P_2 = 6 \text{ bar} = 6 \times 10^5 \text{ N/m}^2$

Initial temperature, $T_1 = 60°\text{C} = 60 + 273 \text{ K} = 333 \text{ K}$

$$C_V = 0.71 \text{ KJ/kg k}, \quad R = 0.287 \text{ KJ/kg k.}$$

As, we know that,

$$P_1V_1 = m.R.T_1 \qquad \text{(gas equation)}$$

$\therefore$ Volume,

$$V_1 = \frac{mRT_1}{P_1} = \frac{1.5 \times 0.287 \times 333 \times 1000}{2 \times 10^5} = 0.71678 \text{ m}^3$$

$$\because \quad V_1 = V_2 \qquad \text{(constant vol. process)}$$

$$\therefore \quad P_2V_2 = m.R.T_2$$

and Final temp.,

$$T_2 = \frac{P_2V_2}{m.R} = \frac{6 \times 10^5 \times 0.71678}{1.5 \times 287} = 998.99 \text{ K}$$

Hence, change in entropy;

$$S_2 - S_1 = m.C_V.\log_e \frac{T_2}{T_1} = 1.5 \times 0.71 \log \frac{998.99}{333} = 1.5 \times 0.71 \times 1.0986 = 1.17 \text{ KJ/k } \textbf{Ans.}$$

Example 3.16. The initial pressure of 1.5 m³ of a gas is 1500 N/m². If this gas is compressed isothermally to a pressure of 7500 N/m² then find out the workdone and the heat transferred during the process.

Solution. Given data :

Initial volume, $V_1 = 1.5 \text{ m}^3$

Initial pressure, $P_1 = 1500 \text{ N/m}^2$

Final pressure, $P_2 = 7500 \text{ N/m}^2$

The process is an isothermal process;

So, Work done, $W = P_1V_1 \log_e \frac{V_2}{V_1}$ ...(i)

Also, $P_1V_1 = P_2V_2$

or $\frac{V_2}{V_1} = \frac{P_1}{P_2}$

Substituting the value of $\frac{V_2}{V_1}$ and eqn. (i), we get

$$\text{Workdone, } W = P_1V_1 \log_e \frac{P_1}{P_2}$$

$$= 1500 \times 1.5 \log_e \frac{1500}{7500}$$

$$= 1500 \times 1.5 \times (-1.61)$$

$$= -3621.24 \text{ Joules } \textbf{Ans.}$$

(Here, it should be noted that work is done on the gas)

Example 3.17. One kg of a gas at 215°C, expands adabatically such that its volume is doubled and temp. falls to 90°C. The work done during the expansion is 90 KJ. Calculate, the sp. heat at const. pressure and sp. heat at constant volume.

Solution. Given data :

Initial temp. $T_1 = 215°C$

$= 215 + 273 \text{ K} = 488 \text{ K}$

Final temp. $T_2 = 90°C$

$= 90 + 273 = 363 \text{ K}$

We know that; in an adiabatic process

Work done is numerically equal to change in int. energy

$\therefore$ Work done, $W = m.C_V(T_1 - T_2)$

or $90 = 1 \times C_V (488 - 363)$

or $C_V = \frac{90}{125} = 0.72$ KJ/kg k **Ans.**

Also, $\frac{T_1}{T_2} = \left(\frac{V_2}{V_1}\right)^{\gamma-1}$

or $\frac{488}{363} = (2)^{\gamma-1}$

$\therefore$ $1.344 = (2)^{\gamma-1}$

Taking log of both sides,

$\log 1.344 = (\gamma - 1) \log 2$

or $0.128 = (\gamma - 1) \times 0.301$

or $0.301\gamma - 0.301 = 0.128$

or $\quad 0.301\gamma = 0.128 + 0.301 = 0.429$

or $\quad \gamma = \dfrac{0.429}{0.301} = 1.425$

But, $\quad \dfrac{C_P}{C_V} = \gamma$

$\therefore \quad C_P = \gamma \times C_V$

$= 1.425 \times 0.72 = 1.026$ KJ/kg k **Ans.**

Example 3.18. 5 kg of air at 450°C expands adabatically. The work done by the air is 240 KJ and the temp. is reduced to 160°C. Find out, both the sp. heats.

Solution. Given data :

Mass of air, m = 5 kg.

Initial temp., T_1 = 450°C = 450 + 273 K = 723 K

Final temp., T_2 = 160˜C = 160 + 273 K = 433 K

Work done by the air, W = 240 KJ

We know that : In an adabatic process;

$$\text{Work done, } W = \frac{P_1V_1 - P_2V_2}{\gamma - 1}$$

$$= \frac{mRT_1 - mRT_2}{\gamma - 1} = \frac{mR(T_1 - T_2)}{\gamma - 1}$$

$$= \frac{m(C_P - C_V)(T_1 - T_2)}{\dfrac{C_P - C_V}{C_V}}$$

$$= \frac{mC_V(C_P - C_V)(T_1 - T_2)}{(C_P - C_V)} = mC_V(T_1 - T_2)$$

$\therefore \quad 240 = 5 \times C_V (723 - 433)$

or $\quad 240 = C_V . 1450$

or $\quad C_V = \dfrac{240}{1450} = 0.1655$ KJ/kg k **Ans.**

$\because$ For air $\gamma = 1.4$

$\therefore$ From relation $\quad \gamma = \dfrac{C_P}{C_V}$

$C_P = \gamma \times C_V = 1.4 \times 0.1655 = 0.232$ KJ/kg k **Ans.**

Example 3.19. 1.5 m^3 of a gas at 100 bar pressure is expanded polytropically. Find the polytropic index 'n' if the final volume and pressure of the gas are 4 m^3 and 5 bar respectively.

Solution. Given data;

Initial volume, V_1 = 1.5 m^3,

Final volume, V_2 = 4 m^3

Initial pressure, P_1 = 100×10^5 N/m^2

Final pressure, P_2 = 5 bar = 5×10^5 N/m^2

We know that : In a polytropic process

$$PV^n = \text{Constant}$$

or
$$P_1V_1^n = P_2V_2^n$$

or
$$\frac{P_1}{P_2} = \left(\frac{V_2}{V_1}\right)^n$$

or
$$\left(\frac{100 \times 10^{15}}{5 \times 10^5}\right) = \left(\frac{4}{1.5}\right)^n$$

or
$$20 = (2.67)^n$$

Now, taking log of both sides, we get

$$\log 20 = \log (2.67)^n$$

or
$$\log 20 = n \log 2.67$$

or
$$1.301 = n \times 0.4265$$

$\therefore$ Polytropic index, $n = \dfrac{1.301}{0.4265} = 3.05$ **Ans.**

Example 3.20. A gas at 160 KPa and 25°C in a closed vessel is compressed at 1050 KPa. After compression its temp. becomes 200°C. If the compression follows the law PV^n = constant, find the value of polytropic index, n.

Solution. Given data;

Initial pressure, $P_1 = 160 \text{ KPa} = 160 \times 10^3 \text{ N/m}^2$

Initial temp. $T_1 = 25°\text{C} = (25 + 273) \text{ K} = 298 \text{ K}$

Final pressure, $P_2 = 1050 \text{ KPa} = 1050 \times 10^3 \text{ N/m}^2$

Final temp. $T_2 = 200°\text{C} = (200 + 273) \text{ K} = 473 \text{ K}$

We know that :

In apolytropic process

$$\left(\frac{T_2}{T_1}\right) = \left(\frac{P_2}{P_1}\right)^{\frac{n-1}{n}}$$

or
$$\frac{473}{298} = \left(\frac{1050 \times 10^3}{160 \times 10^3}\right)^{\frac{n-1}{n}}$$

or
$$1.587 = (0.563)^{\frac{n-1}{n}}$$

Now, taking log of both sides;

$$\log 1.587 = \log (6.563)^{\frac{n-1}{n}}$$

or
$$\log 1.587 = \left(\frac{n-1}{n}\right) \log (6.563)$$

or
$$0.201 = \left(\frac{n-1}{n}\right) \times 0.817$$

or
$$0.201\, n = 0.817\, n - 0.817$$

or $\quad 0.817n - 0.201n = 0.817$

or $\quad 0.616\, n = 0.817$

or $\quad n = \dfrac{0.817}{0.616} = \mathbf{1.326\ Ans.}$

ASSIGNMENT

Objective Type Questions :

(A) Fill in the Blanks :

1. The degree of hotness or coldness is known as ________.
2. Zeroth law provides the basis for comparing the ________.
3. Heat can flow from a body at ________ temperature to a body at ________ temperature.
4. During constant volume process work done is ________.
5. The change in internal energy is ________ during an isothermal process.
6. Hyperbolic and isothermal processes are identical for ________ gases.
7. During an isentropic process ________ is constant.
8. Work done during a polytropic process is ________.
9. Heat transferred during a constant pressure process is ________.
10. During throttling heat transferred is ________.
11. The value of molar volume at N.T.P. is ________.
12. The relation between the two specific heats of a gas and its characteristic gas constant is ________.
13. The value of 'J' in S.I. system of units is ________.
14. Work done during a compression process is ________.
15. The unit of entropy in S.I. system of units is ________.
16. In case of free expansion process, the change in internal energy is ________.

(B) State True or False :

1. Zeroth law determines temperature.
2. The polytropic index is unity for a constant temperature process.
3. The entropy of universe tends to be maximum.
4. A gas can never be liquified above critical temperature.
5. The slope of constant volume line on temp-entropy diagram is more than that of constant pressure line.
6. During constant pressure process, change in internal energy is zero.
7. During an isothermal process work done is $pV \log_e r$.
8. Reversible adiabatic process is the same as adiabatic process.
9. For vapour, hyperbolic and isothermal processes are same.
10. During a polytropic process, heat transferred is not zero.
11. Work done during a constant volume process is zero.
12. Internal energy remains constant during an isothermal process.
13. The internal energy of a gas depends upon its temperature only and is independent of other properties like pressure and volume.
14. During an isothermal process the heat interchange will be equal to work done.

15. In a constant pressure process, the specific heat at constant pressure is to be used for calculating heat transferred.
16. Work done during a throttling process is not zero.
17. The change in internal energy in a free expansion process is zero.

(C) Multiple Choice Questions :

1. Identify the false statement in the context of heat and work.
   (*a*) they exist only during interaction between the system and the boundary.
   (*b*) they are path functions and exact differential.
   (*c*) they may or may not occur simultaneously.
   (*d*) they occur due to some unbalanced potential which tends to promote a change of state.
2. The absolute zero temperature corresponds to the condition when
   (*a*) all the substances exist only as solids.
   (*b*) volume of a gas reduces to zero.
   (*c*) kinetic energy of gas molecules becomes zero.
   (*d*) no pressure is exerted by the gas.
3. The pressure exerted on the walls of container by a gas is due to the fact that gas molecules :
   (*a*) stick to the walls of the container.
   (*b*) lose their kinetic energy.
   (*c*) get accelerated towards the wall.
   (*d*) change their momenta due to collision with the wall.
4. In thermodynamic analysis, a pure substance is that which;
   (*a*) has no dissolved impurities.
   (*b*) consists of only a single species.
   (*c*) may have a number of chemical species but composition remains constant.
   (*d*) behaves as a perfect gas.
5. The expression $\int_1^2 pdv$ gives the measure of work done during :
   (*a*) non-flow reversible process.
   (*b*) steady flow reversible process.
   (*c*) open system and any process.
   (*d*) any system and any process.
6. The work done equals $\int pdv$ for :
   (*a*) an irreversible process.
   (*b*) unrestricted expansion.
   (*c*) non-flow quasi-static process.
   (*d*) the arrangement where work is done on the gas as the paddle wheel turns by a falling weight.
7. The internal energy of an ideal gas is a function of;
   (*a*) pressure only.
   (*b*) absolute temperature only.
   (*c*) pressure and volume.
   (*d*) pressure, volume and temperature.

8. The internal energy of an ideal gas is a function of its absolute temperature only. This statement refers to:
   (*a*) Avogadro's law. (*b*) Maxwell law.
   (*c*) Joule's law. (*d*) Regnault's law.

9. The identity $\delta Q = p\, dv + du$ is true, for;
   (*a*) any process and open system.
   (*b*) any process and closed system.
   (*c*) irreversible process and closed system.
   (*d*) reversible process and closed system.

10. Choose the wrong statement;
   (*a*) if a closed system traces a cycle of operation, then $\oint dp$ is zero.
   (*b*) $\delta Q = \delta W + du$ is true only for reversible process and closed system.
   (*c*) internal energy is a property of the system.
   (*d*) when a gas undergoes a free or unrestricted expansion, then $W \neq \int p.dv$.

11. Heat is being supplied to air in a cylinder fitted with a frictionless piston held by a constant weight. The process is an;
   (*a*) isothermal (*b*) adiabatic
   (*c*) isobaric (*d*) isochoric.

12. Under what conditions the change in the enthalpy of a system equals the heat supplied ?
   (*a*) constant volume
   (*b*) constant pressure
   (*c*) constant temperature
   (*d*) standard, temperature-pressure conditions.

13. Which parameter remains constant during a reversible isothermal process ?
   (*a*) rate of heat exchange (*b*) internal energy
   (*c*) enthalpy (*d*) entropy.

14. During an adiabatic process;
   (*a*) temperature of gas is varied but the pressure is kept constant.
   (*b*) total kinetic energy of the gas molecules remains constant.
   (*c*) gas is compressed at constant temperature and there is no change of entropy.
   (*d*) expansion and compression of a gas occur without any exchange of heat with the surroundings.

15. For an adiabatic process, the pressure and volume are related as pv^{γ} = Constant. The exponent 'γ' represents the ratio of specific heat at constant pressure to specific heat at constant volume and for air, its value equals :
   (*a*) 0.87 (*b*) 1.02
   (*c*) 1.4 (*d*) 1.67.

16. During a general polytropic expansion process characterised by pV^n = Constant the work done is equal to :
   (*a*) $p_1V_1 - p_2V_2$ (*b*) $p_1V_1 \log_e \dfrac{V_2}{V_1}$
   (*c*) $\dfrac{p_1V_1 - p_2V_2}{n+1}$ (*d*) $\dfrac{p_1V_1 - p_2V_2}{n-1}$.

ANSWERS

(A) Fill in the Blanks :

1.	temperature	**2.**	temperature of two bodies
3.	higher, lower	**4.**	zero
5.	zero	**6.**	perfect
7.	entropy	**8.**	$\frac{p_1V_1 - p_2V_2}{n-1}$
9.	change in enthalpy	**10.**	zero
11.	22.41 m^3	**12.**	$C_p - C_v = R$
13.	4.1868 KJ/Kcal	**14.**	Negative
15.	KJ/kg k	**16.**	zero.

(B) State True or False :

1. True	**2.** True	**3.** True	**4.** True	**5.** True
6. False	**7.** True	**8.** False	**9.** False	**10.** True
11. True	**12.** True	**13.** True	**14.** True	**15.** True
16. False	**17.** True.			

(C) Multiple Choice Questions :

1. (*a*)	**2.** (*b*)	**3.** (*a*)	**4.** (*a*)	**5.** (*c*)
6. (*d*)	**7.** (*b*)	**8.** (*d*)	**9.** (*c*)	**10.** (*b*)
11. (*b*)	**12.** (*b*)	**13.** (*d*)	**14.** (*c*)	**15.** (*a*)
16. (*d*).				

(D) Short Answer Type Questions :

Q. 1. Define an 'Isochoric process'.

Ans. The process in which the volume of the given mass of the gas is not allowed to change, is known as an 'Isochoric Process' or 'Constant Volume Process'.

Q. 2. Define an 'Isobaric process'.

Ans. The process in which the pressure of the given mass of a gas is not allowed to change in known as an 'Isobaric Process' or 'Constant Pressure Process'.

Q. 3. Define an 'Isothermal process'.

Ans. The process in which, the temperature of the given mass of a gas is not allowed to change, is known as an 'Isothermal process' or 'Constant-temperature process'.

Q. 4. What is a 'Hyperbolic Process'?

Ans. The process in which the product of the pressure of the gas and the volume remains constant is known as the 'Hyperbolic Process'. So, pV = Constant in this process.

Q. 5. What is an 'Adiabatic Process'?

Ans. The process in which, no heat is added or withdrawn from the system under consideration is known as an 'Adiabatic Process'.

Q. 6. What is an 'Isentropic Process'?

Ans. The process such as reversible adiabatic process, in which the entropy of the system remains constant is known as an 'Isentropic process'.

Q. 7. What is a 'Polytropic Process'?

Ans. The process in which the relation pV^n = Constant, holds good is known as 'Polytropic Process'.

Q. 8. Define the terms 'Throttling' and 'Free Expansion'.

Ans. When a gas or vapour is expanded through an aperture of a minute (narrow) dimensions like a valve, which is very slightly opened, the process is known as 'Throttling Process'.

In 'free expansion' process the fluid expands suddently into a vacuum chamber through on orifice of large dimension.

Q. 9. What is the formula for change in 'Internal Energy' for a constant volume process ?

Ans. In a Constant volume process :

Change in internal energy = $\Delta U = m \cdot C_v (T_2 - T_1)$

where 'm' is the mass, 'C_v' the sp. heat at constant volume and $(T_2 - T_1)$ change in temperature.

Q. 10. What is the formula for heat transferred in a constant volume process ?

Ans. Heat transferred, $Q = m \cdot C_v (T_2 - T_1)$ with usual meanings.

Q. 11. Write the formula for 'Work done' in a constant pressure process.

Ans. Work done, $W = p(V_2 - V_1)$.

where 'p' is the applied pressure and $(V_2 - V_1)$ the change in volume.

Q. 12. What is the formula for change in internal energy in a constant pressure process ?

Ans. Change in internal energy $\Delta U = m \cdot C_v (T_2 - T_1)$.

Q. 13. What is the formula for the 'heat transferred' in a constant pressure process ?

Ans. Heat transferred $Q = m \cdot C_p (T_2 - T_1)$.

Q. 14. What is the formula for 'Work done' in an isothermal process ?

Ans. Work done, $W = p_1 V_1 \log_e r$.

Q. 15. What is the formula for 'Heat Transferred' in an isothermal process ?

Ans. Heat Transferred, $Q = p_1 V_1 \log_e r$.

Q. 16. What is the formula for work done in an 'Adiabatic Process' ?

Ans. Work done, $W = \dfrac{p_1V_1 - p_2V_2}{\gamma - 1}$.

Q. 17. What is the formula for change in internal energy in an adiabatic process ?

Ans. Change in internal energy, $\Delta U = m \cdot C_v (T_2 - T_1)$.

Q. 18. What is the formula for 'work done' in a 'polytropic process' ?

Ans. Work done, $W = \dfrac{p_1V_1 - p_2V_2}{n - 1}$

where 'n' is the polytropic index.

Q. 19. What is the formula for 'change in internal energy in a polytropic process ?

Ans. Change in internal energy $\Delta U = m \cdot C_v (T_2 - T_1)$.

Q. 20. What is the formula for 'Heat Transferred' in a polytropic process ?

Ans. Heat transferred, $Q = \dfrac{\gamma - n}{\gamma - 1} \dfrac{mR(T_2 - T_1)}{(1 - n)}$.

REVIEW QUESTIONS

1. Explain Zeroth law of thermodynamics. **(SBTE Pb., 1981)**
2. What is the concept of Enthalpy and Internal energy ?
3. What is the concept of Entropy ?
4. Briefly describe the various thermodynamic processes.

5. Derive an expression for work done, change in internal energy and heat transferred during a constant pressure process.
6. Derive the expressions for work done during the following processes :
   (*i*) An isothermal expansion.
   (*ii*) An adiabatic expansion. **(SBTE Pb., 1981)**
7. What is a hyperbolic process ?
8. Derive an expression for work done during the polytropic expansion of a gas. **(SBTE Pb., 1978)**
9. Distinguish between an adiabatic and an isothermal expansion process for a gas. Draw P-V diagram for each of them and explain them briefly. **(SBTE Haryana, 1979)**
10. What is an isentropic process ?
11. Distinguish between free expansion and throttling process. **(SBTE Pb., Dec., 1999)**
12. Differentiate between Adiabatic and Polytropic process. **(SBTE Pb., Dec., 1999)**

PROBLEMS FOR PRACTICE

1. A quantity of air at a temperature of 20 °C and pressure of 2.0 bar abs. is found to occupy 0.20 m^3. It is supplied heat energy equivalent to 14720 Joule at constant volume. Find the final pressure and temperature of air. Take $C_v = 0.7128$ KJ/kg k. Also determine the change in internal energy per kg of air. Take $R = 287$ J/kg k. **[Ans.** 4.065 bar, 40.65 °C, 5.926 KJ]
2. When one kg of a gas at an initial temperature of 92 °C is heated to twice its original volume at a constant pressure of 8 bar, what fraction of heat supplied to it converted into work during the change ? Take $C_p = 1.0$, $C_v = 0.7128$ KJ/kg k for the gas. **[Ans.** 0.2872]
3. A gas has $C_v = 0.7128$ KJ/kg k and $C_p = 1.0$ KJ/kg k. Calculate the volume of 0.227 kg of this gas at 21 °C and 1.4 bar absolute. Determine also the quantity of heat added and external work done by this gas if it is heated at a constant pressure of 1.4 bar till its volume is 0.142 m^3. **[Ans.** 0.137 m^3, 2.436 KJ, 1.736 KJ]
4. 0.4 kg of air at a pressure of 1.0 bar and a temperature of 20 °C is compressed isothermally until its volume is 0.08 Cu metre. Determine the work done on the air during this compression. Take $R = 287$ J/kg k. **[Ans.** 482.2 KJ]
5. Determine the volume and internal energy of one kg of air at a pressure of 5.0 bar absolute and temperature 100 °C, the internal energy being reckoned from 0 °C. The air is allowed to expand four time to its initial volume at constant pressure and find the final values of pressure, temperature and work done in each case. **[Ans.** 0.214 m^3, 7.28 KJ, 5 bar, 1219 °C, 2000 KJ]
6. One kg of a gas occupies 0.8 m^3 at 15 °C and 1.05 bar pressure. If after adiabatic compression to 0.256 m^3, the pressure is 4.9 bar. Find the values of (*i*) characteristic gas constant, R; (*ii*) ratio of Sp. heats, γ; (*iii*) C_p; (*iv*) C_v and (*v*) work done during compression. **[Ans.** 291.67 J/kg k, 1.35, 1125 J/kg k; 833 J/kg k; –11840 J]
7. Find the difference in work done when one kg of air is compressed from 1.5 bar abs. and 20 °C to 9 bar abs.
   (*i*) isothermally
   (*ii*) adiabatically.
   Take $C_p = 1.0$ KJ/kg k $C_v = 0.7128$ KJ/kg k and $\gamma = 1.4$. **[Ans.** –813.35 J]

8. The cylinder of an engine contains 0.02 m^3 of air at a pressure of 10 bar and temperature of 25 °C. Calculate the volume and temperature at the end of expansion. Also, calculate the work done during the process of expansion when the pressure at the end of expansion is 1.0 bar abs. The expansion of air takes place polytropically with $n = 1.3$. **[Ans.** 0.117 m^3, 27.67 KJ]

9. Calculate the work done when 0.075 m^3 of air at 16 °C and 7 bar abs. expands to 0.35 m^3 according to the law $pV^{1\cdot2} = C$. **[Ans.** 69.6 KJ] **(SBTE Pb., 1978)**

10. One kg of air at a temperature of 240 °C is allowed to expand adiabatically to double its volume. Find the work done in the process. Take $\gamma = 1.4$, $R =$ 287 KJ/kg k. **[Ans.** 89.1 KJ]

11. Air at 30 °C and 100 KPa is compressed reversibly and polytropically from 5 m^3 to 1 m^3. Calculate the final temperature, pressure, work done and heat transferred if the index of compression is 1.4. Assume $C_p = 1.005$ KJ/kg k and $C_v = 0.7178$ KJ/kg k. **[Ans.** 303.8 °C, 951.8 KPa, 1129.5 KJ, 0]
(SBTE Haryana, MPECS May 1995)

12. The values of specific heat of a gas at constant pressure and at constant volume are 1.0 KJ/kg k and 0.714 KJ/kg k respectively. Determine the charactesitics gas constant and isentropic index of the gas. If 5 kg of this gas is heated at constant pressure from 30 °C to 250 °C. Determine the heat added, the work done and the change in internal energy.
[Ans. 1100 KJ, 400.4 KJ, 699.6 KJ]
(SBTE Haryana, MPECS (Auto Engg.), Nov. 1985)

13. 2500 KJ of heat is added to 2.8 m^3 of a gas at 20 °C and at atm. pressure. If during the process of heat addition the volume remains constant, find the final temperature and pressure. Take R for the gas as 287 J/kg k and $C_v =$ 0.714 KJ/kg k. **[Ans.** 1057.8 °C, 4.6 bar]

14. 0.12 m^3 of air at a temperature of 20 °C and pressure of 1.1 bar is compressed to 20 bar as per the law $pV^{1.3} = C$. Determine, the temperature and volume after compression.

15. 0.15 m^3 of air at a pressure of 1.06 bar is compressed to a volume of 0.008 m^3 at 36 bar. If the compression follows the law $pV^{1.3} = C$ and 'γ' for air is 1.41 find;
    (*i*) quantity of heat added or rejected, and
    (*ii*) change of internal energy. **[Ans.** –11.53 KJ, 31.47 KJ]

16. 1.5 kg of a gas at a pressure of 9 bar is expanded adiabatically to a pressure of 2 bar. If the initial temperature of the gas is 350 °C. What will be the work done during expansion and the change in internal energy of the gas ? Take, 'C_p' for gas as 1.005 KJ/kg k and $C_v = 0.714$ KJ/kg k.
[Ans. 234.6 KJ, –220.4 KJ]

17. A gas having volume 0.23 m^3 and a pressure of 2 bar is compressed polytropically to a volume of 0.013 m^3 and pressure of 40 bar. What is the amount of heat added or rejected.
Take, 'γ' for the gas as 1.4 and the polytropic index 'n' as 1.2.
[Ans. –15 KJ, Work is done on the gas]

❑❑❑

4

CHAPTER

Laws of Thermodynamics

INSIDE THIS CHAPTER

4.1. INTRODUCTION

'Thermodynamics' is based upon the experimental observations and laws like all other branches of science. The dealing of problems in thermodynamics is done wholly on the basis of these laws. The results of this thermodynamic study are generally highly accurate. Therefore, the science of thermodynamics has got an important place in 'Engineering'. The performance of the various thermodynamic processes is measured with the help of thermodynamic laws.

4.2. LAW OF CONSERVATION OF ENERGY

This law states that "Energy can neither be created nor destroyed, but can be transformed from one form to the another". This means the total energy possessed by a body remains constant.

This law when applied to gases means that the total heat supplied or rejected in a system must be equal to the work done plus the change in internal energy.

Thus, $$Q = W + \Delta U$$

where Q = Total heat supplied or rejected.
W = Work done in heat units.
ΔU = Change in internal energy.

In the differential form, the above equation may be written as;

$$dQ = dW + dU$$

where dQ = small amount of heat added or rejected
dW = small amount of work done
dU = small change in internal energy.

It must be remembered that this equation is derived from an application of the first law of thermodynamics to a closed system. The application of first law to such a system shows that out of the energy entering, some is changed in the form of stored energy while some appears as work.

4.3. FIRST LAW OF THERMODYNAMICS

It states that "*Heat and mechanical work are interconvertible*".

4.3.1. Explanation

When mechanical work is produced from heat, a definite quantity of heat goes out of existence for every unit of work done and conversely, when heat is produced by the expenditure of mechanical work the same definite quantity of heat comes into existence for every unit of work spent.

If W is the amount of mechanical work converted from heat energy H then;

$$W \propto H$$

or
$$W = JH \qquad ...(4.1)$$

where 'J' is a constant of proportionality called '*Joule's mechanical equivalent of heat*'.

From eqn. (4.1),

Joule's Mechanical equivalent of heat,

$$J = \frac{W}{H} \qquad ...(4.2)$$

So, the mechanical equivalent of heat is the work done per unit of heat expended.

The value of mechanical equivalent of heat, J

(in MKS system) = 426.934 Kgf-m/kcal

(Which is generally taken as 427 kgf-m/kcal.)

(in S.I. System) = 4.1868 KJ/Kcal

(in F.P.S. system) = 778.16 ft Ib/Btu.

In other words the first law of thermodynamics states that "*when a thermodynamic system executes a cyclic process, the algebraic sum of the work transfers is proportional to the algebraic sum of the heat transfers*". Thus;

$$\oint dW = J\oint dQ \qquad ...(4.3)$$

$$J\oint dQ = J\oint dW = 0 \qquad ...(4.4)$$

where dQ and dW represent infinitesimal elements of heat and work transfers respectively, with the conventional signs, while $\oint$ stands for adding up all the elements occurring on the cyclic process. $\oint$ is known as the Cyclic integral or 'integral around the cycle'.

There exist some *limitations* in the first law of thermodynamics as stated below :

(*i*) It does not specify the conditions under which the conversion of heat into work is possible neither the direction in which heat transfer can take place.

(*ii*) Although, the work can be conveniently converted into heat but it is impossible to convert the heat into work completely *i.e.*, only a fraction of heat supplied is converted into work. This means that heat and work are not completely interconvertible forms of energies. This fact also is not exploded during the discussion of first law of thermodynamics.

These limitations make it necessary to formulate another law which will carry out the desires of the first law of thermodynamics as much as possible. So, this forms the *basis* for the *'Second law of thermodynamics'*.

4.3.2. First Law of Thermodynamics for a Closed or Non-Flow System Undergoing a Cycle

Let us consider a closed system which consists of a known mass of water contained in an adiabatic vessel having a thermometer and a paddle wheel, as shown in the Fig. 4.1.

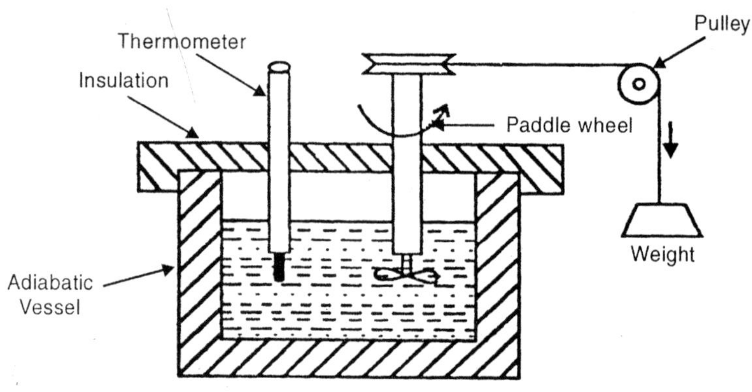

Fig. 4.1. Adiabatic Work.

Let us certain amount of work W_{1-2} be done upon the system by the paddle wheel. The quantity of work can be measured by the fall of weight which drives the paddle wheel through a pulley. The system was initially at temp. t_1 the same as that of atmosphere and after work transfer let the temperature rise to t_2. The pressure is always, one atm. The process 1-2 undergone by the system is shown in the Fig. 4.2 in generalised thermodynamic co-ordinates *X*, *Y*. Let the insulation now be removed. The system and the surroundings interact by heat transfer till the system returns to the original temperature t_1, attaining the condition of thermal equilibrium with the atmosphere. The amount of heat transfer Q_{2-1} from the system during this process 2-1 shown in the Fig. 4.2 can be estimated. The system thus executes a cycle, which consists of a definite amount of work input W_{1-2} to the system followed by the transfer of an amount of heat Q_{2-1} from the system. It has been found that this W_{1-2} is always proportional to the heat Q_{2-1}, and the constant of proportionality is called the **'Joule's mechanical equivalent of heat'**.

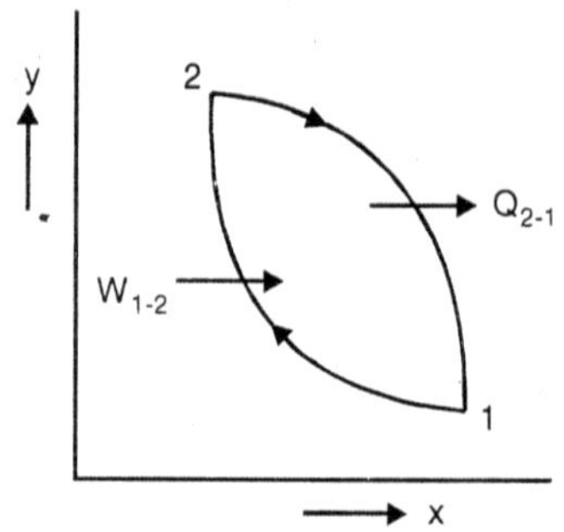

Fig. 4.2.

In the simple example given here, there are only two energy transfer quantities as the system performs a thermodynamic cycle. If the cycle involves many more heat and work quantities, the same result will be found.

Expressed algebraically,

$$\Sigma W \text{ cycle} = J(\Sigma Q) \text{ cycle}$$

where 'J' is the '*Joule's Mechanical equivalent of heat*'.

This is the first law for a closed system undergoing a cycle. In S.I. system of units both heat and work are measured in the derived unit of energy, the Joule. The constant of proportionality J is therefore unity ($J = I$ Nm/J).

4.3.3. First Law of Thermodynamics for an Open System

For any system and in any process, the first law can be written as,

$$Q = \Delta E + W \quad \text{...}(i)$$

where 'E' represents all forms of energy stored in the system. For a pure substance,

$$E = E_K + E_P + U$$

where 'E_k' is the kinetic energy E_P the potential energy and U the residual energy or internal energy stored in the molecular structure of the substance,

$\therefore$ From equation (i)

$$Q = \Delta E_k + \Delta E_p + U + W \quad \text{...(4.5)}$$

When there is 'mass transfer' across the system boundary, the system is called an '*open system*'. Most of the engineering devices are open system involving the flow of fluids through them.

The above equation refers to a system having a particular mass of substance and it is free to move from place to place.

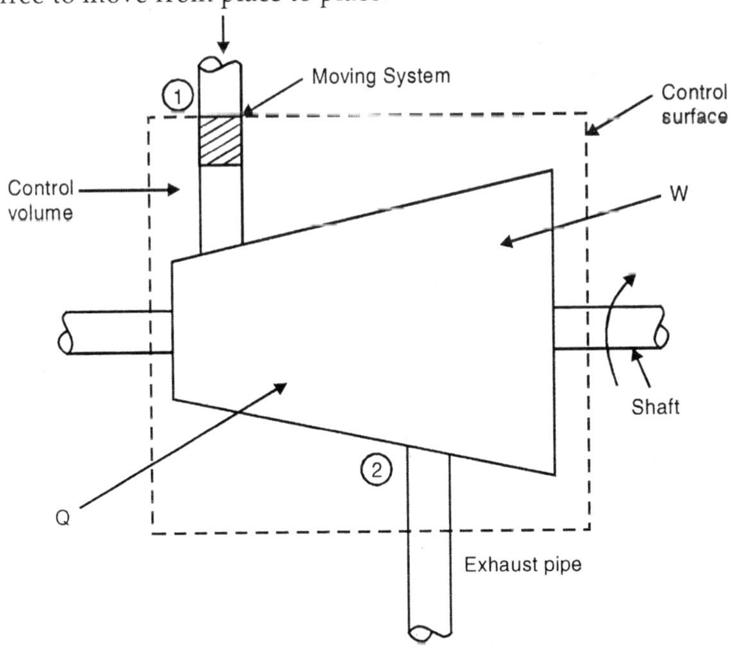

Fig. 4.3.

Consider a steam turbine as shown the Fig. 4.3 in which steam enters at a high pressure, does work upon the turbine rotor and then leaves the turbine at low pressure through the exhaust pipe.

If a certain mass of steam is considered as the thermodynamic system then the energy equation becomes,

$$\therefore \quad Q = \Delta E_K + \Delta Ep + U + W$$

In order to the expansion process in the turbine the moving system is to be followed as it travels through the turbine, taking into account the work and heat interactions all the way through. Sections 1 and 2 allow mass transfer to take place and Q and W are the heat and work interactions respectively.

4.3.4. Application of First Law of Thermodynamics to Non-Flow or Closed Systems

4.3.4.1. Constant-Volume Process or Isochoric Process

The process in which the volume of the given mass of the gas is not allowed to change, is known as 'Constant Volume Process'.

Let us consider a given mass 'm' of a gas be heated in a container of constant volume. As, there is no change in volume, so the process of heating can be represented by a vertical line on the pressure-volume (P-V) diagram, as shown in the Fig. 4.4.

As, the work done, during the process 1-2,

$$W_{12} = \int_1^2 p\, dv$$

or

$$W_{12} = p(V_2 - V_1)$$

As, the change in volume,

$$dv = 0$$

or

$$V_2 - V_1 = 0$$

$$\therefore \text{ Work done, } \quad W_{12} = 0 \quad \text{...(4.6)}$$

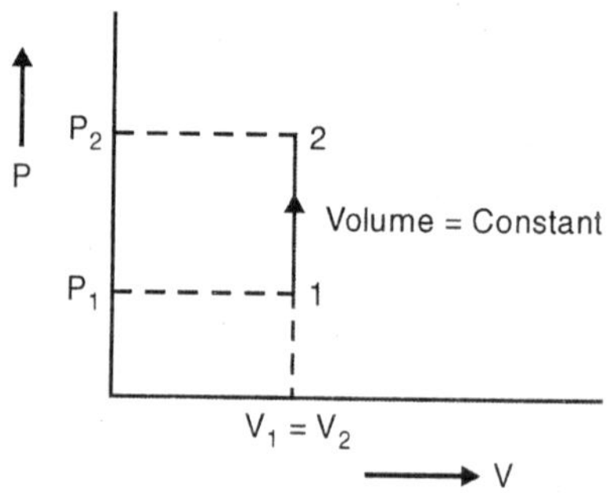

Fig. 4.4.

Now, applying the first law of thermodynamics,

$$Q_{12} - W_{12} = \Delta U_{12}$$

or

$$Q_{12} = \Delta U_{12} \quad (\because W_{12} = 0) \quad \text{...(4.7)}$$

Which means that in constant volume process;

Heat transferred = Change in internal energy

Also, for a perfect gas, we know that

Heat transferred, Q_{12} = Mass × Sp. Heat × Rise in temp.

$$\therefore \quad Q_{12} = m \,.\, C_v \,.\, (T_2 - T_1)$$

$\therefore$ Change in internal energy,

$$\Delta U_{12} = m \,.\, C_v \,.\, (T_2 - T_1) \quad \text{...(4.8)}$$

4.3.4.2. Constant-pressure Process or Isobaric Process

The process in which the pressure of the given mass of a gas is not allowed to change is known as 'Constant pressure process'.

Consider a given mass 'm' of a gas be heated at constant pressure 'p', as shown in Fig. 4.5.

The work done during the process 1-2,

$$W_{12} = \int_1^2 p_1\, dv.$$

or $$W_{12} = p(V_2 - V_1)$$

But, from Gas equation :

$$pV_1 = mRT_1$$

and $$pV_2 = mRT_2$$

or $$p(V_2 - V_1) = mR(T_2 - T_1)$$

$\therefore$ Work done, $W_{12} = m.R(T_2 - T_1)$ ...(4.9)

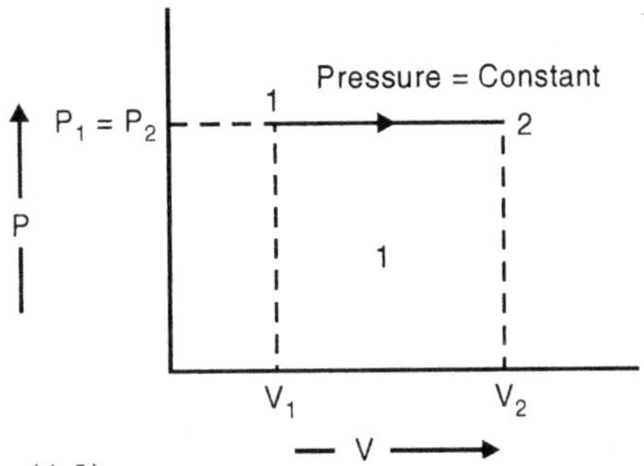

Fig. 4.5.

Now, applying first law of thermodynamics to this process,

$$Q_{12} - W_{12} = \Delta U_{12}$$

or Heat transferred, $Q_{12} = \Delta U_{12} + W_{12}$

or $$Q_{12} = (U_2 - U_1) + p(V_2 - V_1)$$

or $$Q_{12} = (U_2 + pV_2) - (U_1 + pV_1)$$

But, $U + pV$ = Enthalpy, h

$\therefore$ $$Q_{12} = h_2 - h_1$$

or $$Q_{12} = \Delta h_{12} \quad \text{...(4.10)}$$

Which means, that in Constant-Pressure process :

Heat transferred = Change in Enthalpy.

Also, for a perfect gas, we know that :

Heat transferred,

$$Q_{12} = \text{Mass} \times \text{Sp. heat at const. pressure} \times \text{Rise in temp.}$$

or $$Q_{12} = m \,.\, C_p \,.\, (T_2 - T_1)$$

$\therefore$ Change in internal energy,

$$U_{12} = m \,.\, C_p \,.\, (T_2 - T_1) \quad \text{...(4.11)}$$

4.3.4.3. Constant-Temperature or Isothermal Process

The process in which, the temperature of the given mass of a gas is not allowed to change, is known as 'constant-temperature process' or *'isothermal process'.*

In this process, because of constant temperature, the change in thermal energy, which is the function of temperature only remains zero.

$\therefore$ Change in internal energy, $\Delta U = 0$ ...(4.12)

Now, consider a given mass 'm' of a gas be heated from an initial state '1' to state '2' at constant temperature as shown in Fig. 4.6.

The work done during the process,

$$W_{12} = \int_1^2 p_1 dv.$$

Because, in this process the temperature is constant, so applying Boyle's law,

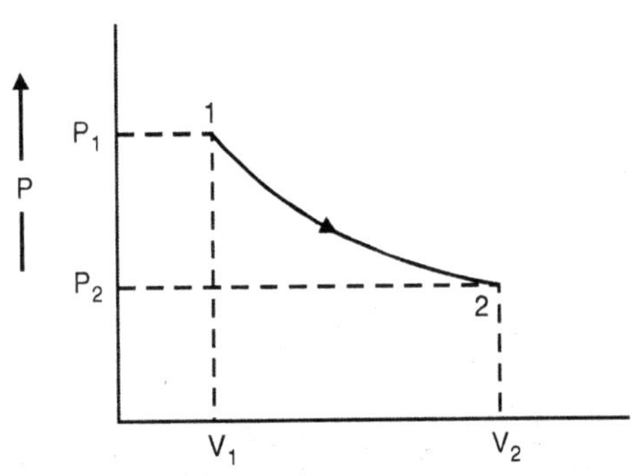

Fig. 4.6.

$$\therefore \quad p_1V_1 = p_2V_2 = pV = \text{Constant.}$$

$$\therefore \quad p = \frac{p_1V_1}{V} = \frac{p_2V_2}{V}$$

and Work done, $$W_{12} = \int_1^2 \frac{p_1V_1}{V}.dv$$

or $$W_{12} = p_1V_1\int_1^2 \frac{dv}{V}$$

or $$W_{12} = p_1V_1 \log_e \frac{V_2}{V_1}$$

or $$W_{12} = p_1V_1\log_e r \quad \left(\because \frac{V_2}{V_1} = r\right) \qquad ...(4.13)$$

or Work done, $$W_{12} = mRT \log_e r$$

Also, as per the first law of thermodynamics,

$$Q_{12} - W_{12} = \Delta U_{12}$$

But, Change in internal energy, $\Delta U_{12} = 0$

$$\therefore \quad Q_{12} = W_{12}$$

or Heat transferred,

$$Q_{12} = mRT \log_e r \qquad ...(4.14)$$

Hyperbolic Process. *The process in which the product of the pressure of the gas and its volume remains constant is known as Hyperbolic Process.* In the above case of isothermal process, it is clear that pV = Constant.

So, the work done and heat transferred during hyperbolic process can be calculated by using the formulae for the above isothermal process.

4.3.4.4. Adiabatic Process

The process in which, no heat is added or withdrawn from the system under consideration is known as an 'Adiabatic Process'. The process is shown in the Fig. 4.7 by the line 1-2.

During this process there is no heat transfer across the boundary of the system.

Therefore, Heat transferred, during the process 1-2,

$$Q_{12} = 0 \qquad ...(4.15)$$

Also, as per first law of thermodynamics,

$$Q_{12} - W_{12} = \Delta U_{12}$$

But, as $$Q_{12} = 0$$

$$\therefore \quad -W_{12} = \Delta U_{12} \qquad ...(4.16)$$

which means, work done, W_{12} = Change in internal energy.

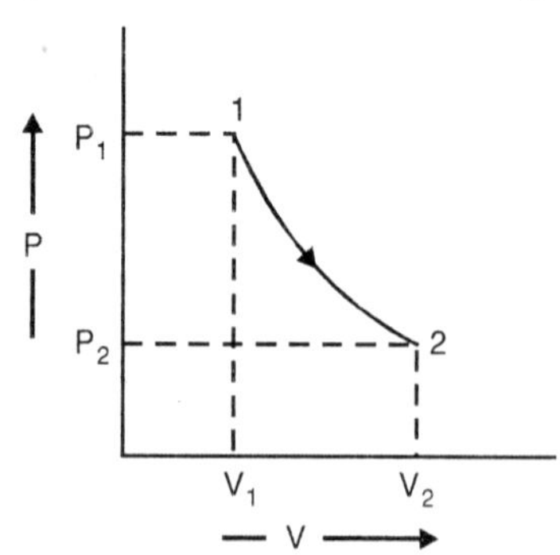

Fig. 4.7.

Also, for any perfect gas, we know that;

$$\Delta U_{12} = \text{Mass} \times \text{Sp. heat} \times \text{Change in temperature}$$

or $$\Delta U_{12} = m.\,C_v.\,(T_2 - T_1) \qquad ...(4.17)$$

The work done during the process, $W_{12} = \int_1^2 p\,dv$

$$\therefore \quad \int_1^2 p\,dv = m.C_V.(T_2 - T_1)$$

or $$p\,dv = -m.C_V.(dT). \qquad (\because dT = T_2 - T_1)$$

But, $$pV = MRT \text{ (Gas Equation)}$$

$$\therefore \quad p = \frac{m.R.T}{V}$$

$$\therefore \quad \frac{m.R.T}{V}.dv = -m.C_v.dT$$

or $$\frac{dT}{T} = -\frac{R}{C_v}.\frac{dv}{V}$$

Also, as $$\frac{R}{C_v} = \frac{C_P - C_v}{C_v} \qquad (\therefore C_P - C_v = R)$$

or $$= \frac{C_P}{C_v} - 1 \quad \text{or} \quad \gamma - 1$$

$$\therefore \quad \frac{dT}{T} = -(\gamma - 1)\frac{dv}{V} \qquad \left(\because \frac{C_p}{C_v} = \gamma\right)$$

Integrating both sides between the limits, we get

$$\log_e \frac{T_2}{T_1} = -(\gamma - 1)\log_e \frac{V_2}{V_1}$$

or $$\frac{T_2}{T_1} = \left(\frac{V_2}{V_1}\right)^{-\gamma-1}$$

or $$\frac{T_2}{T_1} = \left(\frac{V_1}{V_2}\right)^{\gamma-1}$$

or $$TV^{\gamma-1} = \text{Constant}$$

But, $$\frac{V_1}{V_2} = \frac{p_2.T_1}{p_1.T_2} \qquad \left(\because \frac{p_1V_1}{T_1} = \frac{p_2V_2}{T_2}, \text{ Gas} = n\right)$$

So, from above equation,

$$\left(\frac{T_2}{T_1}\right)^{\frac{1}{\gamma-1}} = \frac{p_2.T_1}{p_1.T_2}$$

or $$\frac{T_2}{T_1} = \left(\frac{p_2}{p_1}\right)^{\frac{\gamma-1}{\gamma}}$$

or $$Tp^{\frac{1-\gamma}{\gamma}} = \text{Constant}$$

From above equations

$$\left(\frac{p_2}{p_1}\right)^{\frac{\gamma-1}{\gamma}} = \left(\frac{V_1}{V_2}\right)^{\gamma-1}$$

or $$\left(\frac{p_2}{p_1}\right) = \left(\frac{V_1}{V_2}\right)^{\gamma}$$

$$pV^{\gamma} = \text{Constant}$$

Now, work done during the process 1-2,

$$W_{12} = \int_1^2 p.dv$$

Also $$pV^{\gamma} = p_1V_1^{\gamma} = p_2V_2^{\gamma}$$

$\therefore$ $$p = \left(\frac{p_1V_1^{\gamma}}{V^{\gamma}}\right)$$

$\therefore$ $$W_{12} = \int_1^2 p_1\left(\frac{V_1}{V}\right)^{\gamma}.dv$$

$$= p_1V_1^{\gamma}.\int_1^2 V^{-\gamma}.dv$$

or $$W_{12} = p_1V_1^{\gamma}\left[\frac{V^{-\gamma+1}}{-\gamma+1}\right]_1^2$$

$$= \frac{p_1V_1^{\gamma}}{1-\gamma}\left[V_2^{1-\gamma} - V_1^{1-\gamma}\right]$$

$$= \frac{p_1V_1^{\gamma}.V_2^{1-\gamma} - p_1.V_1^{\gamma}.V_1^{1-\gamma}}{1-\gamma}$$

$$= \frac{p_2V_2 - p_1V_1}{1-\gamma} \qquad (\because p_1V_1^{\gamma} = p_2V_2^{\gamma})$$

or $$W_{12} = \frac{p_1V_1 - p_2V_2}{\gamma - 1} \qquad ...(4.18)$$

or $$W_{12} = \frac{m.R(T_1 - T_2)}{\gamma - 1}$$

Isentropic Process

The process, such as reversible adiabatic process, in which the entropy of the system remains constant is known as an *'isentropic process'*.

Here are some of the *conditions* which must be fulfilled to make the adiabatic process reversible *i.e.* isentropic.

(*i*) The process should be frictionless.

(*ii*) No heat should be transferred to or from the system.

(*iii*) Work should be done by the gas or on the gas.

Here, it is important to note that an isentropic process is not possible in practice, because the friction cannot be eliminated completely from the system. Moreover, some heat is transferred between the gas and the cylinder walls, so the heat transfer from the system or to the system is also unavoidable.

4.3.4.5. Polytropic Process

The process in which the relation pV^n = Constant holds good, is known as 'Polytropic Process', where 'p' is the pressure, 'V' the volume of the gas and 'n' is the polytropic exponent, the value of which varies from zero to infinity for various gases, as shown in the Fig. 4.8.

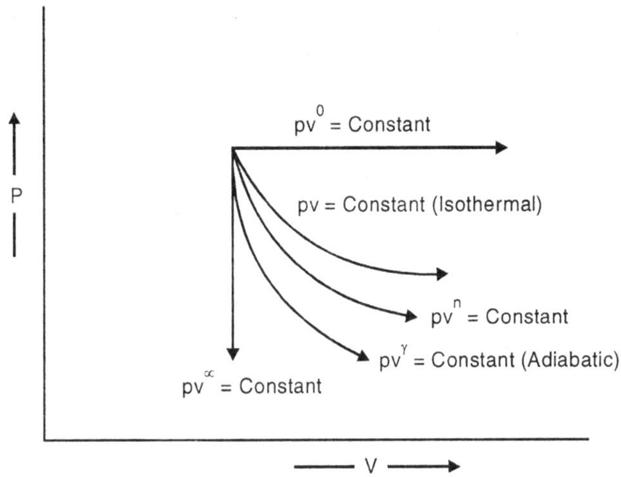

Fig. 4.8.

For constant volume process, $n = \infty$

For constant pressure process, $n = 0$

For an isothermal or constant temperature process

$$n = 1$$

Also, from 1st law of thermodynamics, we have

$$Q_{12} - W_{12} = \Delta U_{12}$$

But work done during process 1-2

$$W_{12} = \int_1^2 pdv$$

For polytropic process.

$$pV^n = p_1 V_1^n$$

or

$$p = \frac{p_1 V_1^n}{V^n}$$

$\therefore$ Work done during the process 1-2,

$$W_{12} = p_1 V_1^n \int_1^2 \frac{dv}{V^n}$$

Which gives,

$$W_{12} = \frac{p_2 V_2 - p_1 V_1}{1 - n}$$

or

$$W_{12} = \frac{p_1 V_1 - p_2 V_2}{n - 1} \qquad ...(4.19)$$

Also,

$$W_{12} = \frac{m.R(T_1 - T_2)}{n - 1}$$

Change in internal energy, during process 1-2.

$$\Delta U_{12} = \text{Mass} \times \text{Sp. heat} \times \text{Change in temperature}$$

or

$$\Delta U_{12} = M \,.\, C_v \,.\, (T_2 - T_1) \qquad ...(4.20)$$

$$\therefore \quad Q_{12} - \frac{m.R(T_2 - T_1)}{(1-n)} = m \,.\, C_v \,.\, (T_2 - T_1)$$

or Heat transferred,

$$\therefore \quad Q_{12} = m \,.\, C_v \,.\, (T_2 - T_1) + \frac{m.R(T_2 - T_1)}{(1-n)}$$

or $$Q_{12} = m.\frac{R}{(\gamma - 1)}.(T_2 - T_1) + \frac{mR(T_2 - T_1)}{(1-n)} \qquad \left(\because \frac{R}{C_v} = \gamma - 1\right)$$

or $$= \frac{m.R(T_2 - T_1)}{(\gamma - 1)}\left(1 + \frac{\gamma - 1}{1 - n}\right)$$

$$= \frac{m.R(T_2 - T_1)}{(\gamma - 1)}\left(\frac{1 - n + \gamma - 1}{1 - n}\right)$$

$$= \frac{(\gamma - n)}{(1-n)(\gamma - 1)} \,.\, m.R(T_2 - T_1) = \frac{(\gamma - n)}{\gamma - 1} \cdot \frac{m.R(T_2 - T_1)}{(1-n)}$$

or $$Q_{12} = \frac{\gamma - n}{\gamma - 1}.W_{12} \qquad ...(4.21)$$

4.3.4.6. Throttling Process

When a gas or vapour is expanded through an aperture of minute (narrow) dimensions like a valve, which is very slightly opened, the process is known as 'throttling process'.

The fluid is forced through the narrow aperture under pressure, but the velocity of the fluid flowing out is reduced up to a negligible amount, by the high resistance offered by the friction, between the fluid and the walls of the aperture. Hence, the kinetic energy of the flowing fluid is very small.

In fact the kinetic energy of the fluid is converted into heat energy, which warms up the fluid to its initial temperature. Thus, no heat is supplied from or rejected to an external source and there is no external work done. Also, there is no change in temp. in the case of a perfect gas. So, in the throttling process;

Work done, $W = 0$ ...(4.22)

Heat supplied, $Q = 0$ ...(4.23)

and change in internal energy,

$\Delta U = 0$ ...(4.24)

4.3.4.7. Free Expansion

In this process *the fluid expands suddenly into a vacuum chamber, through an orifice of large dimension.* No heat is supplied or rejected, there is no work done and during the process the temperature remains constant. So, in free expansion process, also;

Work done $W = 0$ ...(4.25)

Heat supplied, $Q = 0$ ...(4.26)

and change in internal energy,

$\Delta U = 0$ ...(4.27)

4.3.4.8. Summary of Various Methods of Heating and Expanding Gases and Vapours

Table 4.1.

S. No.	Process	Workdone, W (N-m) or J	Change in Internal Energy, ΔU(J)	Heat transferred, Q(J)	Relation between p, V and T
1.	Constant Volume *i.e.* ; (V = Constant)	Zero	$m.C_v.(T_2 - T_1)$	$m.C_v.(T_2 - T_1)$	$\frac{p_1}{T_1} = \frac{p_2}{T_2}$
2.	Constant Pressure *i.e.* ; p = Constant	$p(V_2 - V_1)$	$m.C_v.(T_2 - T_1)$	$m.C_p.(T_2 - T_1)$	$\frac{V_1}{T_1} = \frac{V_2}{T_2}$
3.	Constant Temp. or isothermal *i.e.*, PV = Constant	$p_1 V_1 \log_e r$	Zero	$p_1 V_1 \log_e r$	$p_1V_1 = p_2V_2$
4.	Adiabatic pV^γ = Constant	$\frac{p_1V_1 - p_2V_2}{\gamma - 1}$	$m.C_v.(T_2 - T_1)$	Zero	$p_1V_1^\gamma = p_2V_2^\gamma$ $T_1V_1^{\gamma-1} = T_2V_2^{\gamma-1}$ $T_1p_1^{\frac{1-\gamma}{\gamma}} = T_2p_2^{\frac{1-\gamma}{\gamma}}$
5.	Polytropic pV^n = Constant	$\frac{p_1V_1 - p_2V_2}{n - 1}$	$m.C_v.(T_2 - T_1)$	$\frac{\gamma - n}{\gamma - 1} \cdot \frac{mR(T_2 - T_1)}{(1 - n)}$	$p_1V_1^n = p_2V_2^n$ $T_1V_1^{n-1} = T_2V_2^{n-1}$ $T_1p_1^{\frac{1-n}{n}} = T_2p_2^{\frac{1-n}{n}}$
6.	Throttling	Zero	Zero	Zero	—
7.	Free Expansion	Zero	Zero	Zero	—

4.3.5. Steady Flow Energy Equation (SFEE)

In Fig. 4.9 a steady flow system has been shown in which, one stream of fluid enters and one stream leaves the control volume. Before going for the steady flow energy equation, let us first discuss the conditions, which must be satisfied for the steady flow process :

(*i*) The mass flow rate through the system remains constant.

(*ii*) The state of the fluid, at any point in the system is constant, at all times.

(*iii*) The rate of heat-transfer and work-transfer remains constant.

(*iv*) The rate of energy transfer, at the boundaries is constant and the energy contained by the fluid at any point remains constant.

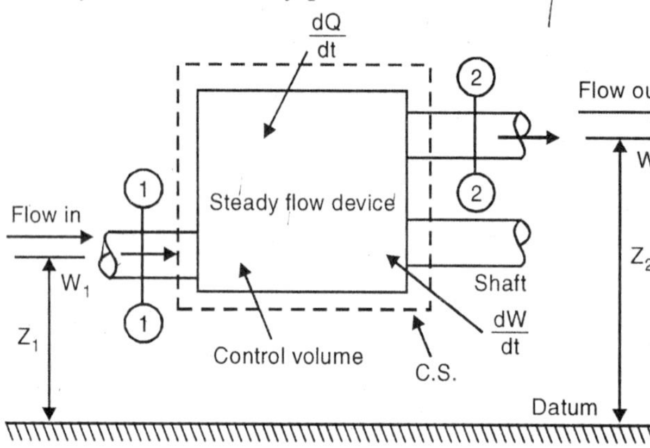

Fig. 4.9. Steady Flow Process.

Consider an open system, in which the working substance flows at a steady rate. Let the mass i.e., the working substance enters the system at section 1-1 and leaves the system at section 2-2.

Let, m = Mass flow rate of the working substance in kg/s at the entrance or exit.

P_1, P_2 = Absolute pressure of the working substance in N/m^2 at the entrance and exit.

V_1, V_2 = Velocity of the working substance in m/s at the entrance and exit.

v_1, v_2 = Specific volume in m^3/kg at the entrance and exit.

z_1, z_2 = Elevation above the datum in 'm' at the entrance and exit.

U_1, U_2 = Specific internal energy in J/kg at the entrance and exit.

Q = Quantity of heat introduced into the system in J.

W = Amount of work done in J.

We know, from the law of conservation of energy.

Total energy at the entrance = Total energy at the exit

As the total energy at the entrance is equal to the sum of internal energy, kinetic energy, potential energy, flow energy, and heat energy supplied to the system, whereas total energy at the exit is equal to the sum of internal energy, kinetic energy, potential energy, flow energy and the work done.

So, $m\left[U_1 + \frac{V_1^2}{2} + gZ_1 + P_1 v_1\right] + Q = m\left[U_2 + \frac{V_2^2}{2} + gZ_2 + P_2 v_2\right] + W$

Also, from the definition of enthalpy,

Enthalpy, h = Internal energy + Pressure – Vol. product

or $h = U + Pv$

Thus, the above equation becomes;

$$m\left[h_1 + \frac{V_1^2}{2} + gZ_1\right] + Q = m\left[h_2 + \frac{V_2^2}{2} + gZ_2\right] + W$$

This is the 'Steady Flow Energy Equation'.

4.3.5.1. Application of Steady-Flow-Energy-Equation to Turbines, Pump, Boilers, Compressors, Nozzles Evaporators

(*i*) Turbines :

(*a*) *Water turbine.* In a water turbine as shown in Fig. 4.10 the water is supplied to the turbine rotor blades from a high head. The potential energy of this water is converted into kinetic energy, when it enters inside the turbine. A part of this kinetic energy is converted into useful work and ultimately to generate electricity.

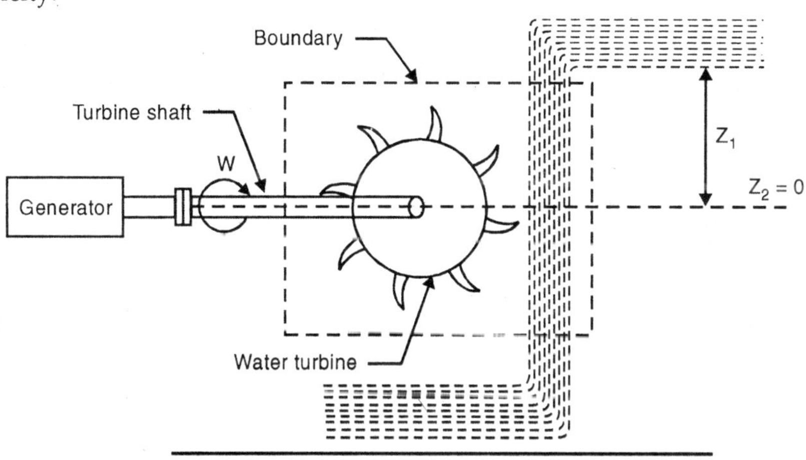

Fig. 4.10. Water turbine.

Consider centre of the turbine shaft as datum or reference. The energy equation can be written as follows:

$$u_1 + p_1V_1 + \frac{v_1^2}{2} + Z_1 g + \frac{dQ}{dm} = u_2 + p_2.V_2 + \frac{v_2^2}{2} + Z_2 g + \frac{dW_x}{dm}$$

Here; $Q = 0$

$\Delta u = u_2 - u_1 = 0$

$V_1 = V_2 = V,$

$Z_2 = 0$

$$\therefore \left[p_1V + Z_1 g + \frac{V_1^2}{2}\right] = \left[p_2V + Z_2 g + \frac{V_2^2}{2}\right] + \frac{dW_x}{dm} \qquad \ldots\left(\frac{53}{4}\times 25\right)$$

$\frac{dW_x}{dm}$ is positive because work is done by the system *i.e.*, work comes out of the boundary of the system.

(*b*) *Steam/Gas Turbine.* In a steam/gas turbine the working fluid is passed through the turbine. A part of its energy is converted into work by the turbine, which further runs a generator to ultimately produce the required electricity as shown in the Fig. 4.11 below. The temp. and pressure of the steam/gas leaving the turbine is decreased :

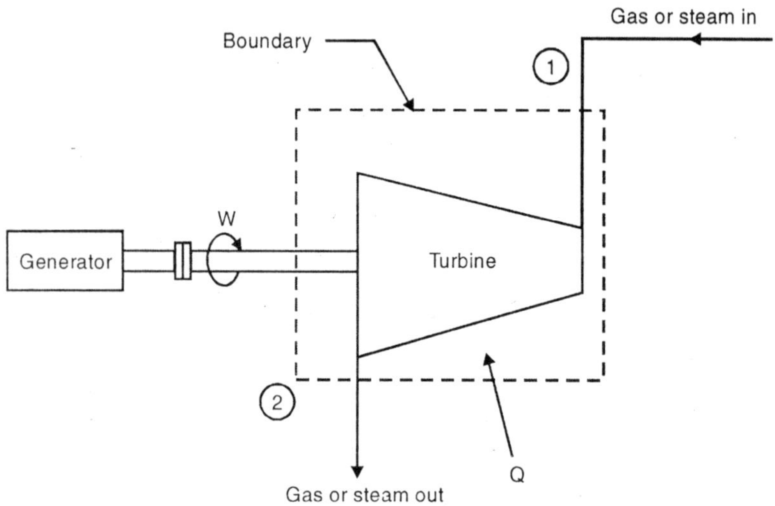

Fig. 4.11. Steam or gas turbine.

Applying S.F.E.E. to this system;
where $Z_1 = Z_2$ or $Z_1 - Z_2 = \Delta Z = 0$

$$h_1 + \frac{v_1^2}{2} - \frac{dQ}{dm} = h_2 + \frac{v_2^2}{2} + \frac{dW_x}{dm}$$

The –ve sign of $\frac{dQ}{dm}$ is because the heat is rejected by the system.

And the +ve sign of $\frac{dW_x}{dm}$ is because the work is done by the system.

(*ii*) **Pump.** Here the case of centrifugal water pump as shown in the Fig. 4.12. is being taken for study.

A centrifugal water pump sucks water from a lower level and deliver it to a higher level. Some work is required to spend to run the pump which is supplied by some external source (may be an electric motor or an I.C. engine).

Here, $\frac{dQ}{dm} = 0$ and $\Delta U = 0$, as there is no temperature gradient

$$V_1 = V_2 = V$$

So, applying S.F.E.E. to this system;

$$p_1V_1 + Z_1g + \frac{V_1^2}{2} = p_2V_2 + Z_2g + \frac{V_2^2}{2} - \frac{dW}{dm}$$

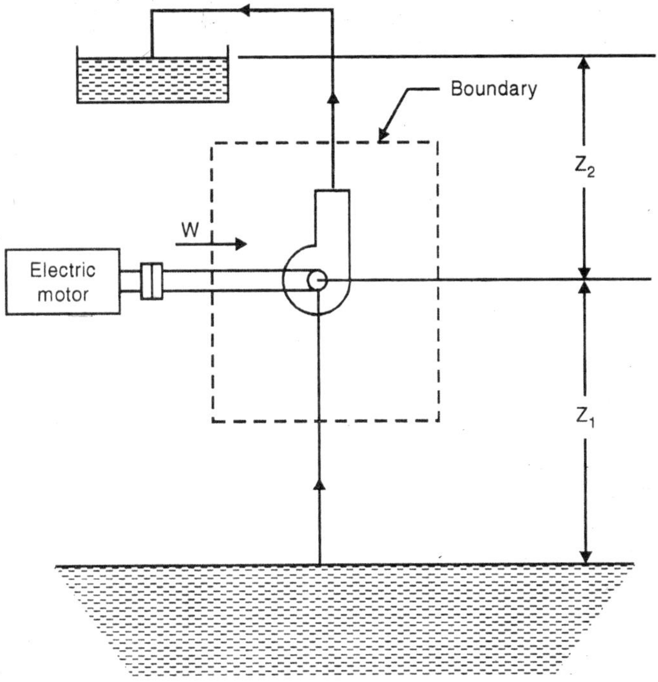

Fig. 4.12. Centrifugal water pump.

The –ve sign of $\frac{dW}{dm}$ indicates that the work is spent on the system.

***(iii)* Boiler.** A boiler or steam generator, transfers the heat of combustion of fuel to the incoming water and converts this water into high temperature/superheated steam as shown in the Fig. 4.13.

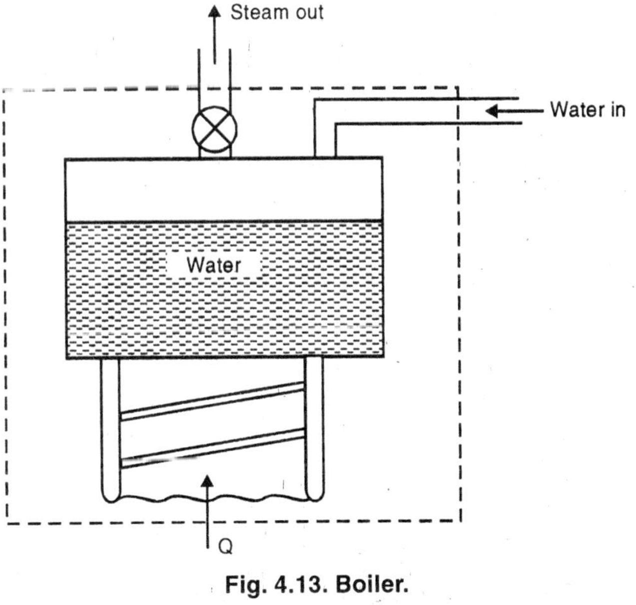

Fig. 4.13. Boiler.

Here, $\Delta z = 0, \Delta \frac{V_2^2}{2} = 0$

$\frac{dW}{dm} = 0$, (Since neither any work is done by the system nor on the system).

So, applying S.F.E.E. to the system

$$h_1 + \frac{dQ}{dm} = V_2$$

***(iv)* Compressor.** As shown in the Fig. 4.14, a reciprocating air compressor draws air from the atmosphere and supplies it at a considerably higher pressure, for various applications. This system can be considered as steady flow system when the control volume includes the receiver, which reduces the fluctuations in the air flow considerably.

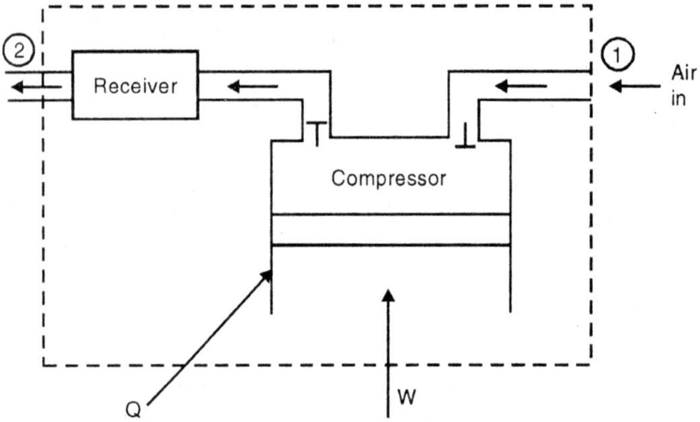

Fig. 4.14. Reciprocating compressor.

Now, applying S.F.E.E. to this system;
where change in P.E. = 0, and change in K.E. = 0

$$\therefore \quad h_1 - \frac{dQ}{dm} = h_2 - \frac{dW}{dm}$$

***(v)* Nozzles.** Refer Fig. 4.15 showing a convergent-divergent nozzle. Here the enthalpy of the working fluid decreases and the pressure, drops whereas simultaneously the flow of working fluid is accelerated. This is generally used to convert a part of the energy of steam into the kinetic energy of steam which is to be supplied to the turbine.

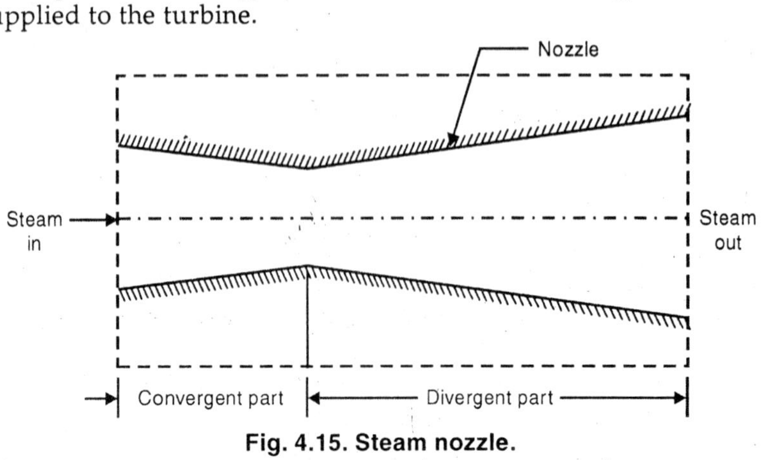

Fig. 4.15. Steam nozzle.

In this case; change in P.E. = 0

$$\frac{dW}{dm} = 0$$

$$\frac{dQ}{dm} = 0$$

So, applying S.F.E.E. to this system;

$$h_1 + \frac{v_1^2}{2} = h_2 + \frac{v_2^2}{2}$$

or $$\frac{v_2^2}{2} - \frac{v_1^2}{2} = h_1 - h_2$$

or $$v_2^2 - v_1^2 = 2(h_1 - h_2)$$

or $$v_2^2 = v_1{}^2 + 2\,(h_1 - h_2)$$

$\therefore$ $$v_2 = \sqrt{v_1^2 - 2(h_1 - h_2)}$$

where the velocity 'V' is in m/s and enthalpy 'h' in joules.

If $v_1 << v_2$ then

$$v_2 = \sqrt{2(h_1 - h_2)}$$

Hence, $$v_2 = \sqrt{2\,\Delta\, h}$$

(vi) **Evaporator.** In an evaporator as shown in Fig. 4.16, the heat is extracted from the system. It is generally used in refrigeration plant to carry heat from the refrigerator and thus, to maintain a low temperature. In this the working fluid or liquid refrigerant is passed through the evaporator and it comes out as vapour after absorbing its latent heat from the surroundings of the evaporator.

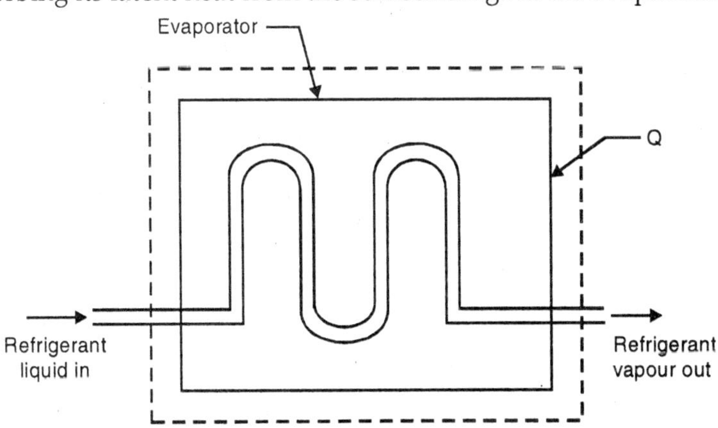

Fig. 4.16. Evaporator.

Here, Change in P.E. = 0

Change in K.E. = 0

Also, $\frac{dW}{dm} = 0$ ($\because$ No work is absorbed or supplied by the system.)

So, applying the S.F.E.E. to this system.

$$h_1 + \frac{dQ}{dm} = h_2$$

$\frac{dQ}{dm}$ is taken as +ve because heat flows from the surroundings to the system as the temperature in the system is lower than the surroundings.

4.3.6. Second Law of Thermodynamics

4.3.6.1. Kelvin-Plank Statement

It states "*It is impossible to construct an engine working on cyclic process whose sole purpose is to convert all the heat energy supplied to it into an equivalent amount of mechanical work*".

In other words no heat engine converts or can convert more than a small fraction of the heat supplied to it, into mechanical work, a large part is necessarily rejected as heat as shown in the Fig. 4.17.

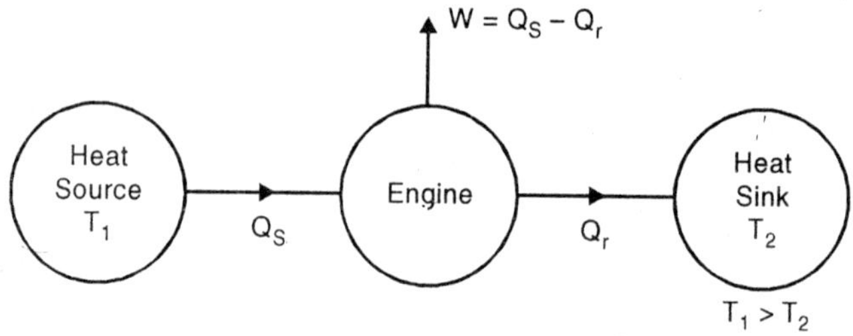

Fig. 4.17.

where Q_S = Heat supplied

Q_r = Heat rejected

and W = Work developed by the engine.

The ratio,

$$\frac{\text{Heat Converted into a work}}{\text{Heat Energy taken in by the engine}} = \frac{W}{Q_S}$$

is thus a fraction always much less than unity. This fraction is called the '*thermal efficiency*' of the engine.

4.3.6.2. Clausius Statement

Clausius stated that "*Heat flows from a hot body to a cold body unaided, but it is impossible for heat to flow from a cold body to a hot body without the aid of external work*".

It means that a body can only absorb heat from a source which is at a higher temperature than the body itself. If it is required to lower the temperature of a body below that of its surroundings, mechanical work must be expended, such as in a refrigerating system.

For further illustration of this statement refer the Fig. 4.18.

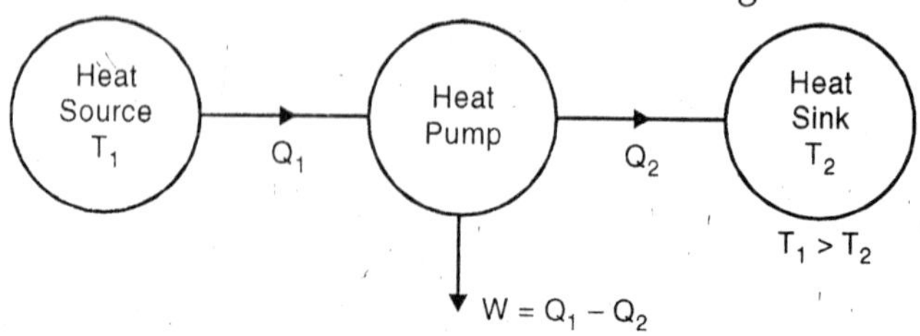

Fig. 4.18.

Thus, if heat source is at a temperature T_1 and a heat sink (a heat reservoir) is at a temperature T_2 with T_1 being more than T_2 then according to Clausius statement heat cannot flow from heat sink to heat source without the aid of an external agency like a heat pump (which takes mechanical work and delivers heat continuously.

4.3.6.3. Equivalence of Statements

Consider a reservoir at higher temperature 'T_1' and another reservoir at a lower temperature 'T_2'; as shown in the Fig. 4.18.

Assume a heat pump which requires no work and transfers heat energy by an amount of Q_2 from a low temperature to a high temperature reservoir, which is in violation to the clausius statement.

Now, let heat energy of an amount 'Q_1' which is more than 'Q_2' be transferred from the high temperature reservoir to the heat engine which ultimately develops a net work 'W' equal to $(Q_1 - Q_2)$ and rejects heat energy 'Q_2' to the low temperature reservoir.

Since, there is no heat energy interaction with the low temperature so it can be eliminated.

The combined system of the heat engine and heat pump acts like a heat engine *i.e.*, it becomes a perpetual motion machine of the second kind which is in violation of the Kelvin-Plank Statement. Hence, our assumption that heat can flow of its own accord from cold reservoir to hot reservoir is wrong. So, both the statements Clausius and Kelvin-Plank's for second law of thermodynamics are equivalent.

4.3.7. Perpetual Motion Machine of the First Kind-PMM 1

The 'First law of Thermodynamics' states the general principle of the 'Law of Conservation of Energy' *i.e.*, "Energy can neither be created nor destroyed but can be transformed from one form to the another."

There can be no machine which would continuously supply mechanical work without some form of energy disappearing simultaneously. Such type of fietritus machine is called a 'Perpetual Motion Machine of First kind'-PMM 1. So, a PMM 1 is impossible.

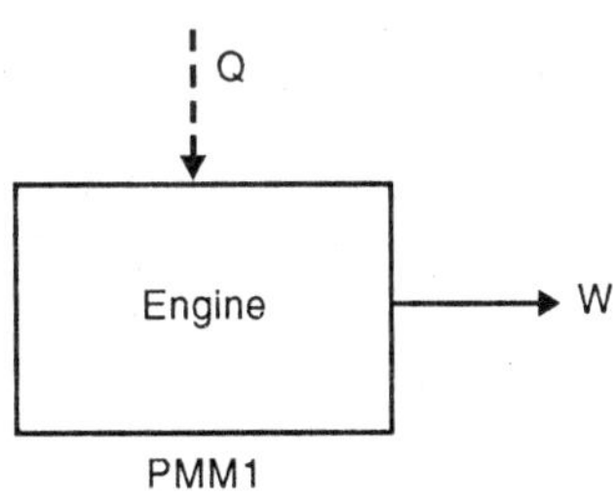

Fig. 4.19.

4.3.8. Perpetual Motion Machine of Second Kind-PMM 2

An imaginary machine which would continuously absorb heat from a single hot reservoir and convert all its heat, completely into work *i.e.*, with 100% efficiency—is called 'Perpetual Motion Machine of Second kind'-PMM 2.

The Fig. 4.20 shows the PMM 2 which obviously violates the 'Second Law of thermodynamics'.

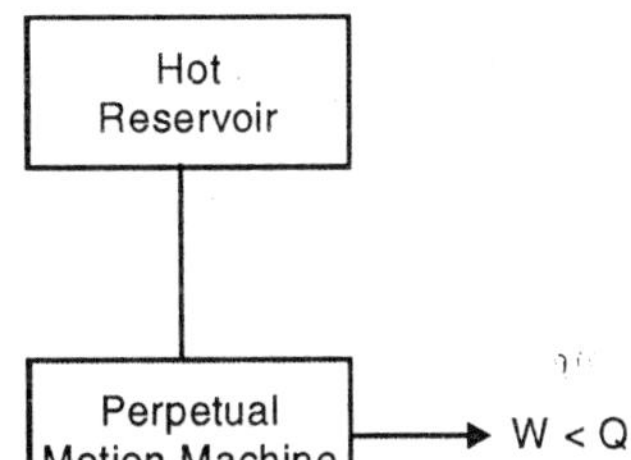

Fig. 4.20. Perpetual motion machine of second kind (PMM 2)

4.3.9. Carnot Engine

An engine working on 'Carnot Cycle' and having max. possible efficiency is called 'a Carnot Engine'

4.3.10. Introduction to Third Law of Thermodynamics

It states that 'When a system is at zero absolute temperature, the entropy of the system is zero".

4.3.11. Entropy

'Entropy is a function of the quantity of heat which shows the possibility of conversion of that heat into useful work'.

The increase in entropy is small when heat is added at a high temperature and is greater when the heat addition is made at a lower temperature. Thus, for maximum entropy there is minimum availability for conversion into work and for minimum entropy there is maximum availability for conversion into work.

NUMERICAL EXAMPLES

Example 4.1. During an expansion of a gas, it is observed that 10,000 N-m of work is done on the piston and 7.0 KJ of heat is absorbed. Calculate the work required to compress the gas to its initial condition if 2.0 KJ of heat is rejected to the surroundings.

Solution. We know

$$Q_{12} - W_{12} = \Delta U_{12} \text{ or } U_2 - U_1$$

During the expansion process (say 1–2),

Work done, $W_{12} = +10{,}000$ N–m $= 10{,}000$ Joule (∵ 1 Nm = 1 Joule)

or $$W_{12} = \frac{10{,}000}{1{,}000} = 10 \text{ KJ}$$

Heat absorbed, $Q_{12} = +7$ KJ

∴ $U_2 - U_1$ *i.e.*; Change in internal energy $= 7 - 10 = -3$ KJ

–ve sign shows that there is loss of internal energy during the expansion process.

Also, during the compression process (2–1)

$$Q_{21} - W_{21} = \Delta U_{21} \text{ or } U_1 - U_2$$

or Work done, $$W_{21} = -(U_1 - U_2) + Q_{21} = -(-3) + 2$$

$$= 3 + 2 = 5 \text{ KJ}$$

or Work done, $W_{21} = 5 \times 1000 = 5{,}000$ J or N-m **(Ans.)**

Example 4.2. How much heat must be supplied to 1.5 kg of a gas to raise its temperature from 90°C to 225°C at constant pressure? Find also the external work done during the supply of heat. Take $C_P = 1.0$ KJ/kg K, $C_v = 0.7128$ KJ/kg K.

Solution. Given data :

Mass, $m = 1.5$ kg.

$$T_1 = 90 + 273 = 363 \text{ K}$$
$$T_2 = 225 + 273 = 498 \text{ K}$$
$$C_p = 1 \text{ and } C_v = 0.7128 \text{ KJ/Kg K}$$

Total heat supplied,

$$Q = m \,.\, C_p.\, (T_2 - T_1) = 1.5 \times 1 \times (498 - 363)$$
$$= 1.5 \times 135 = 202.5 \text{ KJ } \textbf{(Ans.)}$$

Change in internal energy,

$$\Delta U = m \,.\, C_v \,.\, (T_2 - T_1) = 1.5 \times 0.7128 \times (498 - 363)$$
$$= 144.34 \text{ KJ}$$

Now, $Q = W + \Delta U$

$\therefore$ External work done,

$$W = Q - \Delta U = 202.5 - 144.34$$
$$= 58.16 \text{ KJ } \textbf{(Ans.)}$$

Example 4.3. A quantity of a gas occupying 0.08 m^3 at 8.4 bar abs. pressure receives heat at constant pressure until the volume increases to 0.16 m^3. The initial temperature is 200°C and $C_p = 1$, $C_v = 0.7128$ KJ/Kg °k. Find the change in internal energy and external work done during the heat supply.

Solution. Given : $V_1 = 0.08 \text{ m}^3$

$$p_1 = 8.4 \text{ bar} = 8.4 \times 10^5 \text{ N/m}^2 \text{ or Pa.}$$
$$V_2 = 0.16 \text{ m}^3$$
$$C_p = 1,\ C_v = 0.7128 \text{ KJ/Kg k}$$

$\therefore$ $$C_p - C_v = R$$
$$1 - 0.7128 = R$$

or $$R = 0.287 \text{ KJ/Kg k}$$

or $$R = 287 \text{ J/Kg k}$$

Now, $$p_1V_1 = m.R.T_1$$

So, $$m = \frac{p_1V_1}{RT_1} = \frac{8.4 \times 10^5 \times 0.08}{287 \times 473}$$

or $$m = 0.495 \text{ Kg.}$$

Now, applying Charle's law:

$$\frac{V_1}{T_1} = \frac{V_2}{T_2}$$

$\therefore$ $$T_2 = \frac{V_2}{V_1} \times T_1 = \frac{0.16}{0.08} \times 473 = 946 \text{ °K}$$

Change in internal energy,

$$\Delta U = m \,.\, C_v \,.\, (T_2 - T_1)$$

or $$\Delta U = 0.495 \times 0.7128\ (946 - 473)$$
$$= 0.495 \times 0.7128 \times 473$$

or $$\Delta U = 166.89 \text{ KJ } \textbf{(Ans.)}$$

Total heat supplied,

$$Q = m.\, C_p \,.\, (T_2 - T_1) = 0.495 \times 1 \times (946 - 473)$$
$$= 0.495 \times 473$$

or $Q = 234.13$ KJ

Now, $Q = W + \Delta U$

∴ External work done,

$$W = Q - \Delta U$$
$$= 234.13 - 166.89 = 67.24 \text{ (Ans.)}$$

Example 4.4. A closed vessel contains 2.0 kg of a gas at 20°C and 0.7 bar abs. pressure. Heat is supplied to the vessel till the gas acquires a pressure of 1.4 bar abs. Find out;

(*i*) Final temperature of gas

(*ii*) Work done

(*iii*) Heat added and

(*iv*) Change in internal energy.

Take $C_v = 0.7128$ **KJ/Kg k.**

Solution. Given : $p_1 = 0.7$ bar abs.

$T_1 = 20 + 273 = 293$ K

$p_2 = 1.4$ bar abs.

$T_2 = ?$

As the gas is heated in a closed vessel.

∴ $V_1 = V_2$

(*i*) During constant volume process

$$\frac{p_1}{T_1} = \frac{p_2}{T_2}$$

or
$$T_2 = \frac{p_2}{p_1} \times T_1 = \frac{1.4}{0.7} \times 293$$

or
$$T_2 = 586 \text{ K}$$

∴
$$T_2 = 586 - 273 = 313 \text{ °C}$$

(*ii*) Work done, $W = 0$ (as the change in volume, $dv = 0$)

(*iii*) Heat added, $Q_{12} = m \cdot C_v \cdot (T_2 - T_1)$

$$= 2 \times 0.7128 (586 - 293)$$

or
$$Q_{12} = 417.7 \text{ KJ (Ans.)}$$

(*iv*) Change in internal energy

$$\Delta U_{12} = Q_{12}$$

or
$$\Delta U_{12} = 417.7 \text{ KJ (Ans.)}$$

Example 4.5. The value of C_p and C_v for an ideal gas are 1.0 and 0.7128 KJ/Kg K respectively. Find the values of R and ratio of specific heats, γ for the gas. If one Kg of this gas is heated at constant pressure from 25 °C to 200 °C, estimate the heat added, ideal work done and change in internal energy. Also, calculate the pressure and final volume if initial volume is 2 m^3.

Solution. Given : $C_p = 1$

and $C_v = 0.7128$ KJ/Kg k

$V_1 = 2 \text{ m}^3$

$T_1 = 25 + 273 = 298$ k

$$T_2 = 200 + 273 = 473 \text{ k}$$

We know : $C_p - C_v = R$

$\therefore \quad 1 - 0.7128 = R$

or $\quad R = 0.2872$ KJ/Kg k

or $\quad R = 0.287$ J/Kg k **(Ans.)**

Again : $\quad \gamma = \dfrac{C_p}{C_v} \dfrac{1}{0.7128} = 1.4$ **(Ans.)**

During the constant pressure process :

Heat transferred, $Q_{12} = C_p (T_2 - T_1) = 1 (473 - 298)$

or $\quad Q_{12} = 175$ KJ **(Ans.)**

Change in internal energy,

$$\Delta U_{12} = C_v . (T_2 - T_1) = 0.7128 (473 - 298)$$

or $\quad \Delta U_{12} = 124.74$ KJ **(Ans.)**

We know : $Q_{12} - W_{12} = \Delta U_{12}$

$\therefore$ Work done, $W_{12} = Q_{12} - \Delta U_{12} = 175 - 124.74$

$= 50.26$ KJ **(Ans.)**

Now, from characteristics gas equation,

$$10^5 . p_1 V_1 = m . RT_1$$

$$\therefore \quad p_1 = \frac{mRT_1}{10^5 \times V_1} = \frac{1 \times 287 \times 298}{10^5 \times 2}$$

or $\quad p_1 = 0.4276$ bar **(Ans.)**

Final pressure, $p_2 = p_1$ as it is cons. pressure process,

Further during const. pressure process.

$$V_2 = \frac{T_2}{T_1} \times V_1 = \frac{473}{298} \times 2 = 3.1744 \text{ m}^3/\text{Kg}$$ **(Ans.)**

Example 4.6. A gas occupies 0.35 m³ at a pressure of 1 bar abs. Find the work done on the gas if it is compressed isothermally to a pressure of 16 bar abs.

Solution. Given : $p_1 = 1$ bar

$p_2 = 16$ bar

$V_1 = 0.35$ m^3

We know, during an isothermal process

Work done, $\quad W = p_1 V_1 \log_e (r)$

Also, $\quad p_1 V_1 = p_2 V_2$

or $\quad = \dfrac{V_2}{V_1} = \dfrac{p_1}{p_2} = $ 'r' *i.e.* Compression ratio $= \dfrac{16}{1} = 16$

$\therefore$ Work done, $W = 10^5 \times 1 \times 0.35 \log_e 16 \qquad (\because 1 \text{ bar} = 10^5 \text{ N/m}^2)$

$= 10^5 \times 0.35 \times 2.3 \log 16$

$= 96931.66$ Joule or N-m

$= 0.969$ KJ **(Ans.)**

Example 4.7. One kg of a gas at 240°C, expands adiabatically, so that its volume is doubled and the temperature falls to 115°C. The work done during expansion is 90.0 KJ. Calculate the two specific heats.

Solution. Given : $T_1 = 240 + 273 = 513$ k

$T_2 = 115 + 273 = 388$ k

During and adiabatic process the work done is numerically equal to change in internal energy.

$\therefore$ Work done, $W = C_v \cdot (T_1 - T_2)$ $(\because m = 1\text{kg})$

or $90 = C_v (513 - 388) = C_v \times 125$

$$\therefore \quad C_v = \frac{90}{125} = 0.72 \text{ KJ/kg k } \textbf{(Ans.)}$$

Now,
$$\frac{T_1}{T_2} = \left(\frac{V_2}{V_1}\right)^{\gamma - 1}$$

or
$$\frac{513}{388} = (2)^{\gamma-1} \quad (\because V_2 = 2V_1)$$

$$\therefore \quad 1.32 = (2)^{\gamma-1}$$

Taking log of both sides

$$\log 1.32 = (\gamma - 1) \log 2$$

or $0.1206 = (\gamma - 1) \times 0.301$

or $0.301\,\gamma - 0.301 = 0.1206$

or
$$\gamma = \frac{(0.1206 + 0.301)}{0.301} = \frac{0.4216}{0.301}$$

or $\gamma = 1.4$

But,
$$\frac{C_p}{C_v} = \gamma$$

$$\therefore \quad C_p = \gamma \cdot C_v = 1.4 \times 0.72 = 1.01 \text{ KJ/Kg k } \textbf{(Ans.)}$$

Example 4.8. Gas at 1.5 bar abs. and 20°C in a closed vessel is compressed to 10 bar abs. Its temperature then becomes 180°C. If the compression follows the law $pV^n = C$, Find the value of 'n'.

Solution. Given : $p_1 = 1.5$ bar abs.

$T_1 = 20 + 273 = 293$ k

$p_2 = 10$ bar abs.

$T_2 = 180 + 273 = 453$ k

We know, during polytropic process,

$$\frac{T_2}{T_1} = \left(\frac{p_2}{p_1}\right)^{\frac{n-1}{n}}$$

or
$$\frac{453}{293} = \left(\frac{10}{1.5}\right)^{\frac{n-1}{n}}$$

or
$$1.546 = (6.67)^{\frac{n-1}{n}}$$

Taking log of both sides.

$$\log 1.546 = \frac{n-1}{n}.\log 6.67$$

or $$\frac{n-1}{n} = \frac{\log 1.546}{\log 6.67} = \frac{0.1892}{0.8241}$$

or $$\frac{n-1}{n} = 0.2295$$

or $$n-1 = 0.2295\,n$$

or $$n = \frac{1}{1-0.2295}$$

$$n = 1.297 \textbf{ (Ans.)}$$

Example 4.9. One kg of an ideal gas is heated from 18.3 °C to 93.4 °C. Assuming $R = 0.287$ KJ/Kg k and $\gamma = 1.38$ for the gas, find out

(*i*) Specific heats,

(*ii*) Change in internal energy,

(*iii*) Change in enthalpy.

Solution. $R = 0.287$ KJ/Kg k

$\gamma = 1.38$

(*i*) **Specific heats :**

We know, $$C_p - C_v = R$$

or $$R = C_v\left(\frac{C_p}{C_v} - 1\right) = C_v(\gamma - 1)$$

$$\therefore \quad C_v = \frac{R}{(\gamma-1)} = \frac{0.287}{(1.38-1)}$$

$$= \mathbf{0.755 \text{ KJ/Kg k (Ans.)}}$$

$$\therefore \quad C_p = C_v \times \gamma$$

$$\therefore \quad C_p = 1.59 \times 1.18$$

$$= \mathbf{1.04 \text{ KJ/Kg k (Ans.)}}$$

(*ii*) **Change in internal energy :**

We know, $$\Delta U_{12} = m\,.\,C_v\,.\,(T_2 - T_1)$$

$$T_1 = 18.3 + 276 = 291.3 \text{ K}$$

$$T_2 = 93.4 + 273 = 366.4 \text{ K}$$

$\therefore$ For unit mass, $m = 1$

$$\Delta U_{12} = 0.755\,(366.4 - 291.3)$$

$$= 56.70 \text{ KJ/Kg } \textbf{(Ans.)}$$

(*iii*) **Change in enthalpy :**

$$\Delta h = C_p\,(T_2 - T_1) = 1.04\,(75.1)$$

or $$\Delta h = \mathbf{78.1 \text{ KJ/Kg (Ans.)}}$$

Example 4.10. 2 kg of air at a pressure of 7 bar abs. occupies a volume of 0.28 m^3, this air then expanded to a volume of 1.4 m^3 either isothermally or by law $PV^{1.2} = C$. Assume, R = 287 J/Kg k C_p = 1.0, C_v = 0.7128 KJ/Kg k, γ = 1.4 Find,

(a) the final temperatures,

(b) the work done,

(c) heat absorbed or rejected for both the cases mentioned above.

Solution. Given :

$$m = 2 \text{ Kg.}$$
$$p_1 = 7 \text{ bar abs.} = 7 \times 10^5 \text{ N/m}^2$$
$$V_1 = 0.28 \text{ m}^3$$
$$V_2 = 1.4 \text{ m}^3$$

(i) When the gas is expanded isothermally

(a) $\therefore$ Final temp. t_2 = initial temp. t_1

$\therefore$ Applying characteristic gas equation;

$$p_1 V_1 = m \,.\, R \,.\, T_1$$

or
$$T_1 = \frac{p_1 V_1}{m.R} = \frac{7 \times 10^5 \times 0.28}{2 \times 287}$$

or
$$T_1 = 341.5 \text{ k}$$

or
$$= 68.46 \text{ °C}$$

$\therefore$
$$T_2 = T_1 = 68.46 \text{ °C } \textbf{(Ans.)}$$

(b) Work done during isothermal process.

$$W = p_1 . V_1 \log_e r = 7 \times 10^5 \times 0.28 \log_e^{(1.4/0.28)}$$
$$= 7 \times 10^5 \times 0.28 \times 1.61 = 315449.83 \text{ Joules}$$
$$= 315.45 \text{ KJ } \textbf{(Ans.)}$$

(c) Heat absorbed during isothermal expansion is equal to the work done, as change in internal energy is zero.

i.e.,
$$Q = W = 315.45 \text{ KJ } \textbf{(Ans.)}$$

(ii) When the gas is expanded according to $pV^{1.2} = C$

(a) We know :
$$\frac{T_2}{T_1} = \left(\frac{V_1}{V_2}\right)^{n-1} = \left(\frac{0.28}{14}\right)^{1.2-1} = 0.7246$$

or
$$T_2 = 341.5 \times 0.7246 = 247.45 \text{ K}$$
$$= -25.55 \text{ °C } \textbf{(Ans.)}$$

(b) Work done during polytropic process.

$$W = \frac{p_1 V_1 - p_2 V_2}{n - 1} = \frac{mR(T_1 - T_2)}{(n-1)} \text{ N-m}$$

$$= \frac{2 \times 287(341.5 - 247.45)}{(1.2 - 1)}$$

$$= 269923.5 \text{ N-m or J} = 269.9. \text{ KJ } \textbf{(Ans.)}$$

(*c*) Heat transferred during the polytropic process;

$$Q = W \times \frac{\gamma - n}{\gamma - 1} = 269.9 \times \frac{(1.4 - 1.2)}{(1.4 - 1)}$$

$$= 269.9 \times \frac{0.2}{0.4} = 134.95 \text{ KJ } \textbf{(Ans.)}$$

Example 4.11. The ineternal energy of a gas, in a flow process decreases by 100 KJ/kg of gas and flow work increases by 4000 KJ/5 kg of the gas. Determine the, change in specific enthalpy and in total enthalpy of the gas.

Solution. Given data :

Decrease in internal energy of the gas = 100 KJ/kg of gas

and Increase in flow work = 4000 KJ/5 kg of the gas

Now, applying the steady flow energy equation; to the initial and final condition of the gas, we get;

$$h_1 = U_1 + P_1V_1 \text{ and } h_2 = U_2 + P_2V_2$$

∴ Change in specific enthalpy ;

$$h_1 - h_1 = (U_2 - U_1) + (P_2V_2 - P_1V_1)$$

But, change or decrease in specific internal energy ;

$$U_2 - U_1 = -100 \text{ KJ/kg of gas}$$

and increase in flow work,

$$(P_2V_2 - P_1V_1) = \frac{4000}{5} = 800 \text{ KJ/kg of gas}$$

Also, change in specific enthalpy

$$= -100 + 800 = 700 \text{ KJ/kg } \textbf{Ans.}$$

Hence, change in total enthalpy

$$= 700 \times 5 = 3500 \text{ KJ } \textbf{Ans.}$$

Example 4.12. In a gas turbine, the flow rate of air is 4 kg per second. The velocity and enthalpy of air at the entrance are 200 m/sec and 6000 KJ/kg respectively. At the exit the velocity is 100 m/sec and enthalpy is 5000 KJ/kg. A loss of heat of 40 KJ/kg occurs while the air passes through the turbine. Determine, the power developed by the turbine.

Solution. Given data :

Flow rate of air = 4 kg/sec

Velocity of air at the entrance = 200 m/sec

Enthalpy of air at the entrance = 6000 KJ/kg

Velocity of air at the exit = 100 m/sec

and Enthalpy of air at the exit = 5000 KJ/kg

Loss of heat energy, during the flow = 40 KJ/kg

Applying the steady flow energy equation; at the entrance and the exit; we get;

$$h_1 + \frac{v_1^2}{2} + Z_1 = h_2 + \frac{v_2^2}{2} + Z_2 + W$$

But neglecting the change in potential energy *i.e.* Z_1 and Z_2 and considering, the loss of heat energy q; the equation becomes

$$h_1 + \frac{v_1^2}{2} - Q = h_2 + \frac{v_2^2}{2} + W$$

or $$W = (h_1 - h_2) + \frac{v_1^2 - v_2^2}{2} - Q$$

$$= (6000 - 5000) + \frac{(200)^2 - (100)^2}{2 \times 1000} - 40$$

$$= 1000 + \left(\frac{40000 - 10000}{2 \times 1000}\right) - 40$$

$$= 1000 + \frac{30,000}{2 \times 1000} - 40$$

$$= 1000 + 15 - 40 = 975 \text{ KJ/sec}$$

$\therefore$ Total work done $= 975 \times 4 = 3900$ KJ/sec.

Hence, Power developed by the turbine = 3900 KW. **Ans.**

Example 4.13. In a nozzle the steam enters with enthalpy of 3000 KJ/kg, at a velocity of 40 m/sec. During the flow through the nozzle the enthalpy is reduced to 1500 KJ/kg. Determine, the velocity of steam at the exit nozzle. Assume, the nozzle to be hozizontal with no heat-loss.

Solution. Given data :

Enthalpy of steam, at the entrance = 3000 KJ/kg

Velocity of steam, at the entrance = 40 m/sec

Enthaply of steam, at the exit = 1500 KJ/kg

The nozzle is horizontal *i.e.*, $Z_1 = Z_2$

Also, the heat loss equal to zero.

So, applying steady flow energy equation, at the entrance and exit; we get;

$$h_1 + \frac{v_1^2}{2} = h_2 + \frac{v_2^2}{2}$$

or $$h_1 - h_2 = \frac{v_2^2}{2} - \frac{v_1^2}{2} = \frac{v_2^2 - v_1^2}{2}$$

or $$3000 - 1500 = \frac{v_2^2 - (40)^2}{2 \times 1000}$$

or $$1500 = \frac{v_2^2 - 1600}{2000}$$

or $$v_2^2 - 1600 = 30,00000$$

or $$v_2^2 = 30,00,000 + 1600 = 3001600$$

Hence, velocity of steam at exit, of nozzle,

$$v_2 = \sqrt{3001600} = 1732.5 \text{ m/sec}$$ **Ans.**

Example 4.14. A steam-generator produces, 500 kg of steam per hour, from feed water at 30°C. Neglecting the kinetic and potential energy changes, determine the rate of heat-transferred. The enthalpy of the steam is 2000 KJ/kg.

Solution. Given data :

Quantity of steam produced = 500 kg/hour

Feed water temperature = 30°C

Therefore, Enthalpy of feed water = 30 × 4.186 = 125.580

Kinetic and potential energy changes are zero.

Enthalpy of steam = 2000 KJ/kg

Applying the steady flow energy equation, we get

$$q_1 = h_2 - h_1$$

or $$q_1 = 2000 - 125.58 = 1874.42 \text{ KJ/hour}$$

Hence, rate of heat transferred = 1874.42 × 500 = 937210 KJ/hour **Ans.**

Example 4.15. A reversible Carnot engine converts one sixth of the heat input into work. When, the temperature of the sink is reduced to 62°C, its efficiency is doubled. Find, the Temperature of the source and the sink.

Solution. Given data :

(*i*) The Carnot engine converts one sixth of heat input into work.

(*ii*) When temp. of sink is reduced to 62°C, η is doubled.

(*i*) Efficiency of the Carnot engine,

$$\eta_{carnot} = \frac{1}{6}$$

i.e. $$\frac{T_1 - T_2}{T_1} = \frac{1}{6}$$

where, T_1 = Abs. temp. of hot source in K

and T_2 = Abs. temp. of the sink in K.

(*ii*) Temp. of the sink = $(T_2 - 62)$ K

$$\therefore \quad 2 \times \frac{1}{6} = \frac{T_1 - (T_2 - 62)}{T_1}$$

or $$\frac{1}{3} = \frac{T_1 - T_2 + 62}{T_1}$$

or $$\frac{T_1 - T_2}{T_1} + \frac{62}{T_1} = \frac{1}{3}$$

or $$\frac{1}{6} + \frac{62}{T_1} = \frac{1}{3} \qquad \left(\because \frac{T_1 - T_2}{T_1} = \frac{1}{6}\right)$$

or $$\frac{62}{T_1} = \frac{1}{3} - \frac{1}{6} = \frac{2-1}{6} = \frac{1}{6}$$

or $$T_1 = 372 \text{ K} = 99°\text{C}$$

Now, $$\frac{T_1 - T_2}{T_1} = \frac{1}{6} \text{ or } T_1 - T_2 = \frac{T_1}{6}$$

or $$T_2 = T_1 - \frac{T_1}{6} = \frac{5T_1}{6} = \frac{5 \times 372}{6}$$

$$= 5 \times 62 = 310 \text{ K or } = 37°\text{C}$$

Hence,

Temp. of source, $T_1 = 372$ K or 99°C **Ans.**

and Temp. of sink, $T_2 = 310$ K or 37°C. **Ans.**

ASSIGNMENT

Objective Type Questions :

(A) Fill in the Blanks :

1. The degree of hotness or coldness is known as ______.
2. Zeroth law provides the basis for comparing the _____.
3. First law of thermodynamics is the law of _____ of energy.
4. Heat can flow from a body at ______ temperature to a body at ______ temperature.
5. As per first law of thermodynamics "Heat and mechanical work are ______".
6. The relation between heat and work is ______.
7. The internal energy of a substance is the energy possessed by it, due to its ______ and ______.
8. During constant volume process work done is ______.
9. The Change in internal energy is ______ during an isothermal process.
10. Hyperbolic and isothermal processes are identical for ______ gases.
11. During an isentropic process ______ is constant.
12. Work done during a polytropic process is ______.
13. Heat transferred during a constant pressure process is ______.
14. During throttling heat transferred is ______.
15. The value of molar volume at N.T.P. is ______.
16. The relation between the two specific heats of a gas and its characteristics gas constant is ______.
17. The value of 'J' in S.I. system of units is ______.
18. Work done during a compression process is ______.
19. The unit of entropy in S.I. system of units is ______.
20. In case of free expansion process, the change in internal energy is ______.

(B) State True or False :

1. Zeroth law determines temperature.
2. First law of thermodynamics is the law of conservation of energy.
3. Second law of thermodynamics establishes the concept of entropy.
4. The polytropic index is unity for a constant temperature process.
5. The entropy of universe tends to be maximum.
6. A gas can never be liquified above critical temperature.
7. The slope of constant volume line on temp-entropy diagram is more than that of constant pressure line.
8. Heat can flow from a body at a lower temperature to a body at higher temperature.
9. During constant pressure process, change in internal energy is zero.
10. During an isothermal process work done is equal to $pV \log_e r$.
11. Reversible adiabatic process is the same as adiabatic process.
12. For vapour, hyperbolic and isothermal process are same.

13. During a polytropic process, heat transferred is not zero.
14. Work done during a constant volume process is zero.
15. Internal energy remains constant during an isothermal process.
16. The internal energy of a gas depends upon its temperature only and is independent of other properties like pressure and volume.
17. During an isothermal process the heat interchange will be equal to work done.
18. In a constant pressure process, the specific heat at constant pressure is to be used for calculating heat transferred.
19. Work done during a throttling process is not zero.
20. The change in internal energy in a free expansion process is zero.

(C) Multiple Choice Questions :

1. Identify the false statement in the context of heat and work.
   (*a*) They exit only during interaction between the system and the boundary.
   (*b*) They are path functions and exact differential.
   (*c*) They may or may not occur simultaneously.
   (*d*) They occur due to some unbalanced potential which tends to promote a change of state.
2. The absolute zero temperature corresponds to the condition when
   (*a*) All the substances exist only as solids.
   (*b*) Volume of a gas reduces to zero.
   (*c*) Kinetic energy of gas molecules becomes zero.
   (*d*) No pressure is exerted by the gas.
3. The pressure exerted on the walls of container by a gas is due to the fact that gas molecules :
   (*a*) Stick to the walls of the container.
   (*b*) Loose their kinetic energy
   (*c*) Get accelerated towards the wall.
   (*d*) Change their momenta due to collision with the wall.
4. In thermodynamic analysis, a pure substance is that which;
   (*a*) Has no dissolved impurities.
   (*b*) Consists of only a single species.
   (*c*) May have a number of chemical species but composition remains constant.
   (*d*) Behaves as a perfect gas.
5. The expression $\int_1^2 pdv$ gives the measure of work done during :
   (*a*) Non-flow reversible process.
   (*b*) Steady flow reversible process.
   (*c*) Open system and any process.
   (*d*) Any system and any process.
6. The work done equals $\int pdv$ for :
   (*a*) An irreversible process.
   (*b*) Unrestricted expansion.
   (*c*) Non-flow quasi-static process.
   (*d*) The arrangement where work is done on the gas as the paddle wheel turns by a falling weight.

7. First law of the thermodynamics refers to conservation of :
   (*a*) Mass (*b*) Momentum
   (*c*) Energy (*d*) Force.
8. Which of the following statements does not represent the essence of first law of thermodynamics ?
   (*a*) Heat input to the system equals the net work done plus change in thermal energy.
   (*b*) The sum of heat and work effects will be zero, if the system undergoes a cycle.
   (*c*) Heat input cannot be more than the work output.
   (*d*) For an isolated system, energy remains constant.
9. The internal energy of an ideal gas is a function of;
   (*a*) Pressure only
   (*b*) Absolute temperature only
   (*c*) Pressure and volume
   (*d*) Pressure, volume and temperature.
10. The internal energy of an ideal gas is a function of its absolute temperature only. This statement refers to :
   (*a*) Avogadro's law (*b*) Maxwell law
   (*c*) Joule's law (*d*) Regnault's law.
11. The identity $\delta Q = pdv + du$ is true, for :
   (*a*) Any process and open system
   (*b*) Any process and closed system
   (*c*) Irreversible process and closed system.
   (*d*) Reversible process and closed system.
12. Choose the wrong statement :
   (*a*) If a closed system traces a cycle of operation, then $\int dp$ is zero.
   (*b*) $\delta Q = \delta W + du$ is true only for reversible process and closed system.
   (*c*) Internal energy is a property of the system.
   (*d*) When a gas undergoes a free or unrestricted expansion, then $W \neq \int p.dv$.
13. Which is the essence of second law of thermodynamics?
   (*a*) The whole of heat supplied to a system can be converted into equivalent amount of mech. work.
   (*b*) No engine can be 100 percent efficient.
   (*c*) A refrigerator can reduce the temp. to absolute zero.
   (*d*) Reversible engines working between the same temperature limits can have different thermal efficiencies.
14. "Heat cannot be transferred from a system at low temperature to another system at high temperature without the aid of an external agency". This statement of second law of thermodynamics is attributed to :
   (*a*) Gay-Lusac (*b*) Clausius
   (*c*) Max-plant (*d*) Joule Thomson.
15. Heat is being supplied to air in a cylinder fitted with a frictionless piston held by a constant weight. The process is an :

(a) Isothermal (b) Adiabatic

(c) Isobaric (d) Isochoric.

16. Under what conditions the change in the enthalpy of a system equals the heat supplied ?

(a) Constant volume (b) Constant pressure

(c) Constant temperature

(d) Standard, temperature-pressure conditions.

17. Which parameter remains constant during a reversible isothermal process ?

(a) Rate of heat exchange (b) Internal energy

(c) Enthalpy (d) Entropy.

18. During an adiabatic process :

(a) Temp. of gas varied but the pressure is kept constant.

(b) Total kinetic energy of the gas molecules remains constant.

(c) Gas is compressed at constant temperature and there is no change of entropy.

(d) Expansion and compression of a gas occur without any exchange of heat with the surroundings.

19. For an adiabatic process, the pressure and volume are related as $pV^{\gamma} =$ Constant. The exponent 'γ' represents the ratio of specific heat at constant pressure to specific heat at constant volume and for air, its value equals :

(a) 0.87 (b) 1.02

(c) 1.4 (d) 1.67

20. During a general polytropic expansion process characterised by $pV^n =$ Constant the work done is equal to :

(a) $p_1V_1 - p_2V_2$ (b) $p_1V_1 \log_e \frac{V_2}{v_1}$

(c) $\frac{p_1V_1 - p_2V_2}{n+1}$ (d) $\frac{p_1V_1 - p_2V_2}{n-1}$

ANSWERS

(A) Fill in the Blanks :

1. Temperature
2. Temperature of two bodies.
3. Conservation
4. Higher, lower
5. Interconvertible
6. $W = JH$
7. Molecular arrangement, activity
8. Zero
9. Zero.
10. Perfect
11. Entropy
12. $\frac{p_1V_1 - p_2V_2}{n-1}$
13. Change in enthalpy
14. Zero
15. 22.41 m^3
16. $C_p - C_v - R$
17. 4.1868 KJ/Kcal
18. Negative
19. KJ/Kg k
20. Zero.

(B) State True or False :

1. True	2. True	3. True	4. True	5. True
6. True	7. True	8. False	9. False	10. True

11. False	**12.** False	**13.** True	**14.** True	**15.** True
16. True	**17.** True	**18.** True	**19.** False	**20.** True

(C) Multiple Choice Questions :

1. (*a*)	**2.** (*b*)	**3.** (*a*)	**4.** (*a*)	**5.** (*c*)
6. (*d*)	**7.** (*c*)	**8.** (*a*)	**9.** (*b*)	**10.** (*d*)
11. (*c*)	**12.** (*b*)	**13.** (*c*)	**14.** (*c*)	**15.** (*b*)
16. (*b*)	**17.** (*d*)	**18.** (*c*)	**19.** (*a*)	**20.** (*d*)

(D) Short Answer Type Questions :

Q. 1. Define the 'Law of conservation of energy'.

Ans. This law states that, "Energy can neither be created nor destroyed but can be transformed from one form to the another".

Q. 2. Define the 'First law of Thermodynamics'.

Ans. It states that, "Heat and mechanical work are interconvertible". So, $W \propto H$, Where 'W' is the mech.-work and 'H' the heat.

Q. 3. What are the limitations of the first law of thermodynamics?

Ans. The limitations in the first law of thermodynamics as states below :

(*i*) It does not specify the conditions under which the conversion of heat into work is possible neither and direction in which heat transfer can take place.

(*ii*) Although, the work can be conveniently converted into heat but it is impossible to convert the heat into work completely, *i.e.*, only a fraction of heat supplied is converted into work. This means that heat and work are not completely inter convertible forms of energies. This fact also is not exploded during the discussion of the first law of thermodynamics.

Q. 4. What do you know by 'positive and negative thermodynamic work'?

Ans. A 'positive work' is obtained when the work is done by the system whereas, when work is done on the system it is termed as 'Negative work'.

Q. 5. What are the similarities between heat and work?

Ans. The similarities between the heat and work are :

(*i*) Heat and work are both transient phenomena. Systems new possess heat or work, but either may occur when a system undergoes a change of state.

(*ii*) Both heat and work are boundary phenomena. Both are observed at the boundaries of the system and both represent energy crossing the boundary of a system.

(*iii*) Both heat and work are path functions and in exact differentials.

Q. 6. Define the term 'energy'. What are its important types?

Ans. Energy is the capacity of an agent to do work. The energy of the system may be of two types :

(*i*) **Stored Energy.** This is the energy which is contained with the system boundaries. For example, potential energy and internal energy etc.

(*ii*) **Transit Energy.** This is the energy which crosses the system boundaries *e.g.*, heat, work and electrical energy.

Q. 7. Define the 'Second Law of Thermodynamics'.

Ans. **Clausius Statement.** Clausius stated that, "Heat flows from a hot body to a cold body unaided, but it is impossible for heat to flow from a cold body to a hot body without the aid of external work".

Kelvin-Plank Statement. It states, "It is impossible to construct an engine working on cyclic process whose sole purpose is to convert all the heat energy supplied to it into an equivalent amount of mechanical work".

Q. 8. What is 'Entropy' ?

Ans. As per Clausius the entropy may be regarded as "A measure of the unavailability of heat energy for transformation into work".

Q. 9. What is the equation for the 'Law of Conservation of Energy' ?

Ans.

$$Q = W + \Delta U$$

where Q = Total heat supplied or rejected

W = Work done in heat units.

ΔU = Change in internal energy.

Q. 10. What is an 'Isochoric Process' ?

Ans. A thermodynamic process in which the volume of the given mass of the gas is not allowed to change, is known as 'an Isochoric Process'.

Q. 11. What is an 'Isobaric process' ?

Ans. A thermodynamic process in which the pressure of the given mass of a gas is not allowed to change is known as 'Isobaric Process'.

Q. 12. What is an Isothermal Process ?

Ans. A thermodynamic process in which, the temperature of the given mass of a gas is not allowed to change is known as 'Isothermal Process'.

Q. 13. What is a Hyperbolic Process?

Ans. A thermodynamic process in which the product of the pressure of the gas and its volume remains constant is known as 'Hyperbolic Process'.

Q. 14. What is an Adiabatic Process?

Ans. The thermodynamic process in which, no heat is added or withdrawn from the system under consideration is known as an 'Adiabatic Process'.

Q. 15. What is an Isentropic Process?

Ans. A thermodynamic process in which the entropy of the system remains constant is known as an 'Isentropic Process'.

Q. 16. What is a 'Polytropic Process'?

Ans. A thermodynamic process in which the relation 'pV^n = Constant' holds good, is known as a 'Polytropic Process', where 'p' is the pressure and 'V' the volume of the gas whereas 'n' is the polytropic exponent the value of which varies from zero to infinity for various gases.

Q. 17. What is a 'Throttling Process' ?

Ans. A thermodynamic process in which a gas or vapour is expanded through an aperture of minute (narrow) dimensions like a valve, which is very slightly opened, is called a 'throttling Process'.

Q. 18. What is Free-Expansion?

Ans. A thermodynamic process in which the fluid expands suddenly into a vacuum chamber through an orifice of large dimension is called 'Free-Expansion'.

Q. 19. What is the value of work done, heat supplied and change in internal energy in a 'throttling Process" ?

Ans. In a throttling process the work done, heat supplied and the change in internal energy during the process is zero.

Q. 20. What is the value of work done, heat-supplied and change in internal energy in a Free-Expansion ?

Ans. In free-expansion the value of work done, heat-supplied and the change in internal energy is equal to zero.

REVIEW QUESTIONS

1. Explain Zeroth law of thermodynamics. **(SBTE Pb., 1981)**
2. State and explain the first law of thermodynamics. What are its limitations ? **(SBTE Delhi 1977)**
3. What is the concept of Enthalpy and Internal energy ?
4. What are the Clausius and Kelvin-Planck statements of the second law of thermodynamics? Explain it.
5. What is the concept of Entropy ?
6. Briefly describe the various thermodynamic processes.
7. Derive an expression for work done, change in internal energy and heat transferred during a constant pressure process.
8. Derive the expressions for work done during the following processes;
   (*i*) An isothermal expansion.
   (*ii*) An adiabatic expansion. **(SBTE Pb., 1981)**
9. What is hyperbolic process ?
10. Derive and expression for work done during the polytropic expansion of a gas. **(SBTE Pb., 1978)**
11. Distinguish between an adiabatic and an isothermal expansion process for a gas. Draw P-V diagram for each of them and explain them briefly. **(SBTE Haryana, 1979)**
12. What is an isentropic process ?
13. Distinguish between free expansion and Throttling process. **(SBTE Pb., Dec. 99)**
14. Differentiate between Adiabatic and Polytropic process. **(SBTE Pb., Dec. 99)**

PROBLEMS FOR PRACTICE

1. A quantity of air at a temperature of 20 °C and pressure of 2.0 bar abs. is found to occupy 0.20 m^3. It is supplied heat energy equivalent to 14720 Joule at constant volume. Find the final pressure and temperature of air. Take C_v = 0.7128 KJ/Kg k. Also determine the change in internal energy per kg of air, take R = 287 J/Kg k. [**Ans.** 4.065 bar, 40.65°C, 5.926 KJ]
2. When one kg of a gas at an initial temperature of 92°C is heated to twice its original volume at a constant pressure of 8 bar, what fraction of heat supplied to it is converted into work during the change? Take C_p = 1.0, C_v = 0.7128 KJ/Kg k for the gas. [**Ans.** 0.2872]
3. A gas has C_v = 0.7128 KJ/Kg k and C_p = 1.0 KJ/Kg k. Calculate the volume of 0.227 kg of this gas at 21 °C and 1.4 bar abs. Determine also the quantity of heat added and external work done by this gas if it is heated at a constant pressure of 1.4 bar till its volume is 0.142 m^3. [**Ans.** 0.137 m^3, 2.436 KJ, 1.736 KJ]
4. 0.4 kg of air at a pressure of 1.0 bar and a temperature of 20 °C is compressed isothermally until its volume is 0.08 Cu metre. Determine the work done on the air during this compression. Take R = 287 J/Kg k. [**Ans.** 482.2 KJ]
5. Determine the volume and internal energy of one kg of air at a pressure of 5.0 bar abs. and temperature 100 °C, the internal energy being reckoned from 0 °C. The air is allowed to expand four times to its initial volume at constant pressure. Find the final values of pressure, temperature and work done in each case. [**Ans.** 0.214 m^3, 7.28 KJ, 5 bar, 1219 °C, 2000 KJ]
6. One kg. of a gas occupies 0.8 m^3 at 15 °C and 1.05 bar pressure. If after adiabatic compression to 0.256 m^3, the pressure is 4.9 bar. Find the values of

(*i*) characteristic gas constant, R (*ii*) ratio of sp. heats, ρ (*iii*) C_p (*iv*) C_v and (*v*) work done during compression.

[**Ans.** 291.67 J/Kg K, 1.35, 1125 J/Kg K, 833 J/kg K, –11840 J)

7. Find the difference in work done when one kg of air is compressed from 1.5 bar abs. and 20 °C to 9 bar abs.

   (*i*) Isothermally (*ii*) Adiabatically.

   Take C_p = 1.0 KJ/Kg k C_v = 0.7128 KJ/Kg k and γ = 1.4. [**Ans.** –813.35 J]

8. The cylinder of an engine contains 0.02 m^3 of air at a pressure of 10 bar and temperature of 25°C. Calculate the volume and temperature at the end of expansion. Also, calculate the work done during the process of expansion where the pressure at the end of expansion is 1.0 bar abs. The expansion of air takes place polytropically with n = 1.3. [**Ans.** 0.117 m^3, 27.67 KJ]

9. Calculate the work done when 0.075 m^3 of air at 16 °C and 7 bar abs. expands to 0.35 m^3 according to the law pV^{1-2} = C. [**Ans.** 69.6 KJ] **(SBTE Pb., 1978)**

10. One kg. of air at a temperature of 240°C is allowed to expand adiabatically to double its volume. Find the work done in the process. Take γ = 1.4, R = 287 KJ/Kg k. [**Ans.** 89.1 KJ]

11. Air at 30°C and 100 KPa is compressed reversibly and polytropically from 5 m^3 to 1 m^3. Calculate the final temp., pressure, work done and heat transferred if the index of compression is 1.4. Assume C_p = 1.005 KJ/kg k and C_v = 0.7178 KJ/Kg k. [**Ans.** 303.8°C, 951.8 KPa, 1129.5 KJ, 0]

    (SBTE Haryana, MPECS May 1995)

12. The values of specific heat of a gas at constant pressure and at constant volume are 1.0 KJ/kg k and 0.714 KJ/kg k respectively. Determine the characteristics gas constant and isentropic index of the gas. If 5 kg of this gas is heated at constant pressure from 30 °C to 250 °C. Determine the heat added, the work done and the change in internal energy.

    [**Ans.** 1100 KJ, 400.4 KJ, 699.6 KJ]

    (SBTE Haryana, MPECS (Auto Engg.), Nov. 1995)

13. 2500 KJ of heat is added to 2.8 m^3 of a gas at 20°C and at atm. pressure. If during the process of heat addition the volume remains constant, find the final temperature and pressure. Take R for the gas as 278 J/Kg k and C_v = 0.714 KJ/Kg k. [**Ans.** 1057.8 °C, 4.6 bar]

14. 0.12 m^3 of air at a temperature of 20 °C and pressure of 1.1 bar is compressed to 20 bar as per the law $pV^{1.3} = C$. Determine final temp. of air and amount of heat added.

15. 0.15 m^3 of air at a pressure of 1.06 bar is compressed to a volume of 0.008 m^3 at 36 bar. If the compression follows the law $pV^{1.3} = C$ and 'γ' for air is 1.41. Find :

    (*i*) Quantity of heat added or rejected, and

    (*ii*) Change of internal energy. [**Ans.** – 11.53 KJ, 31.47 KJ]

16. 1.5 kg of gas at a pressure of 9 bar is expanded adiabatically to a pressure of 2 bar. If the initial temperature of the gas is 350 °C. What will be the work done during expansion and the change in internal energy of the gas? Take, 'C_p' for gas as 1.005 KJ/kg k and C_v = 0.714 KJ/Kg k.

    [**Ans.** 234.6 KJ, – 220.4 KJ]

17. A gas having volume 0.23 m^3 and a pressure of 2 bar is compressed polytropically to a volume of 0.013 m^3 and pressure of 40 bar. What is the amount of heat added or rejected ?

    Take, 'γ' for the gas as 1.4 and the polytropic index 'n' as 1.2.

    [**Ans.** – 15 KJ, Work done on the gas]

❑❑❑

5

CHAPTER

Ideal and Real Gases

INSIDE THIS CHAPTER

5.1. INTRODUCTION

An *'ideal gas'* is defined as a *gas having no forces of intermolecular attraction.* The gases which *follow the gas laws at all ranges of pressures and temperatures* are considered as *"ideal gases"*. However, *'real gases' follow these laws at low pressures or high temperatures or both.* This is because the forces of attraction between molecules *tend to be very small* at reduced pressures and elevated temperatures.

An *ideal gas* obeys the law $pv = RT$. The specific heat capacities *are not constant but are functions of temperature.* A *perfect gas* obeys the law $pv = RT$ and has *constant specific heat capacities.*

A perfect gas is well suited to mathematical calculation and is therefore a most useful model to use for analysis of practical machinery which uses real gases as a working substance.

In *reality there is no ideal or perfect gas.* At a very low pressure and at a very high temperature, real gases like hydrogen, oxygen, nitrogen, helium etc. behave nearly the same way as perfect gases. These gases are called *semi-perfect* or *permanent gases.* The term semi-perfect has the implication that the behaviour of the gases are nearly the same as that of a perfect gas. The term *'permanent'* was used for these gases by earlier chemists who thought that these gases do not change their phase (*i.e.*, do not condense to a liquid state). Hence they are called permanent gases. There is no gas which does not change phase, and there is no permanent gas in the real sense. However, these gases can be changed into a liquid phase only if they are subjected to a great decrease in temperature and increase in pressure.

All gases behave in nearly in a similar way, especially at pressures considerably lower than the critical pressure, and at temperatures above the critical temperature. *The relation between the independent properties, such as pressure, specific volume and temperature for a pure substance* is known as the *'equation of state'*. For engineering calculations, the equation of state for perfect gases can be used for real gases so long as the pressures are well below their critical pressure and the temperatures are above the critical temperature.

5.2. THE EQUATION OF STATE FOR A PERFECT GAS

Boyle's law. It states that *volume of a given mass of a perfect gas varies inversely as the absolute pressure when temperature is constant.*

If p is the absolute pressure of the gas and V is the volume occupied by the gas, then

$$V \propto \frac{1}{p}$$

or pV = Constant, so long as the temperature is constant ...(5.1)

Fig. 5.1 shows the graphical representation of Boyle's law. The curves are rectangular hyperbolas asymptotic to the p-v axis. Each curve corresponds to a different temperature. For any two points on the curve,

$$\frac{p_1}{p_2} = \frac{V_2}{V_1} \quad ...(5.2)$$

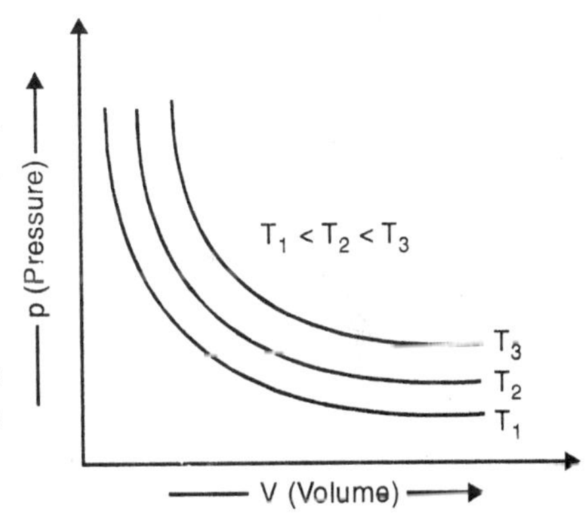

Fig. 5.1. ***p-V*** **relation of a perfect gas at constant temperature.**

Charle's law. It states that *if any gas is heated at constant pressure, its volume changes directly as its absolute temperature.*

In other words, $V \propto T$

or $\dfrac{V}{T}$ = Constant,

so long as pressure is constant ...(5.3)

If a gas changes its volume from V_1 to V_2 and absolute temperature from T_1 to T_2 without any change of pressure, then

$$\frac{V_1}{T_1} = \frac{V_2}{T_2} \quad ...(5.4)$$

Fig. 5.2 gives the graphical representation of Charle's law.

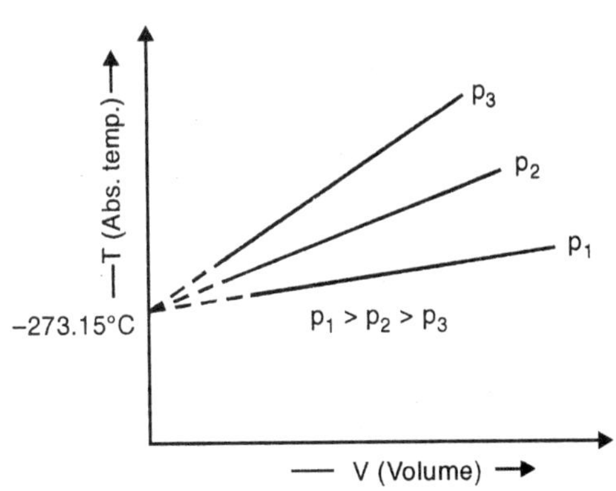

Fig. 5.2. T-v relation of a perfect gas constant pressure.

To derive the *equation of state* for a perfect gas let us consider a *unit mass* of a perfect gas to change its state in the following two successive processes (Fig. 5.3).

(*i*) Process 1-2′ at constant pressure, and

(*ii*) Process 2′-2 at constant temperature.

For the process 1-2′, applying Charle's law

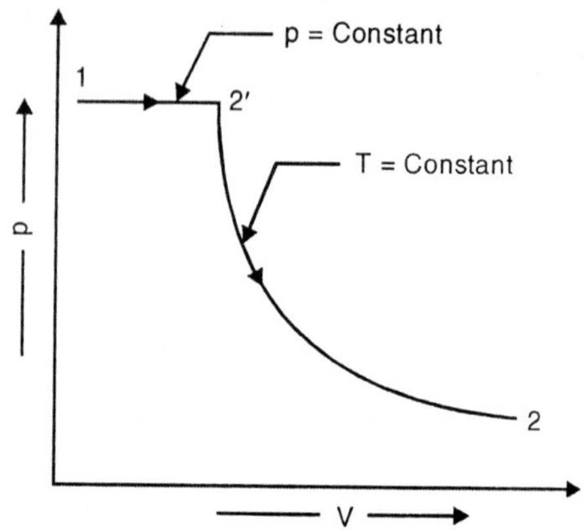

Fig. 5.3. Formulation of equation of state of a perfect gas.

$$\frac{v_1}{T_1} = \frac{v_2'}{T_2'}$$

and since $T_2' = T_2$, we may write

$$\frac{v_1}{T_1} = \frac{v_2'}{T_2'} \quad ...(i)$$

For the process 2′-2, using Boyle's law

$$p_2' v_2' = p_2 v_2$$

and since $p_2' = p_1$

$$\therefore \quad p_1 v_2' = p_2 v_2$$

i.e.,

$$v_2' = \frac{p_2 v_2}{p_1} \quad ...(ii)$$

Substituting the value of v'_2 from eqn. (*ii*) in eqn. (*i*), we get

$$\frac{v_1}{T_1} = \frac{p_2 v_2}{p_1 T_2}$$

or

$$\frac{p_1 v_1}{T_1} = \frac{p_2 v_2}{T_2}$$

i.e.,

$$\frac{pv}{T} = \text{constant} \quad ...(5.5)$$

The magnitude of this constant depends upon the particular gas and it is denoted by *R*, where *R* is called the **specific gas constant.** Then

$$\frac{pv}{T} = R$$

The *equation of the state for a perfect gas* is thus given by the equation

$$pv = RT \quad ...(5.6)$$

or for *m* kg, occupying Vm^3,

$$pV = mRT \quad ...(5.7)$$

If the mass is chosen to be numerically equal to the molecular weight of the gas then 1 mole of the gas has been considered, *i.e.*, 1 kg mole of oxygen is 32 kg oxygen, or 1 kg mole of hydrogen is 2 kg hydrogen.

The equation may be written as

$$pV_0 = MRT \quad ...(5.8)$$

where V_0 = Molar volume, and

M = Molecular weight of the gas.

Avogadro discovered that V_0 *is the same for all gases at the same pressure and temperature* and therefore it may be seen that MR = a constant; R_0 and thus

$$pV_0 = R_0T \qquad ...(5.9)$$

R_0 is called the **molar** or *universal gas constant* and its value is 8.3143 KJ/kg mol K.

If there are n moles present then the ideal gas equation may be written as

$$pV = nR_0T \qquad ...(5.10)$$

where V is the volume occupied by n moles at pressure p and temperature T.

5.3. *p-v-T* SURFACE OF AN IDEAL GAS

The equation of state of an ideal gas is a relationship between the variables pressure (p), volume (V) and temperature (T). On plotting these variables along three mutually perpendiculars axes, we get a surface which represents the equation of state ($pv = RT$). Such a surface is called *p-v-T surface*. These surfaces represent the fundamental properties of a substance and *provide a tool to study the thermodynamic properties and processes of that substance.* Fig. 5.4(a) shows a portion of a *p-v-T* surface for an *ideal gas. Each point on this surface represents a process.* The Fig. 5.4(a) also shows the constant pressure, constant volume and constant temperature lines.

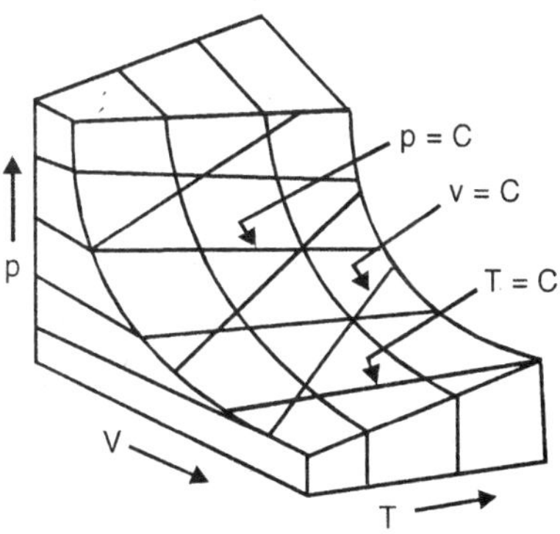

Fig. 5.4(a).

5.3.1. Triple Point

The state change of a pure substance is shown in Fig. 5.4(b) plotted on P-T coordinates. The heating of ice at –10°C to steam at 250°C at the constant pressure of 1 atm is considered. 1-2 is the solid ice heating 2-3 is the melting of ice at 0°C, 3-4 is the liquid heating, 4-5 is the vapourisation of water at 100°C and 5-6 is the heating in the vapour phase. The process will be reversed from state 6 to state 1 upon cooling. The curve passing through the 2, 3 points is called the fusion curve and the curve passing through the 4, 5 points which indicate the vaporisation or condensation at different temperatures and pressures is called the vaporisation curve. If the vapour pressure of a solid is measured at different temperatures and these are plotted, the sublimation curve will be obtained. The fusion curve, the vaporisation curve and the sublimation curve meet at triple point.

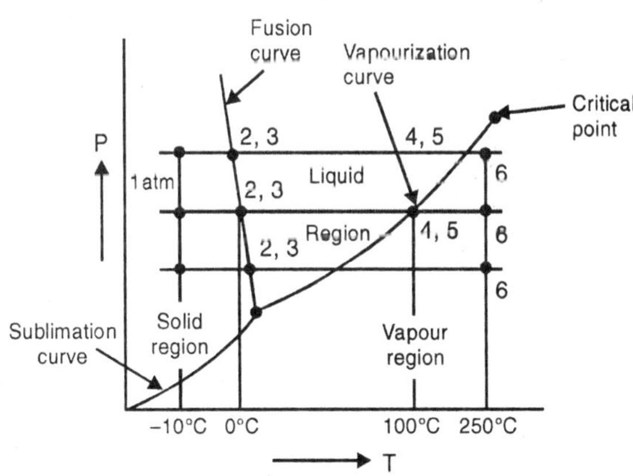

Fig. 5.4(b). Phase equilibrium diagram p-T coordinates triple point.

The triple point of water is 4.58 mm Hg and 273.16 K whereas for CO_2 is at 3885 mm Hg about 5 atm and 216.55 K. Thus when solid CO_2 dry ice is exposed to 1 atm pressure, it gets transformed into vapour directly absorbing the latent heat of sublimation from the surroundings which gets cooled or refrigerated.

5.4. INTERNAL ENERGY AND ENTHALPY OF A PERFECT GAS

Joule's Law. *Joule's law states that the specific internal energy of a gas depends only on the temperature of the gas and is independent of both pressure and volume.*

i.e., $$u = f(T)$$

Joule concluded this result from a series of experiments conducted with an apparatus similar to the one shown in Fig. 5.5.

- Two tanks connected by a valve were submerged in a bath of water.
- Initially one tank was evacuated and the other was filled with air under high pressure.
- A thermometer was placed in the water bath.
- After the tank and water had attained the same temperature, the valve between the two tanks was opened to pass air slowly from high pressure tank to the evacuated tank. Time was allowed for equilibrium to be attained.

Joule observed that there was *no change in temperature of water during or after the process.* Since there was no change in the temperature of water, he concluded that there *was no heat transfer between air and water i.e.,* $\delta Q = 0$. And since there was no work during the process, *i.e.,* $\delta W = 0$, from the first law of thermodynamics, $\delta Q = dE + \delta W$, Joule *concluded that change in internal energy of the air is zero, i.e.,* $dE = 0$.

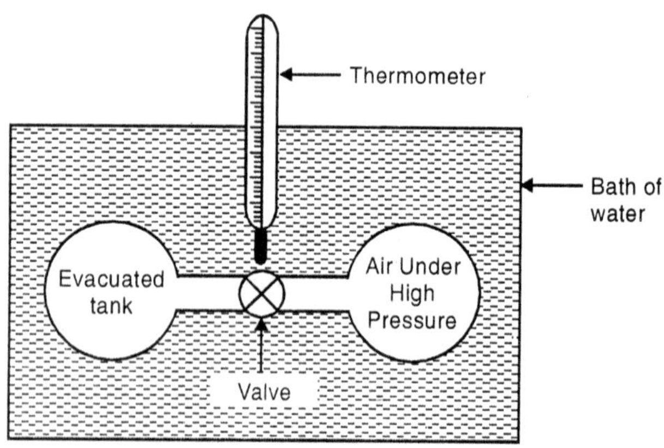

Fig. 5.5. Apparatus for demonstration of Joule's law.

Again, since both pressure and volume changed during the process, he remarked that *internal energy was a function only of temperature; since during the process temperature did not change, the internal energy remained constant.*

Later on when experiments were conducted with more refined instruments, it was found that there was a *very small change in temperature of water,* indicating that *for real gases internal energy was not a function of temperature alone.* However, at low pressure and high temperature where real gases behave like semi-perfect, gases and where the equation of state for a semi-perfect gas, $pv = RT$, is sufficiently accurate, Joule's law holds equally good in that range.

From definition of enthalpy,

$$h = u + pv$$

Also, $$pv = RT$$

$$\therefore \quad h = u + RT \quad \text{...(5.11)}$$

Since u is a function of temperature only, h is a function of temperature,

i.e., $$h = f(T) \quad \text{...(5.12)}$$

5.5. SPECIFIC HEAT CAPACITIES OF AN IDEAL GAS

The specific heat capacity at constant volume of any substance is defined by

$$c_v = \left(\frac{\partial u}{\partial T}\right)_v$$

It may be seen that as Joule's law for an ideal gas states $u = f(T)$, then

$$c_v = \frac{du}{dT} \quad \text{...(5.13)}$$

Since $h = u + pv$, Boyle's law, $pV = f(T)$ and Joule's law $u = f(T)$ together show, $h = f(T)$ and by similar argument to the above it may be seen that :

$$c_p = \frac{dh}{dT} \quad \text{...(5.14)}$$

Further as $h = u + pv$, then $h = u + RT$ and by differentiation

$$\frac{dh}{dT} = \frac{du}{dT} + R$$

Substitution from eqns. (5.13) and (5.14) gives,

$$c_p = c_v + R \quad \text{i.e.,} \quad c_p - c_v = R \quad \text{...(5.15)}$$

If expressed in terms of molar quantities then eqn. (5.15) becomes

$$C_p - C_v = R_0 \quad \text{...(5.16)}$$

where C_p and C_v are *molar specific heat capacities.*

Equations for Specific Heat Capacities of Ideal Gases

Since both u and h are functions of temperature, the equations to c_p and c_v must also be functions of temperature. They are usually expressed in a form :

$$c_p = a + KT + K_1T^2 + K_2T^3 \quad \text{...(5.17)}$$

$$c_v = b + KT + K_1T^2 + K_2T^3 \quad \text{...(5.18)}$$

where a, b, K, K_1 and K_2 are constants. Values of specific enthalpy etc. are then obtained by integration.

5.6. REAL GASES

It has been observed that when experiments are performed at relatively low pressures and temperatures most of the real gases obey Boyle's and Charle's laws quite closely. But the actual behaviour of real gases at elevated pressures and at low temperatures deviates considerably.

The ideal gas equation $pv = RT$ can be derived analytically using the kinetic theory of gases by making the following **assumptions :**

(*i*) A finite volume of gas contains large number of molecules.

(*ii*) The collision of molecules with one another and with the walls of the container are perfectly elastic.

(*iii*) The molecules are separated by large distances compared to their own dimensions.

(*iv*) The molecules do not exert forces on one another except when they collide.

As long as the above assumptions are valid the behaviour of a real gas approaches closely that of an ideal gas.

5.7. VAN DER WAALS' EQUATION

van der Waals' equation (for a real gas) may be written as :

$$\left(p + \frac{a}{v^2}\right)(v - b) = RT \qquad \text{...[5.19 (a)]}$$

The constants a and b are specific constants and depend upon the type of the fluid considered. 'v' represents the volume per unit mass and R is the gas constant.

If the volume of one mole is considered then the above equation can be written as

$$\left(p + \frac{a}{\bar{v}^2}\right)(\bar{v} - b) = R_0T \qquad \text{...[5.19(b)]}$$

The units of p, $\bar{v}$, T, R, a and b are as follows :

$p(N/m^2)$, $\bar{v}$ $(m^3/kg\text{-}mol)$, $T(K)$ and $R = 8314$ Nm/kg mol K, a $[Nm^4/kg\text{-}mol^2]$, $b(m^3/kg\ mol)$.

Table 5.1. Constants of van der Waal's Equation.

S. No.	Substance	a $Nm^4/(kg\text{-}mol)^2$	b $m^3/kg\text{-}mol$
1.	Hydrogen (H_2)	25105	0.0262
2.	Oxygen (O_2)	139250	0.0314
3.	Carbon dioxide (CO_2)	362850	0.0423
4.	Helium (He)	3417620	0.0228
5.	Air	135522	0.0362
6.	Water (H_2O) vapour	551130	0.0300
7.	Mercury (Hg) vapour	2031940	0.0657

van der Waals'' equation was proposed in 1873 for the gaseous and liquid states of a fluid, and accounts qualitatively for many important properties, but quantitatively it fails in many particulars.

The characteristic equation for a perfect gas is obtained by neglecting the finite size of the molecules. If this be taken into account it is obvious that the equation must be modified, for the distance travelled by a molecule between two successive encounters will be less than if the molecules were point spheres. Let the average distance traversed by a molecule between two successive encounters be denoted by λ, the mean free path. In Fig. 5.6 suppose L and M to be the two molecules of diameter 'd' at a distance λ apart. If these molecules were to impinge along the line of centres the path moved over would be less by an amount 'd' than if the molecules were point spheres. Now all the encounters between molecules are not direct, so their mean free paths will be lessened by an amount kd, where k is a fraction. That is, the mean free path is diminished in the ratio $(\lambda - kd) : \lambda$ or $\left(1 - \frac{kd}{\lambda}\right) : 1$.

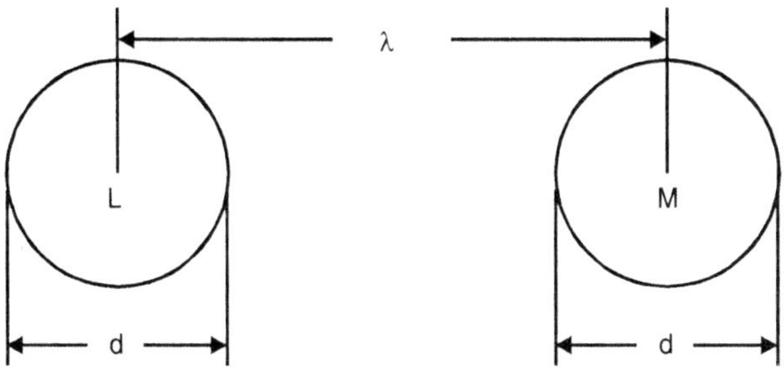

Fig. 5.6.

If the mean free path is lessened in this ratio, the encounters per second will be increased in the ratio $1 : 1 - \frac{kd}{\lambda}$. But the precare of the gas depends upon the encounters per second with the wall of the containing vessel. Hence the new pressure is given by

$$p = \frac{1}{3}\rho\overline{C}^2 \cdot \frac{1}{1 - \frac{kd}{\lambda}} \qquad ...(5.20)$$

where ρ is the density and $\overline{C}$ is the average velocity.

The mean free path is inversely proportional to the density of the gas, for if the volume were halved, *i.e.*, the density doubled, there would be twice as many molecules in the same space, and therefore any molecule would only have to travel approximately half as far before encountering another molecule. Hence writing v for $\frac{1}{\rho}$ and $\frac{b}{v}$ for $\frac{kd}{\lambda}$ in eqn. (5.20), we get

$$pv\left(1 - \frac{b}{v}\right) = \frac{\overline{C}^2}{3} = RT$$

or

$$p(v - b) = RT \qquad ...(5.21)$$

Next consider the *forces of cohesion* which act between a molecule and those surrounding it. When the molecule is sufficiently far removed from the surface of the gas in all directions the resultant of these cohesives forces are equally probable, as the individual forces are varying continuously as the surrounding molecules change their positions. Hence if the resultant is averaged over a sufficient length of time the aggregate force will be nil. *This is not true, however, when the molecule is near the surface.* Let the force from each molecule be resolved into normal and tangential components. All directions for the resultant in the tangential plane are equally likely, but the resultant normal component is most often directed inwards. Averaged over a sufficient length of time the total resultant force will therefore be a normal force always directed inwards. Thus the average effect of the cohesive forces is the same as if there was a permanent field of force acting at and near the surface. This field of force can be regarded as exerting a pressure p_1 over the boundary of the gas. The pressure is proportional to the number of molecules per unit area near the boundary surface and to the normal component of the force. Both of these factors are proportional to the density, so p_1 will be proportional to the square of the density.

i.e., $$p_1 = a\rho^2 \qquad ...(5.22)$$

where a is a constant.

Hence the molecules are not deflected by impact alone on reaching the boundary, but as the total result of their impact and of the action of the supposed field of force. That is, their change of momentum may be supposed to be produced by a total pressure $p + p_1$ instead of by the simple pressure p.

Hence eqn. (5.21) now becomes :

$$(p + p_1)(v - b) = RT,$$

or $$\left(p + \frac{a}{v^2}\right)(v - b) = RT$$

by substitution from p_1 from eqn. (5.22) and replacing ρ^2 by $\frac{1}{v^2}$.

Evaluation of Constants a and b

The general form of the isothermals for carbon dioxide given by van der Waals'' equation is shown in Fig. 5.7. These curves are obtained from the equation,

$$\left(p + \frac{0.00874}{v^2}\right)(v - 0.0023) = \frac{1.00646}{273}T \qquad ...(5.23)$$

where the unit of pressure is the atmosphere, and the unit of volume that of the gas of 0 °C under one atmosphere pressure.

van der Waals' equation being a cubic in v has three roots which may be either all real, or two imaginary and one real, as imaginary roots always occur in pairs. In Fig. 5.7, the 40 °C isothermal corresponds to the first condition, and the other isothermals to the latter. There is one isothermal where there are three real coincident roots at a point of inflexion. All the isothermals for temperatures higher than that corresponding

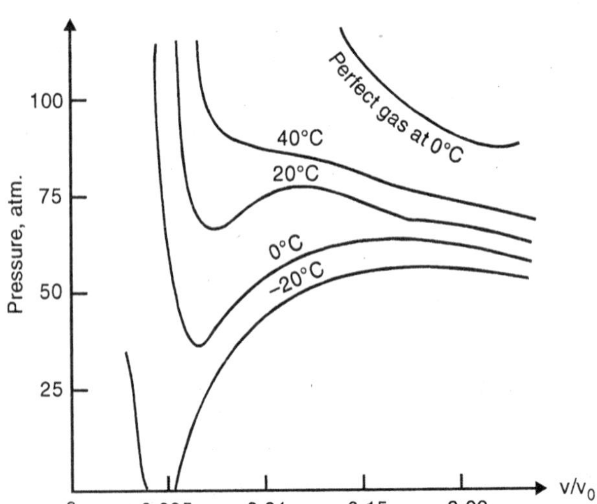

Fig. 5.7. Vander Waals' Isothermal for CO_2.

to the isothermal with three real coincident roots have no horizontal tangent, and all those lower have a maximum and minimum. Consequently this curve is identified with the critical isothermal. The temperature of the critical isothermal is obtained in the following manner. Equation (5.19) may be written

$$v^3 - \left(b + \frac{RT}{p}\right)v^2 + \frac{av}{p} - \frac{ab}{p} = 0$$

$$\left(p + \frac{a}{v^2}\right)(v - b) = RT$$

$$= pv - pb + \frac{a}{v^2} \times v - \frac{a}{v^2} \times b - RT = 0$$

$$= pv - pb + \frac{a}{v} - \frac{ab}{v^2} - RT = 0$$

Multiplying both sides by $\frac{v^2}{p}$, we get

$$pv \times \frac{v^2}{p} - pb \times \frac{v^2}{p} + \frac{a}{v} \times \frac{v^2}{p} - \frac{ab}{v^2} \times \frac{v^2}{p} - \frac{RT}{p}v^2 = 0$$

$$v^3 - \left(b + \frac{RT}{p}\right)v^2 + \frac{av}{p} - \frac{ab}{p} = 0$$

Now at the critical point, as the three roots are equal, the equation must be of the form :

$$(v - v_c)^3 = 0 \quad \text{...(5.25)}$$

where the suffix *c* denotes conditions at the critical point. For the critical point eqn. (5.24) becomes

$$v^3 - \left(b + \frac{RT_c}{p_c}\right)v^2 + \frac{av}{p_c} - \frac{ab}{p_c} = 0 \quad \text{...(5.26)}$$

Equations (5.25) and (5.26) are identical, hence equating coefficients

$$3v_c - b + \frac{RT_c}{p_c}$$

$$3v_c^2 = \frac{a}{p_c},$$

$$v_c^3 = \frac{ab}{p_c},$$

and from these by a simple reduction, we have

$$\left.\begin{aligned} v_c &= 3b \\ p_c &= \frac{a}{27b^2} \\ T_c &= \frac{8}{27} \cdot \frac{a}{bR} \end{aligned}\right\} \quad \text{...(5.27)}$$

From these equations it follows that the *critical volume, pressure and temperature are all completely determined by the constants of eqn.* (5.19).

The eqn. (5.27) indicates the critical constants for a particular gas and leads to the following results :

The values of a and b are also given by

$$a = 3p_c\, v_c^2 = \frac{9}{8} RT_c\, v_c = \frac{27}{64} \cdot \frac{R^2T_c^2}{p_c} \quad ...(i)$$

$$b = \frac{v_c}{3} = \frac{RT_c}{8p_c} \quad ...(ii)$$

and

$$R = \frac{8}{3} \frac{p_c\, v_c}{T_c} \quad ...(iii)$$

Using the values of a, b and R in eqn. (5.23), and substituting in (5.26), we have for carbon dioxide

$$p_c = 61.2 \text{ atmospheres},$$

$$T_c = 305.3 \text{ K or } 32.2\text{ °C}.$$

It is frequently assumed that the approximate agreement between the calculated and experimental values of the critical temperature for carbon dioxide is a sufficient verification of van der Waals' theory, but the constant b cannot be calculated with the required degree of accuracy from Regnault's experiments to make this an adequate test of the theory.

Also from eqns. (5.27), we have

$$\frac{p_c\, v_c}{RT_c} = \frac{3}{8} = 0.375$$

whereas experiment shows that about 0.27 as the average value of this ratio, varying considerably, however, from gas to gas.

The Reduced Equation

When the pressure, volume and temperature of the fluid are expressed as fractions of the critical pressure, volume and temperature the reduced form of van der Waals' equation is obtained thus, writing

$$p = ep_c = \frac{ea}{27b^2},$$

$$v = nv_c = 3nb,$$

$$T = mT_c = \frac{8}{27} \cdot \frac{ma}{bR}$$

and substituting these values in eqn. (5.19), this reduces to

$$\left(e + \frac{3}{n^2}\right)(3n - 1) = 8m$$

In this *"reduced"* equation the three constants which characterised a particular fluid have disappeared. The equation is accordingly true of any substance which satisfies an equation of the van der Waals' type, and the form of the curves connecting e, n and m is the same for all these substances. Thus we see that two substances, the behaviour of each of which is represented by van der Waals' equation, will be in corresponding states when the pressure, volume and temperature are the same multiples of their critical values.

This *theorem of corresponding states*, enunciated by van der Waals', was tested by Amagat and found to be approximately true for a large number of fluids. The theorem of corresponding states is not unique to the equation of van der Waals'. Any equation of state giving a critical point and having not more than three constants will serve equally well to give a reduced equation, in which the constants peculiar to any one fluid disappear, and therefore become the basis of the theorem of corresponding states.

It must be remembered in applying the theorem that the accuracy of results deduced by its aid cannot be greater than the accuracy with which the original equation represents the behaviour of the fluids under consideration.

5.8. AMAGAT'S EXPERIMENTS

As per Amagat's experiments van der Waals' equation accounts for the variation of the product pv with increasing pressure as follows.

Writing equation (5.19) in the form

$$pv = \frac{RT_v}{v-b} - \frac{a}{v},$$

and differentiating with respect to p, keeping T constant, we have

$$\left(\frac{d(pv)}{dp}\right)_T = \left\{\frac{a}{v^2} - \frac{RTb}{(v-b)^2}\right\}\left(\frac{dv}{dp}\right)_T \quad ...(5.28)$$

Since the condition for a minimum on any isothermal is

$$\left(\frac{d(pv)}{dp}\right)_T = 0,$$

The right-hand side of equation (5.28) must vanish at this point. Now $\left(\frac{dv}{dp}\right)_T$ is *never zero*, so we have as the condition for a minimum :

$$\frac{RTb}{(v-b)^2} = \frac{a}{v^2} \quad \text{or} \quad RT.\frac{b}{a} = \left(1-\frac{b}{v}\right)^2 \quad ...(5.29)$$

This equation shows that the *volume at which the minimum value of pv occurs on any isothermal gradually increases as the temperature is raised.*

To find the locus of minima the temperature T must be eliminated from equation (5.29) by substitution from the original equation. Thus from equation (5.19)

$$RT = \left(p + \frac{a}{v^2}\right)(v-b),$$

and substituting this in equation (5.28), we have

$$\frac{b}{a}\left(p + \frac{a}{v^2}\right)(v-b) = \left(1-\frac{b}{v}\right)^2,$$

which reduces to $$v = b\left(\frac{pv^2}{a} + 2\right)$$

Multiply each side of this equation by p, and put $pv = y$ and $p = x$, and we obtain

$$y = b\left(\frac{y^2}{a} + 2x\right) \quad \text{or} \quad y(a - by) = 2abx$$

The above expression gives the locus of minima and is a parabola with axis parallel to the x-axis as shown in Fig. 5.8.

Consider the isothermal which goes through the point A. Here $x = 0$ and $y = \frac{a}{b}$.

Writing van der Waal's equation in terms of x and y, we have

$$\left(1 + \frac{ax}{y^2}\right)(y - bx) = RT,$$

and substituting the coordinates of the point A

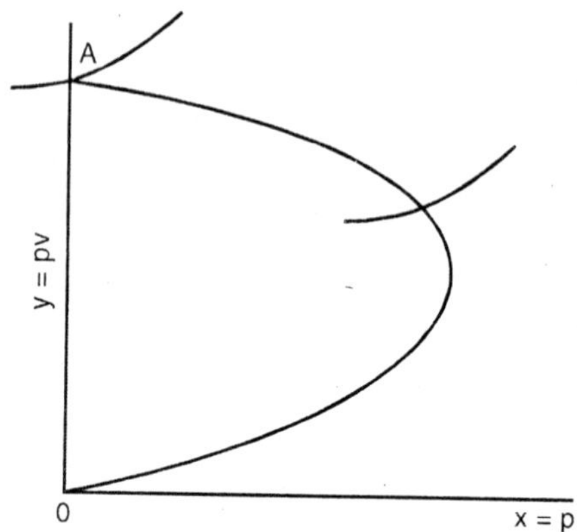

Fig. 5.8.

$$RT = \frac{a}{b}$$

or
$$T = \frac{a}{bR} \qquad \text{...(5.30)}$$

For temperatures above that given by eqn. (5.30) the minima lie in the region of negative pressure, so an Amagat isothermal for a temperature equal to or greater than $\frac{a}{bR}$ will slope upwards along its whole length for increasing values of p, but for a temperature less than $\frac{a}{bR}$ the isothermals first dip to a minimum and then rise.

Using the result from eqn. (5.27)

$$T_c = \frac{8}{27}\frac{a}{bR},$$

We see that the limiting temperature for an isothermal to show a *minimum* is

$$T = \frac{27}{8} . T_c$$

The reason for Amagat finding no dip in the isothermals for hydrogen is now apparent. The critical temperature is 35 K, and therefore the limiting temperature above which minima do not occur is $\frac{27}{8} \times 35 = 118.1$ K or –155 °C, and all Amagat's experiments were conducted between 0 °C and 100 °C.

5.9. THE COOLING EFFECT

The most gases show an inversion of the cooling effect at a certain temperature. The equation of van der Waals' indicates at what temperature this occurs.

We have

$$\left(p + \frac{a}{v^2}\right)(v - b) = RT \qquad \text{[From eqn. (5.15)]}$$

Keeping p constant and differentiating with respect to T, we get

$$\left\{p - \frac{a}{v^2} + \frac{2ab}{v^3}\right\}\left(\frac{dv}{dT}\right)_p = R$$

or

$$\left(\frac{dv}{dT}\right)_p = \frac{R}{\left\{p - \frac{a}{v^2} + \frac{2ab}{v^3}\right\}}$$

Substituting this value of $\left(\frac{dv}{dT}\right)_p$ in the equation $c_p\mu = T\left(\frac{dv}{dT}\right)_p - v$ (where μ is a measure of cooling effect), we get

$$c_p\mu = \frac{RT}{\left\{p - \frac{a}{v^2} + \frac{2ab}{v^3}\right\}} - v$$

and substituting for RT from equation (5.19) this reduces to

$$c_p\mu = \frac{-bp + \frac{2a}{v} - \frac{3ab}{v^2}}{p - \frac{a}{v^2} + \frac{2ab}{v^3}}.$$

The denominator of this expression is *always positive*, since it is $R\left(\frac{dT}{dv}\right)_p$. Hence the cooling effect, μ, is *positive* if

$$bp < \frac{2a}{v} - \frac{3ab}{v^2} \qquad ...(5.31)$$

and *negative* if

$$bp > \frac{2a}{v} - \frac{3ab}{v^2} \qquad ...(5.32)$$

and *inversion occurs* when

$$bp = \frac{2a}{v} - \frac{3ab}{v^2}$$

or

$$p = \frac{a}{b}\left(\frac{2}{v} - \frac{3b}{v^2}\right) \qquad ...(5.33)$$

In order to get the temperature of inversion this equation must be combined with the original equation. Thus

$$\frac{2a}{b}\left(1 - \frac{b}{v}\right)^2 = RT \qquad ...(5.34)$$

Since v is necessarily always greater than b, it will be seen that as v *increases so also does the temperature of inversion.*

The form of curve given by eqn. (5.33) is shown in Fig. 5.9. The pressure is zero when $v = \frac{3b}{2}$, or infinity. These values of v determine the limiting values of the temperature of inversion, as it is only between these limits that p is positive. Substuting these limits of v in equation (5.34) the limiting values of the temperature of inversion are $\frac{2}{9}\frac{a}{bR}$ and $\frac{2a}{bR}$, or from eqn. (5.27), $\frac{3}{4}T_c$ and $\frac{27}{4}T_c$.

The eqn. (5.33) being quadratic there are two values of v for a constant value of p at which inversion occurs, as may also be seen by reference to Fig. 5.9. Consequently by equation (5.34) there are two temperatures for a constant value of p at which inversion occurs. As the temperature increases through the lower of these values the change is from a heating to a cooling effect, and as it increases through the higher of these values the change is from a cooling to a heating effect.

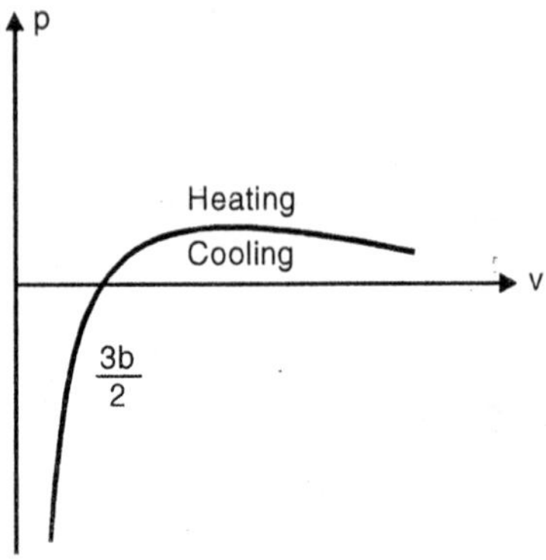

Fig. 5.9.

The inversion will occur when the maximum value of p is $\frac{a}{3b^2}$, when $v = 3b$. For any value of p less than this there is a cooling effect provided the condition of the substance is represented by a point inside the area enclosed by the curve and the axis of volume, Fig. 5.9, and for any greater value of p there is a heating effect as indicated by eqns. (5.31) and (5.32) respectively.

Let us take the case of *hydrogen*. In the experiments of Joule and Thomson the pressure used was 4.7 atmosphere. The critical temperature and pressure are 35 K and 15 atmospheres.

From eqn. (5.33) we can find the values of $\frac{b}{v}$ corresponding to the pressure used by Joule and Thomson, and by substitution in eqn. (5.34) find the two temperatures at which inversion occurs at this pressure. Equation (5.33) can be written as :

$$p = 27p_c\left[2\frac{b}{v} - 3\left(\frac{b}{v}\right)^2\right]$$

Hence
$$\frac{b}{v} = \frac{2 \pm \sqrt{4 - \frac{12p}{27p_c}}}{6} = 0.6608 \text{ or } 0.0058$$

by substitution of the above values for p and p_c.

Writing eqn. (5.34) in the form

$$T = \frac{27}{4}T_c\left(1 - \frac{b}{v}\right)^2,$$

We have by substitution for $\frac{b}{v}$: $T = 233.5$ K or 27.2 K that is, below –245.9 °C there would be a *heating effect*, between –245.9 °C and –39.6 °C *a cooling effect*, and above –39.6 °C a heating effect. Thus **van der Waal's equation qualitatively accounts for the heating effect observed at ordinary temperatures.**

Limitations of van der Waal's Equation

van der Waal's equation under actual condition becomes *invalid* as discussed below :

- The values of a and b (which are assumed to be constant) are found to vary with temperature. Thus the results obtained from the equation are *incorrect when the variation of a and b is large with respect to temperature.*
- The equation is *not accurate enough in the critical region* and it is also obvious from its derivation.

5.10. MASS FRACTION

The ratio of the mass of the constituent to the mass of whole mixture containing it is called 'Mass Fraction'.

5.11. MOLE FRACTION

The ratio of the moles of the constituent to the moles of the mixtures containing it is called mole-fraction. It is also equal to the volume-fraction of the mixture.

5.12. PARTIAL PRESSURE

According to the Dalton's Law, the 'partial pressure' of each constituent is that pressure which the gas would exert if it occupies alone that volume which has been occupied by the mixture at the same temperature.

5.13. COMPRESSIBILITY CHART

The *compressibility factor* (Z) of any gas is a function of only two properties, usually temperature and pressure, to that $Z = f(T_r, p_r)$ except near the critical point. The value of Z for any real gas may be less or more than unity, depending on pressure and temperature conditions of the gas.

The general compressibility chart is plotted with Z versus p_r for various values of T_r. This is constructed by plotting the known data of one or more gases and can be used for any gas. Such a chart is shown in Fig. 5.10. This chart gives best results for the regions well removed from the critical state for all gases.

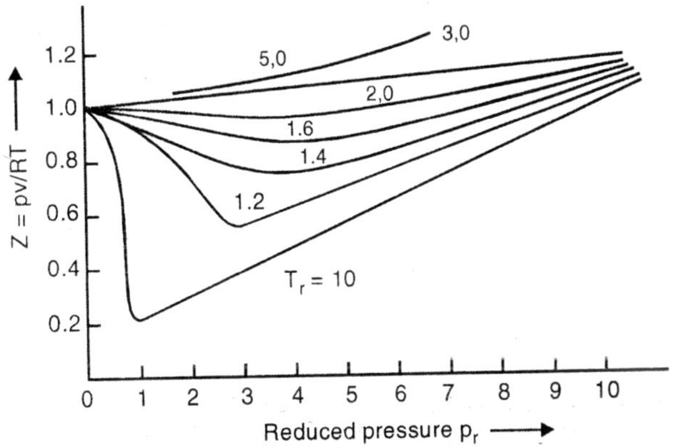

Fig. 5.10. Generalised compressibility chart.

NUMERICAL EXAMPLES

Ideal Gases

Example 5.1. The volume of a high altitude chamber is 40 m³. It is put into operation by reducing pressure from 1 bar to 0.4 bar and temperature from 25 °C to 5 °C.

How many kg of air must be removed from the chamber during the process ? Express this mass a volume measured at 1 bar and 25 °C.

Take R = 287 J/kg K for air.

Solution. Given data,

$V_1 = 40 \text{ m}^3$ $\quad V_2 = 40 \text{ m}^3$

$p_1 = 1 \text{ bar}$ $\quad p_2 = 0.4 \text{ bar}$

$T_1 = 25 + 273 = 298 \text{ K}$ $\quad T_2 = 5 + 273 = 278 \text{ K}$

kg of air to be removed :

Assuming nitrogen to be a perfect gas,

$$p_1V_1 = m_1 RT_1 \qquad \left(\text{or } m_1 = \frac{p_1V_1}{RT_1}\right)$$

$$p_2V_2 = m_2RT_2 \qquad \left(\text{or } m_2 = \frac{p_2V_2}{RT_2}\right)$$

Mass of air removed during the process = $(m_1 - m_2)$ kg

$$(m_1 - m_2) = \frac{p_1V_1}{RT_1} - \frac{p_2V_2}{RT_2}$$

$$= \frac{1}{R}\left(\frac{p_1V_1}{T_1} - \frac{p_2V_2}{T_2}\right) = \frac{1}{287}\left[\frac{(1 \times 10^5) \times 40}{298} - \frac{(0.4 \times 10^5) \times 40}{278}\right]$$

$= 26.71$ kg. **Ans.**

Volume of this mass of gas at 1 bar and 25 °C is given by

$$V = \frac{mRT}{p} = \frac{26.71 \times 287 \times 298}{1 \times 10^5} = 22.84 \text{ m}^3. \textbf{ Ans.}$$

Example 5.2. A steel flask of 0.04 m³ capacity is to be used to store nitrogen at 120 bar, 20 °C. The flask is to be protected against excessive pressure by a fusible plug which will melt and allow the gas to escape if the temperature rises too high.

(*i*) How many kg of nitrogen will the flask hold at the designed conditions ?

(*ii*) At what temperature must the fusible plug melt in order to limit the pressure of a full flask to a maximum of 150 bar ?

Solution. Given data : Capacity of the steel flask, $V = 0.04 \text{ m}^3$

Pressure, $\quad p = 120$ bar

Temperature, $\quad T = 20 + 273 = 293$ K

(*i*) **Kg of nitrogen the flask can hold :**

Now, R for nitrogen (molecular weight, $M = 28$)

$$= \frac{R_0}{M} = \frac{8314}{28} = 296.9 \text{ kJ / kg K.}$$

Assuming nitrogen to be a perfect gas, we get

Mass of nitrogen in the flask at designed condition

$$= m = \frac{pV}{RT} = \frac{120 \times 10^5 \times 0.04}{296.9 \times 293} = 5.51 \text{ kg. } \textbf{Ans.}$$

(*ii*) **Temperature at which fusible plug should melt, *t* :**

When the fusible plug is about to melt

$$p = 150 \text{ bar; } V = 0.04 \text{ m}^3; \; m = 5.51 \text{ kg}$$

Therefore, temperature *t* at which fusible plug must melt is given by

$$T = \frac{pV}{mR} = \frac{15 \times 10^5 \times 0.04}{5.51 \times 296.9} = 366.7 \text{ K}$$

$$\therefore \quad t = 366.7 - 273 = 93.7 \text{ °C. } \textbf{Ans.}$$

Example 5.3. A balloon of spherical shape 6 m in diameter is filled with hydrogen gas at a presssure of 1 bar abs. and 20 °C. At a later time, the pressure of gas is 94 percent of its original pressure at the same temperature :

(*i*) What mass of original gas must have escaped if the dimensions of the balloon is not changed ?

(*ii*) Find the amount of heat to be removed to cause the same drop in pressure at constant volume.

Solution. Given data : Diameter of the spherical balloon = 6 m

Pressure of hydrogen gas, p_1 = 1 bar abs.

Temperature of hydrogen gas, T_1 = 20 °C or 293 K

At a later time pressure of the gas, $p_2 = 0.94\, p_1$ at 293 K.

(*i*) **Mass of original gas escaped :**

$$\Delta m = m_1 - m_2$$

[where m_1 and m_2 are the initial and final masses of the gas]

$$= \frac{p_1V_1}{RT_1} - \frac{p_2V_2}{RT_2} = \frac{V_1}{RT_1}(p_1 - p_2) \qquad [\because V_1 = V_2, T_1 = T_2 \text{ and } p_2 = 0.94\, p_1]$$

$$= \frac{V_1}{RT_1}(p_1 - 0.94\,p_1) = \frac{p_1V_1}{RT_1}(1 - 0.94)$$

$$\therefore \quad \text{\%age mass escaped} = \frac{\Delta m}{m_1} \times 100 = \frac{\dfrac{p_1V_1}{RT_1}(1-0.94)}{\dfrac{p_1V_1}{RT_1}} = 6\%. \; \textbf{Ans.}$$

(*ii*) **Amount of heat to be removed :**

Using the gas equation,

$$\frac{p_1V_1}{T_1} = \frac{p_2V_2}{T_2}$$

or

$$\frac{p_1}{T_1} = \frac{0.94\,p_1}{T_2} \qquad (\because V_1 = V_2 \text{ and } p_2 = 0.94\, p_1)$$

$\therefore \quad T_2 = 0.94\, T_1 = 0.94 \times 293 = 275.4$ K or 2.42 °C

The heat to be removed is given by

$$Q = mc_v\,(T_1 - T_2)$$

where $m = \dfrac{p_1 V_1}{RT_1} = \dfrac{1 \times 10^5 \times \frac{4}{3}\pi \times 3^3}{\frac{8314}{2} \times 293} = 9.28$ kg. $\left[\begin{array}{l} \because \quad MR = 8314 \\ \therefore \quad R = \dfrac{8314}{2} \\ \text{as } M \text{ for } H_2 = 2 \end{array}\right]$

$$c_v = 10400 \text{ J/kg K for } H_2.$$

$\therefore \quad Q$ (heat to be removed) $= 9.28 \times 10400\,(293 - 275.4)$

$= 1.69$ MJ. **Ans.**

Example 5.4. A vessel of capacity 3 m^3 contains 1 kg mole of N_2 at 90 °C.

(*i*) Calculate pressure and the specific volume of the gas.

(*ii*) If the ratio of specific heats is 1.4, evaluate the values of c_p and c_v.

(*iii*) Subsequently, the gas cools to the atmospheric temperature of 20 °C; evaluate the final pressure of gas.

(*iv*) Evaluate the increase in specific internal energy, the increase in specific enthalpy, increase in specific entropy and magnitude and sign of heat transfer.

Solution. Given data : Mass of N_2, $m = 1$ kg mole *i.e.*, 28 kg

Capacity of the vessel, $V = 3$ m^3

Temperature, $T_1 = 90 + 273 = 363$ K

(*i*) **Pressure (p_1) and specific volume (v_1) of the gas :**

Using the relation

$$p_1 V_1 = mRT_1$$

$$p_1 \times 3 = 28 \times \left(\frac{8314}{28}\right) \times 363 \qquad \left[\because R = \frac{R_0}{M} = \frac{8314}{28}\right]$$

$\therefore \quad p_1 = 1005994$ J/m^2 or 10.06 bar. **Ans.**

Specific volume, $v_1 = \dfrac{V_1}{m} = \dfrac{3}{28} = 0.107$ m^3/kg. **Ans.**

(*ii*) $c_p = ?,\ c_v = ?$

$$\frac{c_p}{c_v} = 1.4 \text{ (given)} \qquad ...(i)$$

But $$c_p - c_v = R = \frac{8314}{28} \qquad ...(ii)$$

Solving for c_p and c_v between (*i*) and (*ii*)

$c_p = 1.039$ kJ/kg K; $c_v = 0.742$ kJ/kg K. **Ans.**

(*iii*) **Final pressure of the gas after cooling to 20 °C :**

Initially	After cooling
$p_1 = 10.06$ bar	$p_2 = ?$
$V_1 = 3$ m^3	$V_2 = 3$ m^3

$T_1 = 363\text{ K} \qquad T_2 = 20 + 273 = 293\text{ K}$

Now, $$\frac{p_1V_1}{T_1} = \frac{p_2V_2}{T_2}$$

or $$\frac{p_1}{T_1} = \frac{p_2}{T_2} \qquad \text{(as } V_1 = V_2\text{)}$$

$$p_2 = \frac{p_1T_2}{T_1} = \frac{10.0 \times 293}{363} = 8.12\text{ bar.}$$ **Ans.**

(*iv*) **Δu, Δh, Δs, Q :** For a perfect gas,

Increase in specific internal energy

$$\Delta u = c_v(T_2 - T_1) = 0.742\,(293 - 363) = -51.94\text{ kJ/kg.}$$ **Ans.**

Increase in specific enthalpy,

$$\Delta h = c_p(T_2 - T_1) = 1.039\,(293 - 363) = -72.73\text{ kJ/kg.}$$ **Ans.**

Increase in specific entropy,

$$\Delta s = c_v \log_e\left(\frac{T_2}{T_1}\right) + R\log_e\left(\frac{v_2}{v_1}\right)$$

But $$v_1 = v_2$$

$$\therefore \quad \Delta s = c_v \log_e\left(\frac{T_2}{T_1}\right) = 0.742 \log_e\left(\frac{293}{363}\right) = -0.1589\text{ kJ/kg K.}$$ **Ans.**

Now, $$Q = \Delta u + W$$

Here $W = 0$ as change in volume is zero

$$\therefore \quad Q = \Delta u$$

$\therefore$ Heat transfer, $Q = -51.94\text{ kJ/kg} = -51.94 \times 28 = -1454.32\text{ kJ.}$ **Ans.**

Example 5.5. (*a*) 1 kg of air at a pressure of 8 bar and a temperature of 100 °C undergoes a reversible polytropic process following the law $pv^{1.2}$ = constant. If the final pressure is 1.8 bar determine :

(*i*) The final specific volume, temperature and increase in entropy;

(*ii*) The work done and the heat transfer.

Assume $R = 0.287$ kJ/kg K and $\gamma = 1.4$.

(*b*) Repeat (*a*) assuming the process to be irreversible and adiabatic between end states.

Solution. Given data : (*a*) Mass of air, $m = 1$ kg

Pressure, $p_1 = 8$ bar

Temperature, $T_1 = 100 + 273 = 373$ K

The law followed :

$$pv^{1.2} = \text{Constant}$$

Final pressure, $p_2 = 1.8$ bar

Characteristic gas constant, $R = 0.287$ kJ/kg K

Ratio of specific heats, $\gamma = 1.4$

(*i*) **v_2, T_2 and Δs :**

Assuming air to be a perfect gas,

$$p_1v_1 = RT_1$$

$$\therefore \quad v_1 = \frac{RT_1}{p_1} = \frac{(0.287 \times 1000) \times 373}{8 \times 10^5}$$

$$= 0.1338 \text{ m}^3/\text{kg}$$

Also, $$p_1 v_1^{1.2} = p_2 v_2^{1.2}$$

or $$\frac{v_2}{v_1} = \left(\frac{p_1}{p_2}\right)^{1/1.2}$$

or $$v_2 = v_1 \left(\frac{p_1}{p_2}\right)^{1/1.2} = 0.1338 \left(\frac{8}{1.8}\right)^{1/1.2}$$

$$= 0.4637 \text{ m}^3/\text{kg}$$

i.e., Final specific volume, $v_2 = 0.4637 \text{ m}^3/\text{kg}$. **Ans.**

Again, $$p_2 v_2 = RT_2$$

$$T_2 = \frac{p_2 v_2}{R} = \frac{1.8 \times 10^5 \times 0.4637}{(0.287 \times 1000)} = 290.8 \text{ K}$$

i.e., Final temperature, $t_2 = 290.8 - 273 = 17.8$ °C. **Ans.**

Increase in entropy Δs is given by,

$$\Delta s = c_v \log_e \left(\frac{T_2}{T_1}\right) + R \log_e \left(\frac{v_2}{v_1}\right)$$

But $$\gamma = \frac{c_p}{c_v} = 1.4 \text{ (given)}$$

and $$c_p - c_v = R (= 0.287 \text{ kJ/kg K for air}) \quad ...(ii)$$

Solving for c_v between (*i*) and (*ii*),

$$c_v = 0.717 \text{ kJ/kg K}$$

$$\therefore \quad \Delta s = 0.717 \log_e \left(\frac{290.8}{373}\right) + 0.287 \log_e \left(\frac{0.4637}{0.1338}\right)$$

$$= -0.1785 + 0.3567 = 0.1782 \text{ kJ/kg K}$$

i.e., Increase in entropy, $\Delta s = 0.1782$ kJ/kg K. **Ans.**

(*ii*) **Work done and heat transfer :**

The work done in a polytropic process is given by,

$$W = \frac{p_1 v_1 - p_2 v_2}{n - 1} = \frac{R(T_1 - T_2)}{n - 1}$$

$$= \frac{0.287\,(373 - 290.8)}{(1.2 - 1)} = 117.96 \text{ kj/kg.}$$

i.e., Work done $= 117.96$ kJ/kg. **Ans.**

Heat transfer, $$Q = \Delta u + W$$

where $$\Delta u = c_v (T_2 - T_1)$$

$$= 0.717\,(290.8 - 373) = -58.94 \text{ kJ/kg}$$

$$\therefore \quad Q = -58.94 + 117.96 = 59.02 \text{ kJ/kg}$$

Hence Heat transfer $= 59.02$ kJ/kg. **Ans.**

(*b*) (*i*) Though the process is assumed now to be irreversible and adiabatic, the end states are given to be the same as in (*a*). Therefore, **all the properties at the end of the process are the same as in (*a*). Ans.**

(*ii*) As the process is adiabatic, Q (heat transfer) = 0. **Ans.**

$$\Delta u = \Delta u \text{ in } (a)$$

Applying first law for this process

$$Q = \Delta u + W$$

$$0 = \Delta u + W$$

or $$W = -\Delta u$$

$$= -(-58.94) = 58.94$$

$\therefore$ $$\text{Work done} = 58.94 \text{ kJ/kg. } \textbf{Ans.}$$

Real Gases

Example 5.6. One kg of CO_2 has a volume of 1 m^3 at 100 °C. Compute the pressure by;

(*i*) van der Waals' equation

(*ii*) Perfect gas equation.

Solution. (*i*) Using van der Waals' equation :

Molar specific volume,

$$\bar{v} = 1 \times 44 = 44 \text{ m}^3/\text{kg-mol} \qquad (\because M \text{ for } CO_2 = 44)$$

Temperature, $T = 100 + 273 = 373$ K

The values of a and b for CO_2 (from Table 7.1)

$$a = 362850 \text{ Nm}^4/(\text{kg-mol})^2$$

and $$b = 0.0423 \text{ m}^3/\text{kg-mol}$$

$$R_0 = 8314 \text{ Nm/kg-mol K}$$

van der Waal's equation is written as

$$\left(p + \frac{a}{\bar{v}^2}\right)(\bar{v} - b) = R_0 T$$

or $$p = \left(\frac{R_0 T}{\bar{v} - b} - \frac{a}{\bar{v}^2}\right)$$

Substituting the values in the above equation, we get

$\therefore$ $$p = \frac{8314 \times 373}{44 - 0.0423} - \frac{362850}{44^2}$$

$$= 70548 - 187$$

$$= 70361 \text{ N/m}^2 \text{ or } 0.7036 \text{ bar. } \textbf{Ans.}$$

(*ii*) Using perfect gas equation :

$$p\bar{v} = R_0 T$$

$\therefore$ $$\text{Pressure, } p = \frac{R_0 T}{\bar{v}} = \frac{8314 \times 373}{44}$$

$$= 70480 \text{ N/m}^2 \text{ or } 0.7048 \text{ bar. } \textbf{Ans.}$$

ASSIGNMENT

(A) Multiple Choice Questions :

1. (*a*) A perfect gas does not obey the law $pv = RT$.
   (*b*) A perfect gas obeys the law $pv = RT$ and has constant specific heat.
   (*c*) A perfect gas obeys the law $pv = RT$ but have variable specific heat capacities.
2. Boyle's law states that, when temperature is constant, the volume of a given mass of a perfect gas :
   (*a*) varies directly as the absolute pressure.
   (*b*) varies inversely as the absolute pressure.
   (*c*) varies as square of the absolute pressure.
   (*d*) does not vary with the absolute pressure.
3. Charle's law states that if any gas is heated at constant pressure, its volume :
   (*a*) changes directly as it absolute temperature.
   (*b*) changes inversely as its absolute temperature.
   (*c*) changes as square of the absolute temperature.
   (*d*) does not change with absolute temperature.
4. The equation of the state per kg of a perfect gas is given by :
   (*a*) $p^2v = RT$ (*b*) $pv = RT$
   (*c*) $pv^2 = RT$ (*d*) $p^2v^2 = RT$.
   where p, v, R and T are the pressure, volume characteristic gas constant and temperature of the gas respectively.
5. The equation of state of an ideal gas is a relationship between the variables :
   (*a*) pressure and volume.
   (*b*) pressure and temperature.
   (*c*) pressure, volume and temperature.
   (*d*) none of the above.
6. Joule's law states that the specific internal energy of a gas depends only on :
   (*a*) the pressure of the gas
   (*b*) the volume of the gas
   (*c*) the temperature of the gas
   (*d*) none of the above.
7. Equation for specific heat at constant pressure of an ideal gas is given by :
   (*a*) $c_p = a + KT + K_1T^2 + K_2T^3$
   (*b*) $c_p = a + KT^2 + K_1T^6 + K_2T^4$
   (*c*) $c_p = a + KT^2 + K_1T^4 + K_2T$
   (*d*) $c_p = a + KT^2 + K_1T^6 + K_2T^2$.
   where a, K, K_1 and K_2 are constants.
8. van der Waal's equation may be written as
   (*a*) $\left(p+\frac{a}{v}\right)(v-b) = RT$ (*b*) $\left(p+\frac{a}{v^2}\right)(v-b) = RT$
   (*c*) $\left(p+\frac{a}{v^2}\right)(v^2-b) = RT$ (*d*) $\left(p+\frac{a}{v^2}\right)(v^2-b) = RT^2$.

ANSWERS

(A) Multiple Choice Questions :

1. (*b*) **2.** (*b*) **3.** (*a*) **4.** (*b*) **5.** (*c*)
6. (*c*) **7.** (*a*) **8.** (*b*).

(B) Short Answer Type Questions :

Q. 1. Define an 'Ideal Gas'.

Ans. An 'ideal gas' is defined as a gas having no forces of intermolecular attraction. It may also be defined as a 'gas' which follow the gas laws at all ranges of temperatures and pressures.

Q. 2. What are the 'Real-Gases' ?

Ans. The 'gases' which follow the gas laws at low pressures or high temperatures or both, are called 'Real-gases'.

Q. 3. Define the 'Boyle's law'.

Ans. It states that the volume of the given mass of a perfect gas varies inversely as the absolute pressure when the temperature is kept constant.

Q. 4. Define the 'Charle's law'.

Ans. It states that if any gas is heated at constant pressure, its volume changes directly as its absolute temperature.

Q. 5. Write the equation of the state for a perfect gas.

Ans. The equation of the state for a perfect gas is;

$$pV = RT.$$

Q. 6. Define the Joule's law.

Ans. It states, that the specific internal energy of a gas depends only on the temperature of the gas and is independent of both the pressure and volume.

Q. 7. Write the 'equations for specific heat capacities of ideal gases.

Ans. Since both, internal energy, 'u' and enthalpy 'h' are the functions of temperature the equations to c_p and c_v must also be functions of temperature. They are usually expressed in a form.

$$c_p = a + KT + K_1T^2 + K_2T^3$$

and

$$c_v = b + KT + K_1T^2 + K_2T^3$$

where a, b, K, K_1 and K_2 are the constants.

Values of sp. enthalpy etc. are then obtained by integration.

Q. 8. What are the assumptions for the kinetic theory of gases ?

Ans. (*i*) A finite volume of gas contains large number of molecules.

(*ii*) The collision of molecules with one another and with the walls of the container are perfectly elastic.

(*iii*) The molecules are separated by large distances compared to their own distances.

(*iv*) The molecules do not exert forces on one another except when they collide.

As long as there assumptions are valid the behaviour of a real gas approaches closely that of an ideal gas.

Q. 9. What is the 'van der Waal's Equation' for a real gas ?

Ans. van der Waal's equation for a real gas may be written as;

$$\left(p + \frac{a}{v^2}\right)(v - b) = RT.$$

The constants '*a*' and '*b*' are specific constants and depend upon the type of the fluid considered. v is the volume per unit mass and R the gas constant.

Q. 10. What are the limitations of van der Waal's equation ?

Ans. (*i*) The values of '*a*' and '*b*' (both constants) are found to vary with temperature. Thus, the results obtained from the equation are incorrect when the variation of '*a*' and '*b*' is large w.r.t. temperature.

(*ii*) The equation is not accurate enough in the critical region and it is also obvious from its derivation.

Q. 11. Define 'Mass-Fraction'.

Ans. The ratio of the mass of the constituent to the mass of the whole mixture containing it is called 'Mass-Fraction'.

Q. 12. Define 'Mole-Fraction'.

Ans. The ratio of the moles of the constituent to the moles of the mixtures containing it is called the 'Mole-Fraction'. It is also equal to the volume fraction of the mixture.

Q. 13. Define partial pressure.

Ans. As per Dalton's law, the partial pressure of each constituent is that pressure which the gas would exert if it occupies alone that volume which has been occupied by the mixture at the same temperature.

Q. 14. Is there any perfect-gas existing ?

Ans. No, in reality, there is no ideal or perfect gas but the real gases just approach them upto certain extent.

Q. 15. What are the permanent gases ?

Ans. At a very low pressure and at a very high temperature, real gases like Hydrogen, Oxygen, Nitrogen, Helium etc. behave nearly the same way as the perfect gases, so they are called semi-perfect or permanent gases.

REVIEW QUESTIONS

1. What is an ideal gas ?
2. What is the difference between an ideal and a perfect gas ?
3. What are semi-perfect or permanent gases ?
4. Define 'Equation of state'.
5. State Boyle's and Charle's laws and derive an equation of the state for a perfect gas.
6. What is a *p-v-T* surface ? Draw a portion of a such a surface.
7. Derive the relationship between the two principal specific heats and characteristic gas constant for a perfect gas..
8. Write a short note on van der Waal's equation.

PROBLEMS FOR PRACTICE

Ideal Gases

1. A vessel of 0.03 m^3 capacity contains gas at 3.5 bar pressure and 35 °C temperature. Determine the mass of the gas in the vessel. If the pressure of this gas is increased to 10.5 bar while the volume remains constant, what will be the temperature of the gas ?

   For the gas take R = 290 J/kg K. **[Ans.** 0.118 kg, 650 °C]
2. The tyre of an automobile contains a certain volume of air at a gauge pressure of 2 bar and 20 °C. The barometer reads 75 cm of Hg. The temperature of air

in the tyre rises to 80 °C due to running of automobile for two hours. Find the new pressure in the tyre.
Assume that the air is an ideal gas and tyre does not stretch due to heating.
[Ans. 2.62 bar]

3. A tank made of metal is designed to bear an internal gauge pressure of 7 bar. The tank is filled with a gas at a pressure of 5.5 bar abs., and 15 °C. The temperature in the tank may reach to 50 °C when the tank stands in the sun.
   (*i*) If the tank does not expand with temperature, will the design pressure be exceeded on a day when atmospheric pressure is 1 bar and air in the tank reaches 50 °C when exposed to hot sun ?
   (*ii*) What temperature would have to be reached to raise the air pressure to the design limit ? **[Ans.** (*i*) 6.16 bar, (*ii*) 147 °C]
4. A vessel of spherical shape is 1.5 m in diameter and contains air at 40 °C. It is evacuated till the vacuum inside the vessel is 735 mm of mercury. Determine :
   (*i*) The mass of air pumped out;
   (*ii*) If the tank is then cooled to 10 °C what is the pressure in the tank ?
   The barometer reads 760 mm of mercury. Assume that during evacuation, there is no change in temperature of air. **[Ans.** (*i*) 1.91 kg, (*ii*) 3 kPa]
5. A balloon of spherical shape is 8 m in diameter and is filled with hydrogen at a pressure of a 1 bar abs. and 15 °C. At a later time, the pressure of gas is 95 percent of its original pressure at the same temperature.
   (*i*) What mass of original gas must have escaped if the dimensions of the balloon are not changed ?
   (*ii*) Find the amount of heat to be removed to cause the same drop in pressure at constant volume. **[Ans.** (*i*) 5 percent, (*ii*) 3.26 MJ]
6. Find the molecular weight and gas constant for the gas whose specific heats are as follows :
   c_p = 1.967 kJ/kg K, c_v = 1.507 kJ/kg K. **[Ans.** 180.461 kJ/kg K]
7. A constant volume chamber of 0.3 m^3 capacity contains 1 kg of air at 20 °C. Heat is transferred to the air until its temperature is 200 °C. Find :
   (*i*) Heat transferred;
   (*ii*) Change in entropy and enthalpy.
   [Ans. (*i*) 128.09 kJ, (*ii*) 0.339 kJ/kg K, 180.8 kJ]

Real Gases

8. One kg-mol of oxygen undergoes a reversible non-flow isothermal compression and the volume decreases from 0.15 m^3/kg to 0.06 m^3/kg and the initial temperature is 50 °C. The gas obeys van der Waal's equation during the compression. Find :
   (*i*) The work done during the process;
   (*ii*) The final pressure. **[Ans.** (*i*) –6706500 Nm/kg-mol, (*ii*) 13.825 bar]
9. Determine the compressibility factor for O_2 at (*i*) 100 bar –70 °C and (*ii*) at 5 bar and 30 °C. **[Ans.** (*i*) 0.71, (*ii*) 0.98]
10. Determine the pressure of air at 205 °C having a specific volume of 0.00315 m^3/kg by means of :
    (*i*) Ideal gas equation;
    (*ii*) van der Waal's equation;
    (*iii*) Beattie-Bridgeman equation.
    [Ans. (*i*) 435.7 bar, (*ii*) 557.3 bar, (*iii*) 525.8 bar]

❑❑❑

6

CHAPTER

Properties of Steam

INSIDE THIS CHAPTER

6.1. INTRODUCTION

We have studied that the perfect gases do not exhibit any change of phase during a thermodynamic process *i.e.*; the gas remained in gaseous state even after the end of the process. But there are certain working fluids which when subjected to a thermodynamic process may change their phases. These working fluids are termed as vapours. Their most common example is ammonia and steam.

With vapour, the laws of perfect gases do not apply. So, for their calculations, we use the predetermined values listed in tables.

In this chapter, we shall discuss the formation of steam and its properties, because it is the most important working fluid being used in steam prime-movers.

6.2. FORMATION OF STEAM

Let us consider one kg of ice under a pressure 'p' and at sub zero temperatures (say –4°C) contained in a vessel. Now, if heat is supplied to heat it gradually at constant pressure, the following changes will take place, as shown in Fig. 6.1.

(*i*) The temperature of ice will gradually rise till it just reaches the freezing temperature, corresponding to pressure 'p'. It is represented by point '2' and is equal to 0°C at atmospheric pressure. The process of heating is shown by line 1-2 in Fig. 6.1.

(*ii*) If more heat is supplied beyond the point '2' the ice will start melting at constant temperature, till whole of the ice would be converted into water. The heat added during this period from 2 to 3 is called the *'latent heat of fusion of ice.*

(*iii*) Further addition of more heat beyond the point '3' will lead to a gradual rise in temperature of water till it reaches the boiling point '4'. Its value is different at different pressures and is 100°C *at the atmospheric pressure*. The heat added during process 3-4 is called '*sensible heat*' of liquid, as shown in Fig. 6.1. It is generally denoted by 'h_f'.

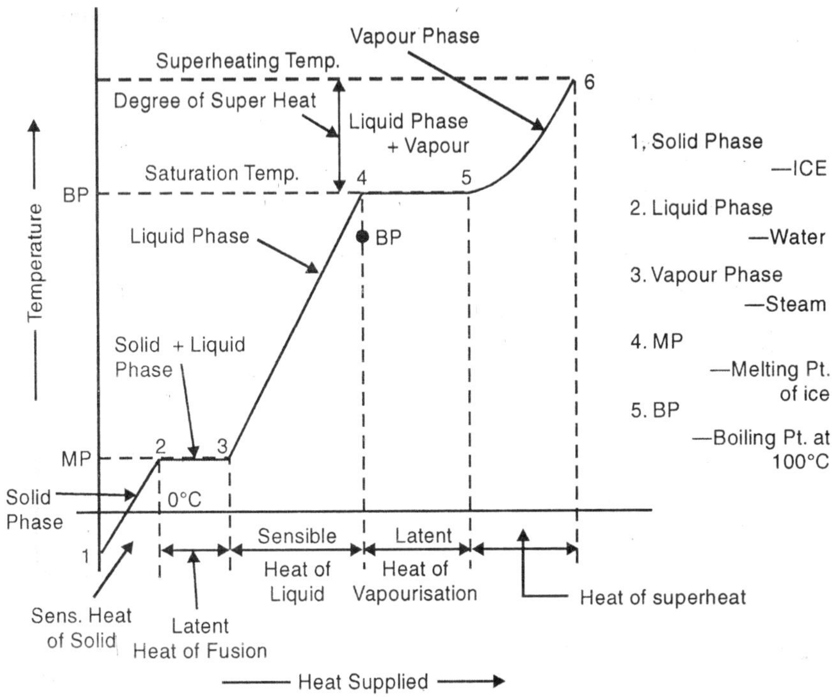

Fig. 6.1.

(*iv*) Beyond point '4' if the heat is gradually supplied, to the given sample at pressure 'p', the water will start to convert into steam but at constant temperature. This steam if still contains some particles of water in suspension is called the '*Wet steam*', which does not contain sufficient heat energy to maintain all of the water in gaseous state. Further, a state will be reached when whole of the water in the vessel is converted into wet steam. On further heating the water particles, which are in suspension will start evaporating gradually and in the end the steam becomes free from these water particles, which is called *dry or dry saturated steam* represented by point '5' on Fig. 6.1. The heat added during the period from point 4 to 5 is called the '*latent heat of vaporisation of steam*, or simply '*latent heat of steam*'.

(*v*) If the dry saturated steam is further heated beyond the point '5' at same pressure 'p' the process is called '*superheating*' and the steam so produced is called *the super heated steam, which follows the gas laws*. The temperature of steam corresponding to point '6' is called *superheating temperature*, generally denoted by t_s°C. The quantity of heat contained in the sample of superheated steam is more than the required to maintain the water in gaseous state by an amount ($t_s - t_5$) °C which is called the '*degree of superheat*'. The heat supplied during the process of super heating from point '5' to '6' is called the '*heat of superheat*'.

6.2.1. Wet-Steam

If saturated steam, contains suspended water particles in it, it is known as 'Wet steam'. It does not contain sufficient heat energy to maintain all water in it, in the gaseous state.

6.2.2. Dry-Steam

If saturated steam does not contain any water particle suspended in it, it is known as 'Dry-steam' or *Dry saturated steam.* It contains just sufficient heat energy to maintain all of the water in it, in a gaseous state.

6.2.3. Saturated Steam

Saturated steam is a vapour at the temperature corresponding to the boiling point of water at the imposed pressure.

6.2.4. Super-Heated Steam

The steam heated beyond the saturation temperature for a given pressure, is termed as *'superheated steam'*.

6.3. DRYNESS FRACTION (x)

The ratio of the weight of actual dry steam in the given sample of wet steam, to the weight of wet steam containing it, is known as *'dryness-fraction*, It is generally denoted by 'x'.

If m_s = Mass of dry steam in kg

and m_w = Mass of water vapour in suspension.

then,
$$x = \frac{m_s}{m_s + m_w}$$

6.4. DEGREE OF SUPER-HEAT

It is the difference between the temperature of the superheated steam and that of the saturated steam at the imposed pressure.

6.5. LATENT-HEAT OF VAPOURISATION (h_{fg})

The quantity of heat required at a constant pressure to convert one kg of water at saturation temperature into dry and saturated steam is known as *'latent heat of vapourisation'* or simply *'latent heat of steam'*. It is generally denoted by h_{fg}.

6.6. ENTHALPY OF STEAM (h)

The quantity of heat required to convert, at constant pressure one kg of water at freezing point, into wet, dry-saturated or superheated steam as the case may be is known as Enthalpy of that steam accordingly. It is also called *'total heat of steam'*, and generally denoted by 'h'.

6.6.1. Total Heat of Wet Steam

The quantity of heat required to convert, at constant pressure one kg of water at freezing point, into wet steam is known as *'total heat of wet steam'*. It is also known as *enthalpy of wet steam* and is generally denoted by 'h'.

In other words, we can say that enthalpy or total heat of wet steam is the sum of sensible heat of water and latent heat of wet steam, as shown in Fig. 6.2.

i.e., $h = h_f + x\, h_{fg}$

where h_f = Sensible heat of water

x = Dryness fraction of wet steam

h_{fg} = latent heat of wet steam.

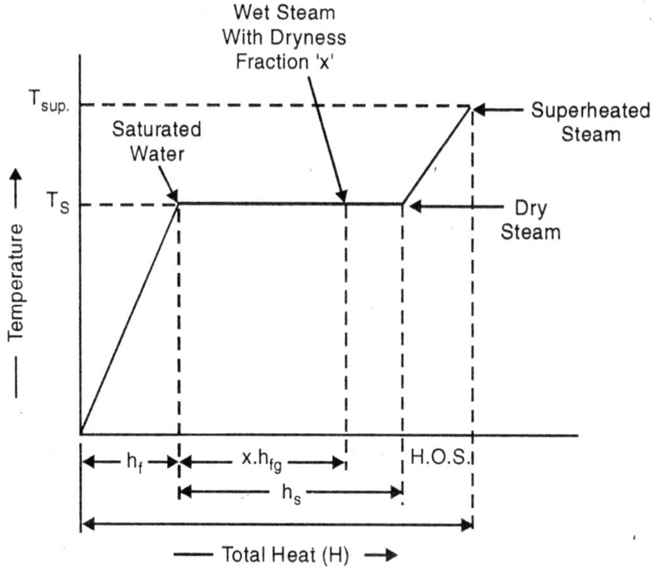

Fig. 6.2.

6.6.2. Total Heat of Dry-Saturated Steam

The quantity of heat required to convert, at constant pressure, one kg of water at freezing point into dry and saturated steam is known as *total heat of dry-saturated steam*. It is also called *enthalpy of dry-saturated steam* and generally denoted by h as shown in Fig. 6.2.

i.e., h = Sensible heat + Latent heat

or $h = h_f + h_{fg}$

6.6.3. Total Heat of Superheated Steam

The quantity of heat required, to convert at constant pressure one kg of water, at freezing point into superheated steam, is known as *'total heat of superheated steam'*. It is also called *'enthalpy of super heated steam* and generally denoted by h_s. Refer to Fig. 6.2.

h_s = Total heat of dry steam + Heat reqd. to superheat the steam.

or $h_s = h_f + h_{fg}$ + Heat of superheat

But, heat of superheat = Mass × Sp. heat × Rise in temp.

$= 1 \times C_p \times (t_s - t) = C_p\,(t_s - t)$

where t_s = Final temp. of superheated steam

t = Saturation temp.

and C_p = Sp. heat of steam at constant pressure.

Hence, $h_s = h_f + h_{fg} + C_p\,(t_s - t)$

6.7. USE OF STEAM TABLES

The various properties of steam *i.e.*, saturation temp., sensible heat, latent heat, total heat, specific volume and entropy etc. corresponding to various pressures can be determined on the basis of experiments. In actual practice it is not convenient to do so. Therefore, *the values of these properties are made available in the tabular form. These tables are generally known as 'steam tables'*, provided at the end of this book.

It is important to note that these properties are corresponding to one kg of dry and saturated steam and these values are reckoned above freezing temperature of water. These values may also be used to determine the corresponding values for the wet steam.

In these tables, it is not convenient to give various properties corresponding to all pressures. So, the value of properties at any pressure (other than given in these tables) can be calculated by interpolation of properties by the two values, one below and one above the actual value.

6.7.1. Specific Volume of Dry Steam

The volume of one kg of dry steam at a given pressure, is known as *'specific volume of dry steam'*.

It is generally abbreviated as v_g. Its value corresponding to different pressures are given in the steam tables.

The density of dry steam, $\rho = \frac{1}{v_g}$ kg/m^3

6.7.2. Specific Volume of Wet Steam

The volume of one kg of wet steam will be equal to the volume of dry portion of steam plus the volume of water in suspension.

i.e., Volume of one kg of wet steam

= Volume of dry steam + Volume of water in suspension.

Let, v_g = Volume of one kg of dry steam.

x = Dryness fraction of steam

v_f = Volume of one kg of water.

As the weight of given sample of steam is one kg. Therefore, the weight of dry steam will be 'x' kg and remaining $(1 - x)$ will be the weight of water in suspension.

$$\therefore \quad v_w = x \times v_g + (1 - x)\, v_f$$

The term $(1 - x)\, v_f$ can be neglected as the value of v_f is very small as compared to v_g.

$$\therefore \quad v_w' = x \times v_g \text{ m}^3/\text{kg}$$

The density of wet steam $= \frac{1}{x \times v_g}$

6.7.3. Volume of Superheated Steam

The volume of the superheated steam can be calculated on the assumption that the dry-steam starts behaving (approximately) like a perfect gas when it is heated.

So, applying Charle's law :

Assuming that steam is superheated at pressure 'p'

$$\frac{v_s}{t_s} = \frac{v_g}{t}$$

or

$$v_s = v_g \frac{t_g}{t}$$

Where v_s = Volume of one kg of superheated steam at pressure 'p'

v_g = Specific volume of dry steam at pressure 'p' (from steam tables)

t_s = Final temperature of superheated steam

t = Saturation temp. at pressure 'p'.

6.7.4. External Work Done During Evaporation

The amount of work done due to large increase in volume when changing from water into steam, is known as '*external work done during evaporation*.

Corresponding to the evaporation of one kg of water into dry saturated steam at pressure 'p' bar we get :

External work done during evaporation,

$$= 10^5 p\,(v_g - v_f) \text{ N–m or Joule}$$

where v_g = Specific volume of dry saturated steam at pressure 'p' in m^3.

and v_f = Volume of one kg of water in m^3

The value of v_f is very small as compared to v_g, therefore, it may be neglected.

∴ External work done per kg of dry steam

$$= 10^5 \,.\, p \,.\, v_g \text{ N-m or Joule}$$
$$= 10^2\, p \,.\, v_g \text{ KJ}$$

If the steam is wet, then the volume of one kg of steam with dryness fraction 'x' will be :

$$= x \times v_g.$$

∴ External work done per kg of wet steam

$$= 10^5 \,.\, p \,.\, x \,.\, v_g \text{ N–m or Joule}$$
$$= 10^2\, p \,.\, x \,.\, v_g \text{ KJ}$$

6.7.5. Internal Latent Heat

The difference between the latent heat and the external work of evaporation is known as '*Internal Latent Heat*'.

The energy for performing the external work is actually obtained from the latent heat absorbed during the evaporation. This means that the total latent heat absorbed consists of two parts as :

(*i*) true latent heat.

(*ii*) external work done of expansion.

i.e., Total latent heat = True latent heat + $10^5\, p \,.\, v_g$

True latent heat = $h_{fg} - 10^5\, p \,.\, v_g$ Joule.

The 'true latent heat' is also known as 'Internal latent heat'

∴ Internal latent heat = $h_{fg} - 10^5\, p.\, v_g$ Joule

6.7.6. Internal Energy of Steam

The actual heat energy stored in the steam is known as *'internal energy of steam'*.

It is generally abbreviated as 'U'

We know that the total heat energy stored in one kg of dry and saturated steam.

$$h = h_f + h_{fg}$$
$$= h_f + (\text{internal latent heat + external work of evaporation})$$

The energy due to external work done of evaporation is not stored in the steam, as it is spent in doing external work.

Actual energy stored in one kg of steam,

$$U = h_f + \text{Internal latent heat.}$$
$$= h_f + (h_{fg} - 10^5 p \,.\, v_g)\text{ J}$$
$$= h - 10^5\, p \,.\, v_g\text{ J.} \qquad (\because h = h_f + h_{fg})$$
$$\therefore \quad U = h - 10^5\, p \,.\, v_g\text{ Joule}$$

If the steam is superheated

$$U = h_s - 10^5\, p\, v_s\text{ J}$$

where

$$h_s = h_f + h_{fg} + C_p(t_s - t)$$

6.8. ADVANTAGES OF SUPERHEATED STEAM

The process of superheating is carried out in the superheater of a boiler. Most of the modern boilers used for power plants generate superheated steam. The advantages of this superheated steam are as follows :

(*i*) The superheated steam can be considerably cooled during expansion in an engine cylinder, before its temperature falls so low as to cause condensation. Therefore, the heat lost as a result of condensation of steam on the cylinder walls in greatly reduced.

(*ii*) The temperature of the superheated steam being higher, it gives a higher thermal efficiency, when used for the working a prime mover.

(*iii*) The steam has an increased capacity to work due to a higher heat content. So, the use of superheated steam results in an economy in the steam consumption of steam engines or steam turbines.

6.9. ENTROPY(S)

Entropy is a measure of unavailability of heat energy, for its transformation into work. It is a calculated property of matter.

6.9.1. Entropy of Water (S_f)

If water is heated from absolute temperature t_0 to t then the increase in entropy per kg of water, at constant pressure will be given by,

$$\text{Increase in entropy} = \log_e \frac{t}{t_0}$$

But, the entropy of steam is always reckoned above the freezing point ol water,

$\therefore \quad t_0 = 273$ k

$\therefore$ Entropy of water above its freezing point,

$$S_f = \log_e \frac{t}{273}$$

6.9.2. Entropy Increase During Evaporation (S_{fg})

We know that change in entropy is given by $\int \frac{dQ}{T}$

During evaporation, at constant pressure 'p' temperature of steam (saturation) 't' remains constant and the heat absorbed during complete evaporation is equal to the latent heat of evaporation of steam 'h_{fg}', at the given pressure.

$\therefore$ Change in entropy during complete evaporation (dry steam) or

Entropy increase during evaporation for drysteam $= \frac{h_{fg}}{t}$

Similarly, entropy increase during incomplete or partial evaporation *i.e.*, for wet steam with dryness fraction, $x = \frac{x.h_{fg}}{t}$.

6.9.3. Entropy of Steam (S_g)

It is the sum of entropy of water and entropy of evaporation, *i.e.*,

$$S_g = S_f + \text{Entropy of evaporation.}$$

where $\quad S_g$ = Entropy of steam

and $\quad S_f$ = Entropy of water.

(*i*) Entropy per kg of wet steam above freezing point

$$= S_f + \frac{x.h_{fg}}{t} = \log_e \frac{t}{273} + \frac{x.h_{fg}}{t}$$

(*ii*) Entropy per kg of dry-steam above freezing point,

$$= S_f + \frac{h_{fg}}{t} - \log_e \frac{t}{273} + \frac{h_{fg}}{t}$$

(*iii*) Entropy of superheated steam.

Let dry saturated steam is superheated to temp. t_s at constant pressure 'p'.

Then increase in entropy per kg of steam during superheating

$$= C_p.\log_e \frac{t_s}{t}$$

where $\quad C_p$ = Specific heat of superheated steam.

$\therefore$ Entropy per kg of superheated steam above freezing point,

$$= S_f + \frac{h_{fg}}{t} + C_p.\log_e \frac{t_s}{t} = \log_e \frac{t}{273} + \frac{h_{fg}}{t} + C_p.\log_e \frac{t_s}{t}$$

6.10. TEMPERATURE-ENTROPY DIAGRAM FOR STEAM

Figure 6.3 shows the temperature-entropy phase diagram for water and steam.

The absolute temperature is plotted vertically and entropy is plotted horizontally. It is very useful in solving engineering problems concerning with steam.

The diagram has three regions. These are separated out by two curves the water curve and the saturation curve as shown in the Fig. 6.3. The water curve represents completely wet steam *i.e.*, having 'x' = 0 and the saturation curve completely dry steam *i.e.*, having dryness fraction $x = 1$. Between these two lines is the wet region and obviously the steam in this region is wet, having *condition*, represented by the constant dryness fraction lines.

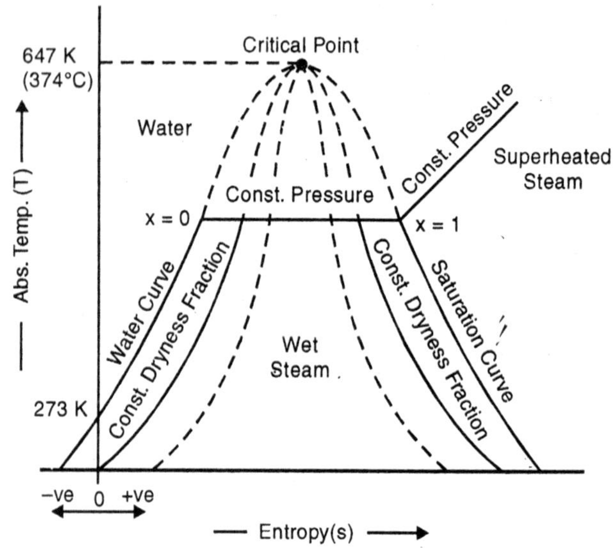

Fig. 6.3.

The water curve and the saturation curve meet at a point which is known as the *'critical point'*. The temp. corresponding to this point is the critical temperature of water having value 374°C, where as the pressure corresponding to this point is called 'critical pressure' having value of 220.65 bar. This point is very unstable and is unaffected by the presence of gases in the water. Since the horizontal distance between these two curves at critical point vanishes, so the latent heat absorbed becomes zero. Thus, *at this temperature the water is converted into dry-steam, without the absorption of any latent heat.*

The isothermal process can be represented by a horizontal line on this diagram, as the temp. remains constant during the process.

Whereas, the adiabatic (isentropic) process is represented by a vertical line on this diagram, as the entropy remains constant during this process.

6.11. MOLLIER DIAGRAM OR ENTHALPY-ENTROPY (H-S) DIAGRAM

It is the diagram representing the relation between specific enthalpy (on vertical line) and specific entropy (on horizontal line) as shown in Fig. 6.4.

The regions above and below the saturation curve (line) represent the superheated and wet conditions of steam respectively. The lines of constant dryness fraction are shown in the wet steam region, while the lines of constant temperature in the region of superheat.

It should be noted that the lines of constant pressure are straight in the wet region but curved in the region of superheat. In order to facilitate taking readings, horizontal and vertical lines are drawn at small intervals.

An adiabatic process will be represented by a vertical line in this diagram and throttling process or the expansion during which the total heat remains constant can be represented by a horizontal line.

One of the outstanding advantage of this diagram is that the drop in the total heat during an adiabatic expansion can be directly read from it.

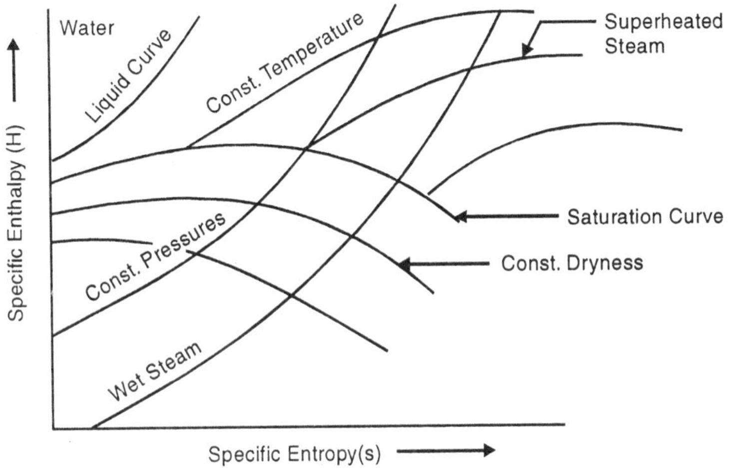

Fig. 6.4.

6.12. EXPANSION OF VAPOURS (STEAM) ACCORDING TO VARIOUS THERMODYNAMIC PROCESSES

The various methods of heating and expanding the vapours (steam) are explained below :

6.12.1. Constant Volume Expansion

The constant volume heating process is represented on p-v and T-S diagrams in Fig. 6.5 in which case steam is wet in initial state at pressure p_1 and it is superheated in final state at pressure p_2. Consider one kg of wet steam at a certain pressure and dryness fraction, to be heated at constant volume and let the steam is wet in the final state.

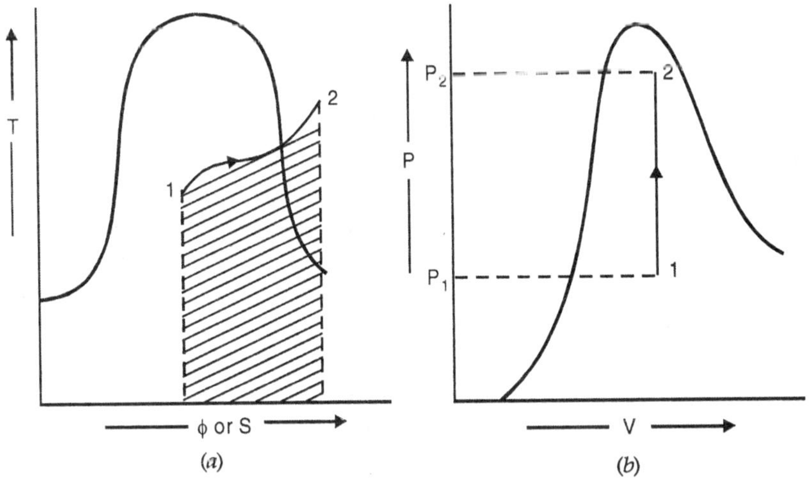

Fig. 6.5.

Let, p_1 = Initial pressure of steam in bar

x_1 = Initial dryness fraction of steam

v_{g1} = Initial specific volume of steam in m^3/kg obtained from steam tables corresponding to pressure 'p_1'

U_1 = Initial energy of one kg of steam at the pressure p_1

h_1 = Total heat of wet steam at pressure 'p_1'

p_2 = Final pressure of steam in bar

x_2 = Final dryness of fraction of steam

v_{g2} = Final specific volume on m^3/kg obtained from steam tables corresponding to the pressure 'p_2'

U_2 = Internal energy of one kg of steam at pressure p_2

h_2 = Total heat of the wet steam at pressure p_2

Now, initial volume of steam = $x_1 . v_{g1}$

Final volume of steam = $x_2 . v_{g2}$

Since, the volume of steam remains constant.

$$\therefore \quad x_1 . v_{g1} = x_2 . v_{g2}$$

or

$$x_2 = \frac{v_{g1}}{v_{g2}} . x_1$$

Thus, x_2 can be calculated.

Now, internal energy of steam before heating

$$U_1 = h_1 - 10^2 p_1 . x_1 . v_{g1} \text{ KJ}$$

Similarly, internal energy of steam after heating

$$U_2 = h_2 - 10^2 . p_2 . x_2 . v_{g2} \text{ KJ}$$

Thus, the values of U_1 and U_2 can be calculated

Since there is no change in volume

So, work done, $W = 0$

By the law of conservation of energy.

Total heat supplied, Q = Work done + Change in internal energy

or $= 0 + (U_2 - U_1)$

$$\therefore \quad Q = U_2 - U_1$$

6.12.2. Constant Pressure Expansion

Now, so long as the steam is wet there is only one temperature corresponding to the given pressure which, in this case, is kept constant, the temperature of wet steam remains constant during constant pressure process. In other words, the constant pressure process is also isothermal process in case of wet steam. Fig. 6.6 shows constant pressure heating process on *p-v* and *T-S* diagrams.

In this case, $p_1 = p_2 = p$ bar (say)

h_{fg} = Latent heat of the steam, which also remains constant since, the pressure is constant.

The specific volumes v_g will be same in this case.

let $v_{g1} = v_{g2} = v_g$

Work done, $W = 10^2 p \times$ change in volume. KJ

$$= 10^2 p(x_2 . v_g - x_1 v_g) \text{ KJ}$$

$$= 10^2 p v_g (x_2 - x_1) \text{ KJ}$$

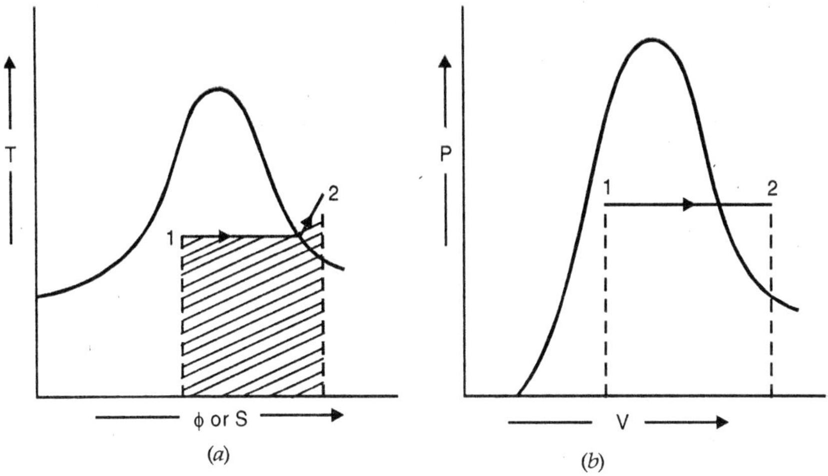

Fig. 6.6.

Internal energy of steam before expansion,

$$U_1 = h_1 - 10^2\, px_1 \,.\, v_g \text{ KJ}$$

Internal energy of steam after expansion

$$U_2 = h_2 - 10^2 p \,.\, x_2\, v_g \text{ KJ}$$

Change in internal energy

$$U_2 - U_1 = (h_2 - 10^2 p \,.\, x_2\, v_g) - (h_1 - 10^2 px_1 \,.\, v_g) \text{ KJ}$$
$$= (h_2 - h_1) - 10^2 pv_g\,(x_2 - x_1) \text{ KJ}$$

By the law of conservation of energy

$$Q = W + (U_2 - U_1) \text{ KJ}$$

or $$= 10^2 p \,.\, v_g(x_2 - x_1) + [(h_2 - h_1) - 10^2 \,.\, pv_g(x_2 - x_1)] \text{ KJ}$$

or $$Q = (h_2 - h_1) \text{ KJ}$$

Now, $$h_1 = h_{f1} + x_1\, h_{fg}$$
$$h_2 = h_{f2} + x_2\, h_{fg}$$
$$Q = h_2 - h_1$$
$$= (h_{f2} + x_2 h_{fg}) - (h_{f1} + x_1 h_{fg})$$
$$= (x_2 h_{fg} - x_1 h_{fg}) \qquad (\because h_{f1} = h_{f2} \text{ since pressure is constant})$$

or $$Q = h_{fg}\,(x_2 - x_1) \text{ for wet steam}$$

In case the steam is superheated.

Q = Total heat of superheated steam – Total heat of saturated steam

$Q = h_s - h$ (for superheated steam)

Note. The expansion will no longer be the isothermal, once the steam is **'superheated.'**

6.12.3. Constant Temperature (Isothermal) Expansion

The isothermal process of wet steam will be same as the constant pressure process as mentioned earlier. But, this fact will not hold good once the steam is superheated in which case it will behave as a gas and will follow Boyle's law (pv = constant). Therefore, the process will be hyperbolic. Thus, *we come to the conclusion that for the superheated steam the isothermal and hyperbolic process will be identical, while for wet steam isothermal and constant pressure process will be identical.* The isothermal expansion is represented on T–S diagram in Fig. 6.7(a).

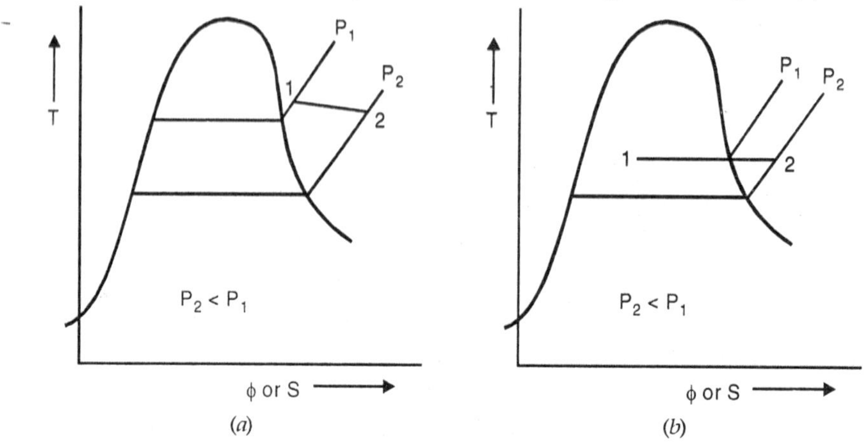

Fig. 6.7.

6.12.4. Hyperbolic Expansion

During a hyperbolic process (pv = constant) the process may be regarded same as that of isothermal but only in case of superheated steam, refer Fig. 6.7(b).

Consider the hyperbolic expansion of one kg of steam at a pressure of p_1 bar and dryness fraction x_1, till its final pressure is p_2 bar and the corresponding dryness fraction x_2. The values of the corresponding specific volumes v_{g1} and v_{g2} may be found from steam tables.

Then, initial volume of steam $= x_1 \cdot v_{g1}$

and Final volume of steam $= x_2 \cdot v_{g2}$

As, pv = Constant (in hyperbolic process)

$$\therefore \quad p_1 x_1 v_{g1} = p_2 \cdot x_2 \cdot v_{g2}$$

or
$$x_2 = \frac{p_1 x_1 v_{g1}}{p_2 . v_{g2}}$$

Thus x_2 can be calculated, if it comes out to be more than unity, it means that the steam is superheated.

For superheated steam

$$\frac{p_2 v_{g2}}{t_2} = \frac{p_2 v_{s2}}{t_s}$$

or
$$t_s = \frac{v_{s2}}{v_{g2}} \times t_2$$

which gives the value of t_s.

Now, $h_1 = h_{f1} + x_1 h_{fg1}$ for the initial condition of steam

$h_2 = h_{f2} + x_2 h_{fg2}$ (when the final condition of steam is wet)

$h_{s2} = h_{f2} + h_{fg2} + C_p(t_s - t)$

(When the final condition of steam is superheated)

Now, $U_1 = h_1 - 10^2\, p_1 x_1 v_{g1}$

and $U_2 = h_2 - 10^2 p_2 . x_2 . v_{g2}$ (if steam is wet)

$U_2 = h_{s2} - 10^2 p_2 v_{s2}$ (if the steam is superheated)

But, as the expansion is hyperbolic

$$p_1 x_1 v_{g1} = p_2 x_2 v_{g2} = p_2 v_{s2}$$

$$\therefore \quad 10^2 p_1 x_1 v_{g1} = 10^2 p_2 x_2 v_{g2} = 10^2 p_2 v_{s2} \text{ KJ}$$

Thus, we conclude that the external work of evaporation is constant during a hyperbolic expansion.

$\therefore$ Change in internal energy,

$$U_2 - U_1 = (h_2 - 10^2 p_2 x_2 v_{g2}) - (h_1 - 10^2 p_1 x_1 v_{g1})$$

$$= (h_2 - h_1) \text{ if the steam is finally wet.}$$

Hence, for hyperbolic expansion of wet steam, the change in internal energy is equal to the change in total heat or enthalpy.

When the steam is superheated,

$$U_2 - U_1 = h_{s2} - h_1$$

In other words for the hyperbolic expansion of superheated steam, change in internal energy is equal to the change in total heat.

Work done, $W = 10^2 p_1 v_1 . \log_e r$ KJ

Applying the Law of Conservation of Energy;

Total heat supplied, $Q = W + (U_2 - U_1)$

or $Q = 10^2 p_1 x_1 v_{g1} \log_e r + (U_2 - U_1)$

Which gives the heat absorbed during hyperbolic expansion.

6.12.5. Polytropic Expansion (PV^n = Constant)

Consider one kg of steam at an initial pressure p_1 bar and dryness fraction x_1 to expand to a pressure p_2 bar and dryness fraction x_2 while following the general law pV^n = constant during the expansion.

Initial volume of steam = $x_1 v_{g1}$

Final volume of steam = $x_2 v_{g2}$

But pV^n = Constant

$$\therefore \quad p_1(x_1 . v_{g1})^n = p_2(x_2 . v_{g2})^n$$

or

$$x_2 = \left(\frac{p_1}{p_2}\right)^{1/n} \times \left(\frac{v_{g1}}{v_{g2}}\right) \times x_1$$

which gives the value of x_2. If it is greater than unity the steam is superheated after expansion and its temperature in that case should be found out by applying Charle's Law as discussed in the case of hyperbolic expansion.

Now, total heat of steam before expansion,

$$h_1 = h_{f1} + x_1 h_{fg1}$$

and total heat of steam after expansion,

$$h_2 = h_{f2} + x_2 h_{fg2}$$

Internal energy of steam before expansion,

$$U_1 = h_1 - 10^2 p_1 . x_1 . v_{g1}$$

Internal energy of steam after expansion,

$$U_2 = h_2 - 10^2 p_2 \cdot x_2 \cdot v_{g2}$$

Work done, $$\dot{W} = \frac{10^2\left(p_1 \cdot x_1 \cdot v_{g1} - p_2 \cdot x_2 \cdot v_{g2}\right)}{n-1} \text{ KJ}$$

Applying the law of conservation of energy,

Heat supplied, Q = Work done + Change in internal energy

or $$Q = \frac{10^2\,(p_1 \cdot x_1 \cdot v_{g1} - p_2 \cdot x_2 \cdot v_{g2})}{n-1} + (U_2 - U_1)$$

6.12.6. Isentropic Expansion of Steam

Isentropic process is the second name of reversible adiabatic process. During this process, there is no heat transfer and the entropy remains constant.

Now, let us consider that one kg of steam at a pressure p_1 and dryness fraction x_1 is expanded isentropically to the pressure p_2. The dryness at the end of expansion 'x_2' can be calculated as follows;

During an isentropic expansion,

Entropy before expansion = Entropy after expansion.

i.e., $$S_{f1} + \frac{x_1 h_{fg1}}{t_1} = S_{f2} + \frac{x_2 h_{fg2}}{t_2}$$

or $$\frac{x_2 h_{fg2}}{t_2} = \left[(S_{f1} - S_{f2}) + \frac{x_1\, h_{fg}}{t_1}\right]$$

or $$x_2 = \frac{t_2}{h_{fg2}}\left[(S_{f1} - S_{f2}) + \frac{x_1 h_{fg1}}{t_1}\right]$$

(*i*) When the steam is initially dry; *i.e.;* if $x_1 = 1$

$\therefore$ $$x_2 = \frac{t_2}{h_{fg2}}\left[(S_{f1} - S_{f2}) + \frac{h_{fg}}{t_1}\right]$$

(*ii*) When the steam is initially superheated :

Let, t_s = Final temp. of the superheated steam.

Then,

$$S_{f1} + \frac{h_{fg1}}{t_1} + C_p \log_e \frac{t_s}{t_1} = S_{f2} + \frac{x_2 h_{fg2}}{t_2}$$

or $$\frac{x_2\, h_{fg2}}{t_2} = \left[(S_{f1} - S_{f2}) + \frac{h_{fg1}}{t_1} + C_p \log_e \frac{t_s}{t_1}\right]$$

or $$x_2 = \frac{t_2}{h_{fg2}} \times \left[(S_{f1} - S_{f2}) + \frac{h_{fg1}}{t_1} + C_p \cdot \log_e \frac{t_s}{t_1}\right]$$

After determining the dryness fraction at the end of expansion, the internal energies at initial and final conditions can be determined.

i.e., $$U_1 = h_1 - 10^5 \cdot p_1 \cdot x \cdot v_{g1}$$

and $$U_2 = h_2 - 10^5 \cdot p_2 \cdot x_2 \cdot v_{g2}$$

Therefore, change in internal energy,

$$\Delta U = U_2 - U_1$$

As the heat transfer during the expansion is zero, therefore from the law of conservation of energy, the work done on piston during expansion can be calculated as follows :

$$Q = W + \Delta U$$

or $$0 = W + \Delta U$$

or $$W = -\Delta U = (U_1 - U_2)$$

The above equation gives the value of work done, but sometimes the approximate value of work done during adiabatic expansion can be calculated by assuming the expansion to follow the law pV^n = constant.

Where, n = 1.13 for wet steam

= 1.3 for superheated steam.

Isentropic process can be represented by a vertical line on the T–S diagram.

6.13. DETERMINATION OF DRYNESS FRACTION OF STEAM

There are following methods of measuring the dryness fraction of steam :

(*i*) Barrel calorimeter.

(*ii*) Separating calorimeter.

(*iii*) Throttling calorimeter.

(*iv*) Combined separating and throttling calorimeter.

(*v*) An electric calorimeter.

6.13.1. Barrel Calorimeter

It consists of a copper calorimeter placed on wooden blocks and provided with a wooden cover as shown in the Fig. 6.8. The water equivalent of the calorimeter is determined experimentally and its value is stamped on it. It contains a known quantity of cold water and is surrounded by an outer vessel, thus providing an air space in between, which acts as an insulator.

A thermometer which is inserted through one of the holes in the wooden cover is provided to measure the temperature. The whole assembly is placed on the platform of a weighing bridge. The steam from the main steam pipe enters through the sampling tube and via the control valve and the pipe, flows into the cold water through fine exit holes of the ring provided as shown in Fig. 6.8. The condensation of steam takes place as it comes in contact with the cold water, with the result that the temperature of the latter rises. The quantity of steam condensed can be known from the difference of readings of the weighing bridge before and after the condensation of steam. The pressure at which steam enters the calorimeter is indicated by the pressure gauge.

Let, W = Weight of the steam condensed

x = Dryness fraction of steam

h_{fg} = Latent heat of steam

t = Temperature of formation of steam (corresponding to the pressure indicated by the pressure gauge)

w = Weight of the cold water in the calorimeter in the beginning

w_c = Water equivalent of the calorimeter
t_1 = Initial temperature of cold water in the calorimeter
t_2 = Final temperature of the mixture (after condensation of steam)

Heat lost by steam = $W [xh_{fg} + (t - t_2)]$

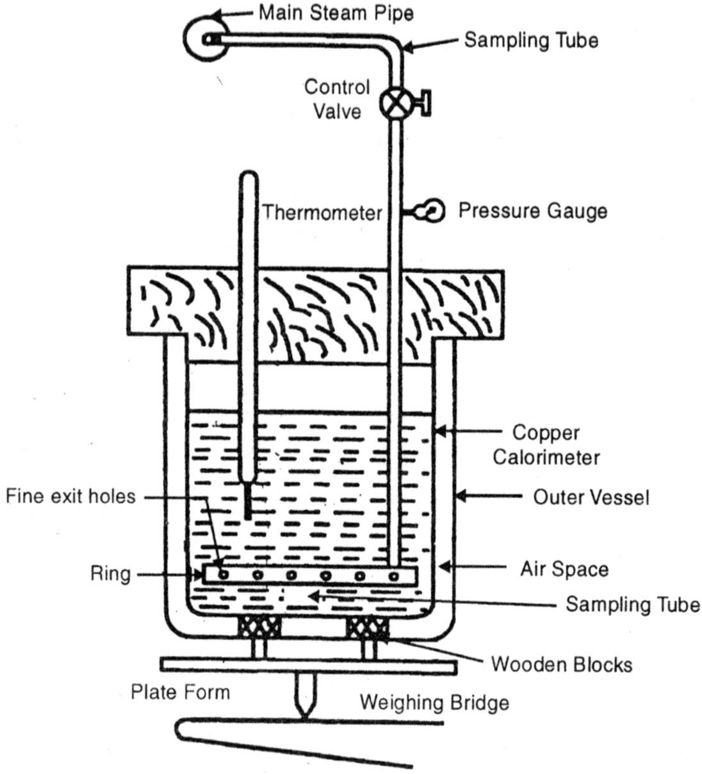

Fig. 6.8. Barrel Calorimeter.

Heat gained by cold water and calorimeter

$$= (t_2 - t_1)(w + w_c)$$

Assuming that the steam enters the calorimeter as vapour and not as a condensate and further that there is no heat lost to the surroundings.

Heat lost by steam = Heat gained by cold water and the calorimeter

or $$W [xh_{fg} + (t - t_2)] = (t_2 - t_1)(w + w_c)$$

or $$x = \frac{(t_2 - t_1)(w + w_c) - (t - t_2) W}{W h_{fg}}$$

or $$x = \frac{\frac{(w + w_c)}{W}(t_2 - t_1) - (t - t_2)}{h_{fg}}$$

In some cases instead of a copper calorimeter or some such vessel, we use a wooden calorimeter or barrel in which w_c can be neglected, in that case the above equation takes the following form *i.e.*, putting $w_c = 0$.

$$x = \frac{(t_2 - t_1) - (t - t_2)}{h_{fg}}$$

A sufficient rise in temperature should be obtained by blowing a large quantity of steam into calorimeter. But, at the same time the final temperature of the mixture should not be too high giving incorrect results due to heat lost by conduction and radiation. Too much opening of the control valve should be avoided otherwise the excessive steam flowing into the cold water would result in the bubbles of steam left uncondensed escaping out of the cold water thus leading to wrong results.

This method of finding the value of the dryness fraction of steam, 'x' by blowing steam into the cold water gives only *approximate results*. So, *the value of 'x' thus calculated is always lower than the actual value.*

6.13.2. Separating - Calorimeter

The Fig. 6.9 shows the schematic line diagram of a separating calorimeter. It consists of two concentric chambers, the inner chamber

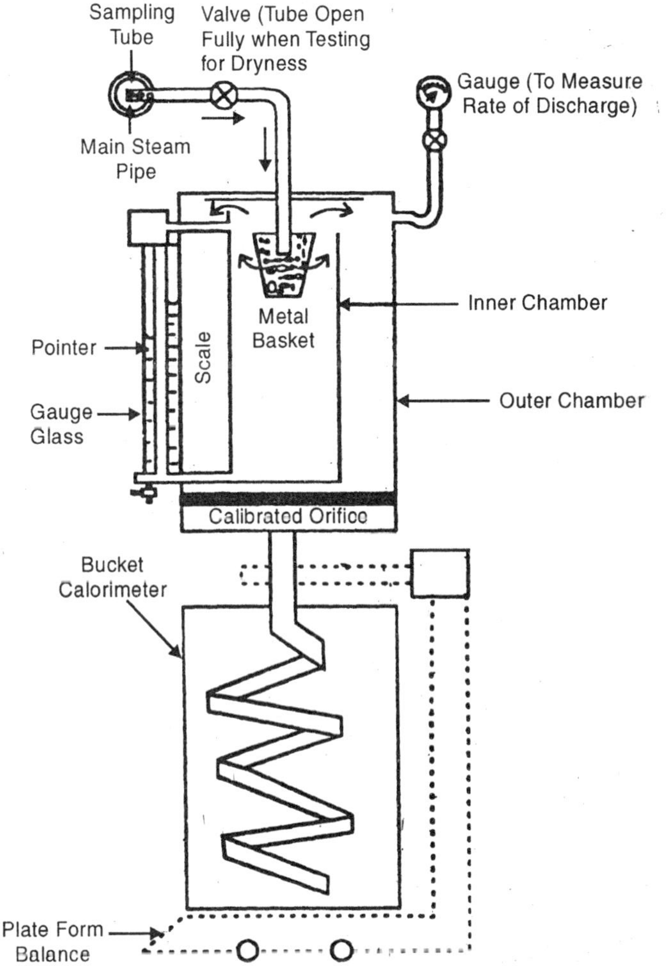

Fig. 6.9. Separating calorimeter.

and the outer chamber, which communicate with each other through an opening at the top. The steam to be tested flows from the main steam pipe through a sampling tube to the calorimeter pipe via the valve which must be fully opened when testing for dryness in order to ensure that there is no wire drawing or throtling taking place. As the steam discharges through the metal basket, which has a large number of performations all around but not at the base, it is made to move through an abuse angle.

The water particles due to their heavier momentum get separated from the steam due to their heavier momentum, and collect into the inner chamber as shown. The quantity water collected is indicated by the level in the gauge glass and the pointer which moves on a scale as shown. The scale is graduated in such a way as to give reading directly in kg. The dry steam in the inner chamber moves up and then down again through the annual space between the two chambers, the gauge fitted as shown has two scales, the inner one indicating the pressure, while the outer on indicates the rate of discharge of dry saturated steam per ten minutes. Afterwards the steam flows through the calibrated orifice to the bucket calorimeter, which is placed on the plateform balance by which the weight of steam discharged can be further confirmed.

Let, w_d = Weight of dry steam condensed

w = Weight of suspended moisture collected

Then, $$x = \frac{w_d}{w_d + w}$$

This method also does not give accurate results as the steam discharged is not absolutely dry.

6.13.3. Throttling Calorimeter

The fact that the total heat of steam during throttling process remains constant is made use of in a throttling calorimeter, as shown in Fig. 6.10. Which consists of the usual sampling tube (ST) through which the steam from the main steam pipe (MSP) flows to be throttling valve (TV). The steam in flowing through this valve becomes superheated (usually 5°C of superheat are kept) and it further flows into the inner chamber (IC). Then the steam first flows down and rises up again to flow out of the inner chamber, into the annular space (AS) between the inner chamber and the outer chamber (OC), which is insulated to avoid heat losses.

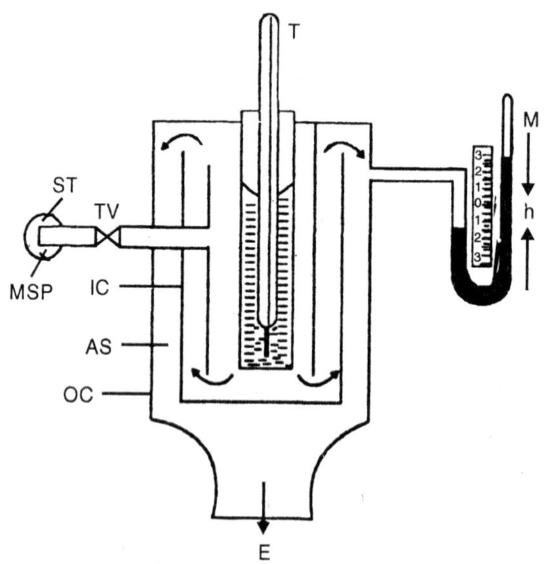

MSP : Main Steam Pipe
ST : Sampling Tube
TV : Throttling Valve
IC : Inner Chamber
AS : Annular Space
T : Thermometer in the Pocket Filled with Cylinder Oil
E : Exhaust
M : Manometer
h : Gauge Pressure after Throttling in mm of Hg.

Fig. 6.10. Throttling calorimeter.

The loss of heat by radiation from the inner chamber is minimised by the hot steam around the outside of the throttled steam is measured by the thermometer (T) placed in the pocket filled with cylinder oil. This method of determining the valve of dryness fraction of steam, 'x' will not give correct results unless the steam issuing from the throttle valve and entering the inner chamber is superheated. In order to ensure it a manometer (M) if attached as shown, to measure the pressure of the steam.

Corresponding to this pressure, say p bar absolute, the temperature of formation of steam is read from the steam tables and if this value of temperature is lower than that indicated by the thermometer (T), the steam is superheated, otherwise not. The steam finally escapes out through the exhaust (E).

Let, p_1 = Initial pressure of wet steam before expansion.

x_1 = Dryness fraction of the wet steam

h_{fg1} = Latent heat of wet steam

t_s = Temperature of superheated steam after expansion (as indicated by the thermometer)

h_{fg2} = Latent heat of steam after expansion.

t_2 = Temperature of formation of steam after expansion

C_p = Specific heat of super heated steam at constant pressure.

Total heat of steam before throttling = $h_{f1} + x_1 h_{fg1}$

Total heat of steam after throttling = $h_{f2} + h_{fg2} + C_p (t_s - t_2)$

Now total heat before throttling = Total heat after throttling

$$\therefore \quad h_{f1} + x_1 h_{fg1} = h_{f2} + h_{fg2} + C_p (t_s - t_2)$$

or
$$x_1 = \frac{h_{f2+} + h_{fg2} + C_p (t_s - t_2) - h_{f1}}{h_{fg1}}$$

$$= \frac{h_2 + C_p (t_s - t_2) - h_{f1}}{h_{fg1}} \qquad (\because h_2 = h_{f2} + h_{fg2})$$

Thus,
$$x_1 = \frac{h_2 - h_1 + C_p (t_s - t_2)}{h_{fg1}}$$

In calculating results from the observations taken with a throttling calorimeter, it is assumed that no heat is lost to the external bodies *i.e.*, surroundings. Also, if the steam supplied to the throttle valve is just dry *i.e.*, $x_1 = 1$, in that case,

$$h_{f1} + h_{fg1} = h_1$$

So,
$$h_1 = h_{f2} + h_{fg2} + C_p (t_s - t_2) = h_2 + C_p (t_s - t_2)$$

or
$$C_p = \frac{h_1 - h_2}{t_s - t_2}$$

Thus, the specific heat of superheated steam, at constant pressure can be calculated with the help of this calorimeter. The throttling calorimeter is also called, *wire drawing calorimeter.*

6.13.4. Combined Separating and Throttling Calorimeter

The steam passing out from a separating calorimeter may still contain some water particles in it. In other words, it may not be absolutely dry. Again in a throttling calorimeter the steam after passing through the throttle valve must be superheated or at least dry saturated. This limits the extent of the dryness fraction that can be measured by a throttling calorimeter, it has been observed that the minimum dryness fraction that can be reliably measured depends upon the pressure of steam in the main steam pipe. Thus, a sample of steam which has dryness fraction less than this valve may be still wet after passing through the throttle valve *i.e.,* it will not be superheated. So, under this condition throttling calorimeter fails to enables us in determining the value of dryness fraction of steam.

To overcome above mentioned difficulties we make use of the combined separately and throttling calorimeter as shown in the Fig. 6.11. The steam is first passed through the separating calorimeter where it losses most of its moisture and becomes comparatively drier, it is then passed through the throttling calorimeter where superheating takes place without change of total heat.

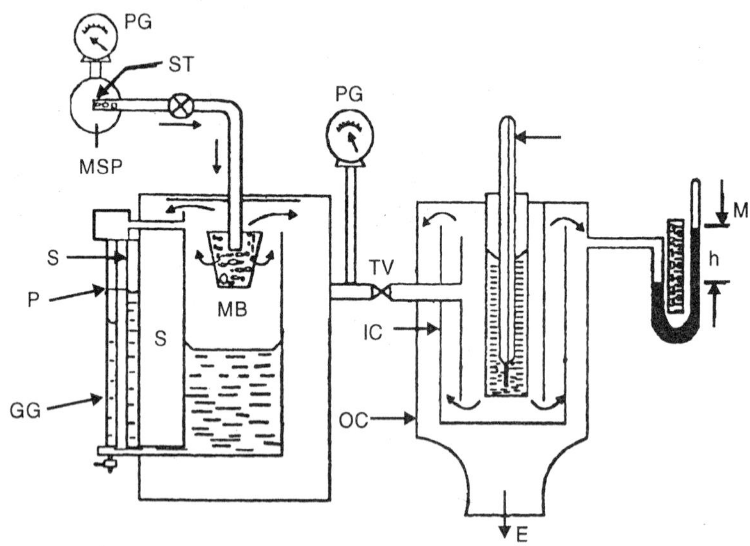

MSP : Main Steam Pipe
ST : Sampling Tube
TV : Throttling Valve
V : Valve
MB : Metal Basket
P : Pointer
GG : Gauge Glass
IC : Inner Chamber
OC : Outer Chamber
TV : Throttle Valve
T : Thermometer in the Pocket Filled with Cylinder Oil
E : Exhaust
M : Manometer
h : Gauge Pressure after Throttling in mm of Hg.
G : Pressure Gauge

Fig. 6.11. Combined separating and throttling calorimeter.

The temperature and pressure of steam after throttling are measured by using a thermometer and pressure gauge respectively.

Let, p_1 = Pressure of wet steam entering the throttling calorimeter

x_1 = Dryness fraction of steam entering the throttling calorimeter

h_{fg1} = Latent heat of wet steam entering the throttling calorimeter

w = Weight of the suspended moisture collected in the separating calorimeter

w_1 = Weight of the steam leaving the separating calorimeter and entering the throttling calorimeter

p_2 = Pressure of steam after throttling

h_{fg2} = Latent heat of steam at pressure p_2.

Since, total heat of steam entering the throttling calorimeter = Total heat of steam leaving the throttling calorimeter.

$$\therefore \quad h_{f1} + x_1 h_{fg1} = (h_{f2} + h_{fg2}) + C_p(t_s - t_2)$$

or

$$x_1 = \frac{(h_2 - h_1) + C_p (t_s - t_2)}{h_{fg1}} \text{ as before}$$

Let,

$$\frac{w_1}{w + w_1} = x_1$$

Now, 'x' may be called 'apparent dryness fraction' of steam.

Let x = Dryness fraction of steam entering the separating calorimeter.

Weight of dry steam entering the whole apparatus = $w_1 x_1$

$$\therefore \quad x = \frac{\text{Total weight of dry steam}}{\text{Total weight of steam drawn through the sampling tube}}$$

(*i.e.*, entering the separating calorimeter)

$$= \frac{w_1 x_1}{(w + w_1)} = \left(\frac{w_1}{(w + w_1)}\right) x_1 = x_1 . x$$

$$\therefore x_1 = x_1' . x$$

Hence, dryness fraction of steam = dryness fraction of steam entering the throttling calorimeter × apparent dryness fraction.

The throttling unit may be coupled directly to the steam pipe in case the steam is comparatively dry. Otherwise both units should be together as in the illustration, support being optained by a light strap or bracket. The manometer (M) should be filled to zero level with water, the average of the readings on the two limbs being taken for determining the pressure of steam after throttling. After erection the instrument should be throughly lagged to minimise heat losses. The thermometer pocket should be half filled with cylinder oil before inserting the thermometer. The commencement of taking the readings should not be hurried even when good lagging is provided. The value v should be opened sufficiently long to allow the proper heating of the unit.

6.13.5. An Electric Calorimeter

The dryness fraction of steam can also be measured by an electric calorimeter as shown in the Fig. 6.12. A sample of the steam from the steam main at a pressure 'p_1' as indicated by the sampling tube. This sample of steam is then passed in steady flow through an electric heater as shown. The steady flow been regulated by the regulating valve provided in the inlet pipe to the heater.

The electrical energy input, 'Q' should be sufficient to take the steam to the superheated region, where pressure 'p_2' and the temperature 't_2' are measured. If 'I' is the current flowing through the heater in amperes and 'V' is the voltage across the electrical heater coil, then at steady state :

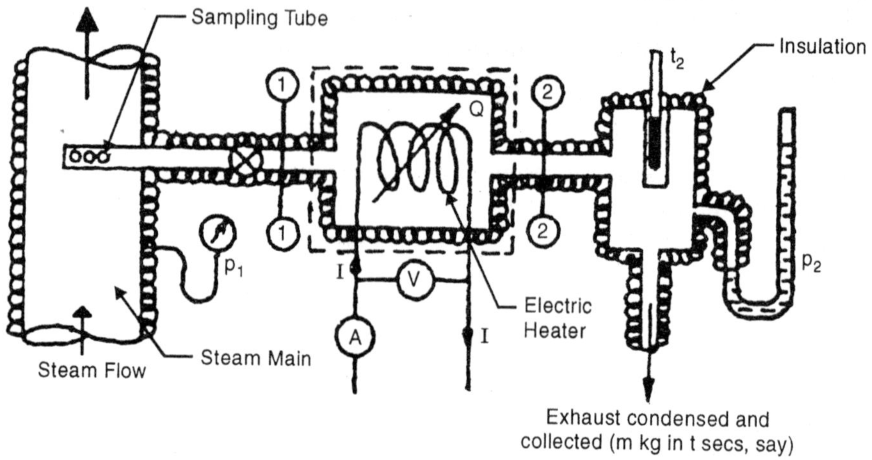

Fig. 6.12. An electric calorimeter.

$$Q = V \cdot I \times 10^{-3} \text{ kw}$$

If '*m*' is the mass of steam taken in '*t*' seconds under steady flow conditions, then the steady flow energy equation for the heater (as control volume) gives :

$$w_1 \cdot h_1 + Q = w_1 \cdot h_2$$

where 'w_1' is the steam flow rate in kg/s *i.e.*, $w_1 = m/t$ kg/s and h_1, h_2 the enthalpies or total heats at point 1 and point 2 respectively.

$$\therefore \qquad h_1 + \frac{Q}{w_1} = h_2$$

Now, with H_2, Q and w, being the known quantities h_1 can be computed

Also, $\qquad h_1 = h_{f1} + xh_{fg}$

Hence, dryness fraction of steam

$$x = \frac{h_1 - h_{f1}}{h_{fg}}$$

***H–S* diagrams.** The quality of steam decreases *i.e.*, the dryness fraction of steam decreases.

6.13.6. Throttling of Steam

During this process, the steam expands from higher pressure to lower pressure through a small hole in such a way that both the heat transfer and work done are zero. This process is also known as '*wire drawing*'.

Let us consider that one kg of steam at pressure 'p_1' and dryness fraction 'x_1' is throttled to pressure 'p_2'. The value of final condition of steam 'x_2' can be determined as follows :

During the throttling expansion,

Total heat before expansion = Total heat after expansion.

$$h_{f1} + x_1 \cdot h_{fg1} = h_{f2} + x_2 \cdot h_{fg2}$$

or

$$x_2 = \frac{1}{h_{fg2}}\left[(h_{f1} - h_{f2}) + x_1 \cdot h_{fg1}\right]$$

It the value of 'x_2' comes out to be greater than one. Then the steam after expansion is superheated. How much the superheating is achieved, can be calculated through the following equation.

i.e., $$h_{f1} + x_1 . h_{fg1} = h_{f2} + h_{fg2} + C_p (t_s - t_2)$$

Throttling process is represented by a horizontal line on the *H–S* diagram. Also, during this process, a part of sensible heat is transferred into the latent heat. And at the end of the throttling expansion, the dryness fraction increases at the cost of decrease in its temperature.

NUMERICAL EXAMPLES

Example 6.1. Calculate the total heat of 5 kg of stream at a pressure of 8 bar (abs.) having dryness fraction of 0.8. Also calculate heat in KJ required to convert this steam into dryness and saturated steam. (Pb. May 2001, 2000)

Solution. We know that;

Total heat of wet steam,

$$h = h_f + x.h_{fg}$$

From steam tables : Against 8 bar (abs.) pressure

$$h_f = 720.9 \text{ KJ/kg}$$
$$x = 0.8$$
$$h_{fg} = 2046.5 \text{ KJ/kg}$$

∴ Total heat of one kg of wet steam

$$h = 720.9 + 0.8 \times 2046.5 = 2358.1 \text{ KJ/kg}$$

∴ Total heat of 5 kg of steam

$$= 2358.1 \times 5 = 11790.5 \text{ KJ}$$

Also, from steam tables :

Total heat of dry and saturated steam at 8 bar pressure

$$= 2767.4 \text{ KJ/kg.}$$

Hence, total heat of dry and saturated steam at 8 bar pressure = 2767.4 × 5 = 13830 KJ

Therefore, heat required to convert 5 kg of wet steam into dry and saturated steam,

$$= 13830 - 11790.5 = 2039.5 \text{ KJ}$$ **Ans.**

Example 6.2. Determine the dryness fraction of a sample of steam weighing 4.7 kg if the total heat of the sample is 11000 KJ. Steam pressure is 5.0 bar (abs.).

Solution. The properties of steam at 5.0 bar pressure.

From steam tables are as follow :

$$h_f = 640.1 \text{ KJ/kg}$$
$$h_{fg} = 2107.4 \text{ KJ/kg.}$$

Let x be the dryness fraction of steam.

∴ Total heat of 4.7 kg steam = 4.7 $(h_f + h_{fg})$

or $$11{,}000 = 4.7(640.1 + x \times 2107.4) \text{ KJ.}$$

Hence, $$x = 0.807$$ **Ans.**

Example 6.3. If 250 KJ/kg are added to 0.8 dry steam at 7.0 bar (abs.) pressure, determine the condition of steam after heat addition.

Solution. From steam tables at 7.0 bar (abs.) pressure.

$$h_f = 697.1 \text{ KJ/kg.}$$
$$h_{fg} = 264.9 \text{ KJ/kg}$$

Total heat of wet steam $= h_f + x \cdot h_{fg}$

$$= 697.1 + 0.8 \times 2064.9 = 2349 \text{ KJ/kg.}$$

Now, by adding 250 KJ/kg, total heat of steam becomes,

$$= 2349 + 250 = 2599 \text{ KJ/kg}$$

Total heat of dry and saturated steam at 7.0 bar pressure

$$= 697.1 + 2064.9 = 2762.0 \text{ KJ/kg.}$$

This shows that given sample of steam is wet even after heat addition.

Let, 'x_2' be the dryness fraction after heat addition.

Then, $\quad 2599 = 697.1 + x_2 \times 2064.9$

or $\quad x_2 = 0.921$ **Ans.**

Example 6.4. Determine if the steam is wet or superheated in the following states :

(*i*) $p = 6.0$ bar, $h = 2690$ KJ/kg.

(*ii*) $p = 8.0$ bar, $v_g = 0.2\ m^3/kg$.

(*iii*) $p = 75.0$ bar, $t = 300°C$.

Also determine dryness fraction in case of wet steam and degree of superheat for superheated steam.

Solution. At 6.0 bar pressure

$$h_f = 670.4 \text{ KJ/kg.}$$
$$h_{fg} = 2085.0 \text{ KJ/kg.}$$

Hence, $\quad h = h_f + h_{fg} = 670.4 + 2085.0$

$$= 2755.4 \text{ KJ/kg.}$$

The given sample has $h = 2690$ KJ/kg so the steam is wet

Let 'x' be the dryness fraction of this steam

Hence, $\quad 2690 = 670.4 + x.\ 2085$

or $\quad x = 0.967.$ **Ans.**

(*ii*) At 8.0 bar pressure, specific volume of steam $v_g = 0.2403\ m^3/kg$

As the volume of given sample is $0.2\ m^3/kg$, hence the sample is wet steam, Let 'x_2' be the dryness fraction.

Then, Volume of steam $x_2 \times 0.2403 = 0.2$

or, $\quad x_2 = 0.832.$ **Ans.**

(*iii*) At 75.0 bar pressure saturation temperature of steam

$$= 290.5°C$$

So, the sample is superheated steam

and Degree of superheat = 300 – 290.5 = 9.5°C. **Ans.**

Example 6.5. 4 kg of 0.5 dry stream at 6.0 bar pressure is heated so that it becomes :

(*i*) 0.95 dry at 6.0 bar pressure.

(*ii*) Dry and saturated at 6.0 bar pressure.

(*iii*) Superheated to 300°C at 6.0 bar pressure.

(*iv*) Superheated to 250°C at 8.0 bar pressure.

Determine in each case the quantity of heat to be supplied.

Solution. From steam tables at 6.0 bar pressure the properties of steam are

$$h_f = 670.4 \text{ KJ/kg.}$$
$$h_{fg} = 2085.0 \text{ KJ/kg.}$$
$$\text{Sat, temp.} = 158.8°$$

Total heat of 4 kg of 0.5 dry steam at 6.0 bar pressure

or, $h_1 = 6851.6$ KJ

(*i*) Total heat of 4 kg of 0.95 dry steam at 6.0 bar pressure

$$= 4\,(670.4 + 0.95 \times 2085.0) = 10604.6 \text{ KJ.}$$

Hence, quantity of heat to be supplied

$$= 10604.6 - 6851.6 = 3753 \text{ KJ.}$$ **Ans.**

(*ii*) Total heat of 4 kg of dry saturated steam at 6.0 bar pressure

$$= 4\,(670.4 + 2085.0) = 11021.4 \text{ KJ}$$

Hence, quantity of heat to be supplied

$$= 11021.4 - 6851.6 = 4169.8 \text{ KJ.}$$ **Ans.**

(*iii*) For steam at 6.0 bar pressure

Saturation temperature = 158.8°C

Temperature of steam = 300°C

Degree of superheat = 300 – 158.8 = 141.2°C

Total heat of superheated steam at 6.0 bar pressure and 300°C

(from steam tables) = 3062.3 KJ/kg

Total heat of 4 kg of superheated steam,

$$= 4 \times 3062.3 = 12249.2 \text{ KJ}$$

Hence, quantity of heat to be supplied

$$= 12249.2 - 6851.6 = 5397.6 \text{ KJ.}$$ **Ans.**

(*iv*) From steam tables total heat of superheated steam, at 8.0 bar pressure and 250°C = 2950.4 KJ

Hence, total heat of 4 kg of superheated steam

$$= 4 \times 2950.4 = 11801.6 \text{ KJ}$$

Hence, quantity of heat to be supplied

$$= 11801.6 - 6851.6 = 4950 \text{ KJ.}$$ **Ans.**

Example 6.6. Calculate volume, enthalpy, internal energy and entropy of 4.73 kg of steam at 7.8 bar pressure under the following conditions.

(*i*) Dry and saturated,

(*ii*) Wet steam having wetness of 37%

(*iii*) Superheated steam at 240°C

Solution. In the steam tables properties of steam are available for 7.5 bar and 8.0 bar pressure.

These properties are as given below :

Pressure in bar	Specific Volume (Vapour) m^3/kg (v_g)	Specific enthalpy		Specific entropy	
		Water KJ/kg (h_r)	Evaporation KJ/kg (h_{fg})	Water KJ/kg (S_r)	Evaporation KJ/kg (S_{fg})
7.5	0.2555	709.3	2055.6	2.009	4.662
8.0	0.2403	720.9	2046.5	2.046	4.614

By interpolation the properties can be evaluated for 7.8 bar pressure.

Specific volume at 7.8 bar

$$= 0.2555 - \left(\frac{0.2555 - 0.2403}{0.5}\right) \times 0.3 = 0.2555 - 0.0091$$

$$= 0.2464 \text{ m}^3/\text{kg}$$

Specific enthalpy of water at 7.8 bar pressure

$$= 709.3 + \frac{(720.9 - 709.3)}{0.5} \times 0.3 = 709.3 + 6.96$$

$$= 716.3 \text{ KJ/kg}$$

Specific enthalpy of evaporation at 7.8 bar pressure,

$$= 2055.6 - \frac{(2055.6 - 2046.5)}{0.5} \times 0.3 = 2055.6 - 5.46$$

$$= 2050.14 \text{ KJ/kg}$$

Specific entropy of water at 7.8 bar pressure,

$$= 2.009 + \frac{(2.046 - 2.009)}{0.5} \times 0.3 = 2.009 + 0.0222$$

$$= 2.0312 \text{ KJ/kg}$$

Specific entropy of steam at 7.8 bar pressure,

$$= 4.662 - \frac{(4.662 - 4.614)}{0.5} \times 0.3$$

$$= 4.662 - 0.0288 = 4.6332 \text{ KJ/kg k}$$

Thus the properties of saturated steam at 7.8 bar pressure are :

Specific volume, v_g = 0.246m^3/kg **Ans.**

Specific heat h_f = 716.3 KJ **Ans.**

Latent heat, h_{fg} = 2050.14 KJ **Ans.**

Specific entropy, water, S_f = 2.0312 KJ/kg k **Ans.**

Specific entropy evaporation, $S_g - S_f$ = 4.6332 KJ/kg k **Ans.**

(*i*) **Dry and saturated steam :**

Volume of 4.73 kg of steam = 4.73 × 0.2462 = 1.1645 m^3 **Ans.**

Internal energy of steam,

$$U = h - pV = 13085 - (7.8 \times 100) \times 1.1645 = 12176.67 \text{ KJ. } \textbf{Ans.}$$

Entropy of steam = 4.73 (2.0312 + 4.6332) = 31.5226 KJ/k. **Ans.**

(*ii*) **Wet steam :**

Wetness of steam = 37%

∴ Dryness of steam = 100 – 37 = 63%

or the dryness fraction of steam, $x = 0.63$.

Volume of steam, $V = m \cdot x \cdot v_g = 4.73 \times 0.63 \times 0.2462$

$= 0.7337 \text{ m}^3$ **Ans.**

Enthalpy of steam $h = m(h_f + xh_{fg}) = 4.73(716.3 + 0.63 \times 2050.14)$

$= 9497.31$ KJ **Ans.**

Internal energy of steam $= h - pV = 9497.31 - (7.8 \times 100)(0.7337)$

$= 8925.02$ KJ **Ans.**

Entropy of steam $= m\,[S_f + x(S_g - S_f)]$

$= 4.73\,(2.0312 + 0.63 \times 4.6332)$

$= 23.414$ KJ/K **Ans.**

(*iii*) **Superheated steam :**

Saturation temperature at 7.8 bar = 169.4°C

From steam tables, the volume of superheated steam at 7.0 bar and 8.0 bar at 200°C and 250°C is as follows :

Pressure bar	Volume of superheated steam (m^3/kg)	
	200°C	250°C
7.0	0.2999	0.3364
8.0	0.2608	0.2932

By interpolation, specific volume of steam at 7.8 bar and 240°C

$= 0.2952 \text{ m}^3/\text{kg}$.

Volume of 4.73 kg of steam = $4.73 \times 0.2952 = 1.3963 \text{ m}^3$ **Ans.**

Specific enthalpy of superheated steam may be calculated by interpolation = 2926.8 KJ/kg.

∴ Enthalpy of 4.73 kg of steam = $4.73 \times 2926.8 = 13844$ KJ **Ans.**

Internal energy, $U = h - pV = 13844 - 7.8 \times 100 \times 1.3963$

$= 12755$ KJ **Ans.**

Similarly, specific entropy of superheated steam at 7.8 bar, 240°C by interpolation = 7.009 KJ/kg k

∴ Entropy of 4.73 kg of steam = $4.73 \times 7.009 = 33.153$ KJ/k **Ans.**

Example 6.7. Steam at a pressure of 7 bar abs. and dryness fraction 0.8 is throttled to a pressure of 4 bar abs. Find the final condition of the steam.

Solution. First to all, locate the point corresponding to 7 bar abs pressure and 0.8 dryness fraction, on the Mollier diagram. From this point, draw a horizontal line to cut the pressure line of 4 bar abs. pressure. Then it will be noticed that the steam even after the throttling process is wet.

The dryness fraction at this point, $x_2 = 0.82$ **Ans.**

To check this result proceed as follows :

As we know that during a throttling process;

Total heat before process = Total heat after process.

$\therefore \quad h_{f1} + x_1 h_{fg1} = h_{f2} + x_2 h_{fg2}$

Now, substituting the values in the above equation

or $697.2 + 0.8 \times 2066.3 = 604.7 + 2133.8 \times x_2$

$$x_2 = 0.818 \textbf{ Ans.}$$

Example 6.8. Steam at a pressure of 8.5 bar abs. and dryness fraction 0.98 is subjected to throttling expansion to a pressure of 1 bar abs. Find the final condition of steam, assuming C_p for the superheated steam as 2.2.

Solution. First of all, locate the point corresponding to a pressure of 8.5 bar abs and 0.98 dryness fraction on the Mollier diagram. Now, from this point, draw a horizontal line cutting the pressure line of 1 bar abs. It will be noticed that the condition of steam after this expansion is superheated.

Also, from the Mollier diagram ;

Temperature of the superheated steam, $t_s = 125°C$ **Ans.**

To check this result proceed as follows :

Corresponding to a pressure of 8.5 bar abs. It is observed from steam tables that;

$$h_{f1} = 732.2 \text{ KJ}$$
$$h_{fg1} = 2039.4 \text{ KJ}$$
$$x_1 = 0.98$$
$$h_{f2} = 417.5 \text{ KJ}$$
$$h_{fg2} = 2258 \text{ KJ}$$
$$t_s = 99.6 \text{ °C}$$

Also, we know that during throttling process :

$$h_{f1} + x_1 h_{fg1} = h_{f2} + h_{fg2} + C_p (t_s - t_2)$$

Substituting the values in the above equation; we get

$$732.2 + 0.98 \times 2039.4 = 417.5 + 2258 + 2.2 (t_s - 99.6)$$

and $\quad t_s = 124.7°C$ **Ans.**

Hence, our result is true.

Example 6.9. Steam at a pressure of 8.5 bar abs. and 200°C is throttled to 3 bar abs. and then expanded adiabatically to 0.5 bar. Find out the change in entropy and enthalpy during these two processes.

Solution. First of all, locate the point 1, where the lines of 8.5 bar abs. pressure and 200°C temp. cut each other on the Mollier diagram. Then from this point 1 draw a horizontal line representing the throttling process cutting the pressure line of 3 bar abs. This is point 2. Now, from the point 2 draw a vertical line representing the adiabatic expansion cutting the pressure line of 0.5 bar abs. This is point 3, as shown in Fig. 6.13.

Now, from the Mollier diagram

$$h_1 = h_2 = 2836.45 \text{ KJ/kg}$$
$$S_1 = 6.723 \text{ KJ/kg k}$$
$$h_3 = 2500 \text{ KJ/kg}$$
$$S_2 = S_3 = 7.15 \text{ KJ/kg k}$$

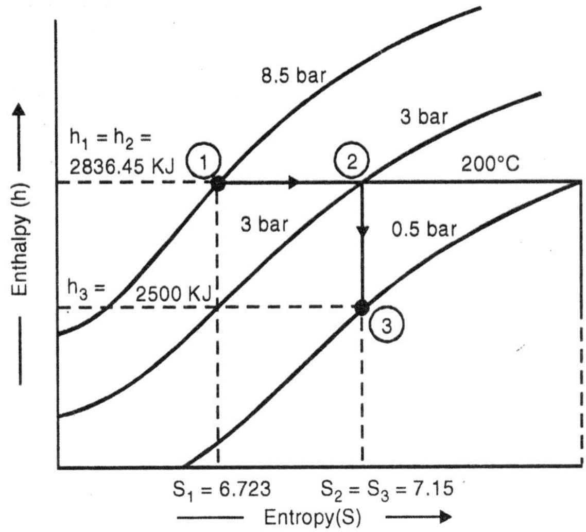

Fig. 6.13.

1. Throttling Process, 1–2 :

(*i*) Change in entropy during the throttling process 1–2,

$$= S_2 - S_1 = 7.15 - 6.723$$
$$= 0.427 \text{ KJ/kg k } \textbf{Ans.}$$

(*ii*) Change in enthalpy during throttling process 1–2 will be zero

As, $h_1 = h_2$ **Ans.**

2. Adiabatic Process, 2–3 :

(*i*) Change in entropy during the adiabatic process, 2–3 will be zero,

As $S_2 = S_3$ **Ans.**

(*ii*) Change in enthalpy during the adiabatic process, 2–3,

$$= h_2 - h_3$$
$$= 2836.45 - 2500 = 336 \text{ KJ/kg k } \textbf{Ans.}$$

Example 6.10. A steam turbine receives dry and saturated steam at a pressure of 10 bar abs. and expands it, adiabatically to a pressure of 0.5 bar abs. If the steam is supplied at the rate of 1.5 kg/s then find the theoretical heat loss and the power developed.

Solution. Because, it is an adiabatic (isentropic) expansion, therefore, using Mollier diagram;

First of all locate the point where 10 bar abs. pressure line cuts the dry and saturated line. This is point 1.

Now, draw horizontal line as shown in Fig. 6.14 the enthalpy

$$h_1 = 2780 \text{ KJ/kg}$$

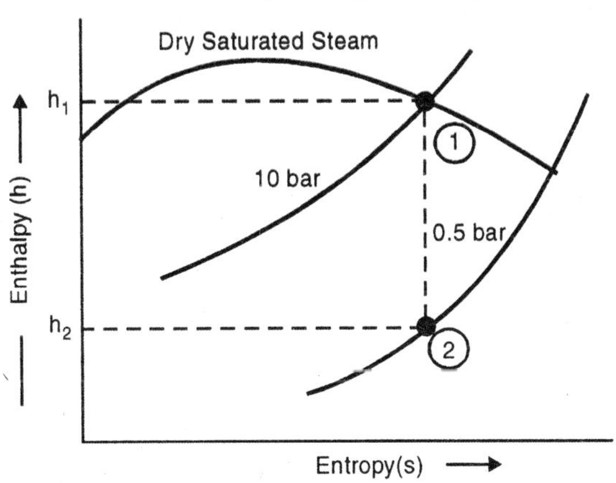

Fig. 6.14.

Now, from this point 1 draw a perpendicular line to cut the pressure line of 0.5 bar abs. This is point 2. Draw a horizontal line again to find the enthalpy at point 2.

$$h_2 = 2290 \text{ KJ/kg}$$

$\therefore$ Heat lost = Mass × Heat drop/kg

$$= m(h_1 - h_2) = 15(2780 - 2290)$$

$$= 7350 \text{ KJ/s } \textbf{Ans.}$$

Example 6.11. Find the total heat of superheated steam at a pressure of 15 bar abs. and temp. of 250°C, assuming that the specific heat of superheated steam is 2.5 KJ/kg K. Find the heat required to raise 3 kg of this steam from steam at 3.5 bar abs. and dryness fraction 0.88.

Solution. Given data :

Abs. pressure of steam = 15 bar

Superheating temp. of steam, T_{sup} = 250°C

Sp. heat of superheated steam = 2.5 KJ/kg K

As, we know that :

$$h_s = h_f + h_{fg} + C_p (T_{sup} - T_s)$$

From steam tables; at a pressure of 15 bar

$$T_{sat} = 198.3°C$$

$$h_{fg} = 1947.3 \text{ KJ/kg K}$$

$$h_f = 844.9 \text{ KJ/kg K}$$

$\therefore$ $$h_s = 844.9 + 1947.3 + 2.5 (250 - 198.3)$$

or $$h_s = 2921.45 \text{ KJ/kg}$$

But, corresponding to 3.5 bar pressure;

$$h_f = 584.3 \text{ KJ/kg}$$

$$h_{fg} = 2148.1 \text{ KJ/kg}$$

So, total heat of this wet steam

$$= 584.3 + 0.88 \times 2148.1$$

$$= 2474.63 \text{ KJ}$$

Hence,

Total heat required = 3(2921.45 – 2474.63)

$$= 1340.46 \text{ KJ } \textbf{Ans.}$$

Example 6.12. Using steam tables, find the amount of heat required to superheat one kg of dry saturated steam at a pressure of 7 bar abs. to a temp. of 300°C. Calculate the specific heat of this steam.

Solution. Given data :

Mass of dry saturated steam = 1 kg

Abs. pressure of steam = 7 bar

Temp. of steam = 300°C

From steam tables, at abs pressure of 7 bar

$$h_g = 2763.5 \text{ KJ/kg}$$

$$h_s = 3059.1 \text{ KJ/kg}$$

$\therefore$ Heat required for superheating = 3059.1 – 2763.5 = 295.6 KJ

Now, we know that ;

$$h_s = h_g + C_p (T_{sup} - T_s)$$

$$\therefore \quad C_p = \frac{h_s - h_g}{T_{sup} - T_s}$$

or $$C_p = \frac{3059.1 - 2763.5}{300 - 165}$$

Hence sp. heat, C_p = 2.19 **Ans.**

Example 6.13. Estimate, the amount of heat required to convert 10 kg of water at 35°C, into steam at a pressure of 16 bar abs. with 130°C superheat. Find, the average value of sp. heat of the steam over this range of superheat and calculate the volume of superheated steam.

Solution. Given data :

Mass of water = 10 kg

Temp. of water = 35°C

Abs. pressure of steam = 16 bar

Superheating temp., T_{sup} = 130°C

From steam tables at 16 bar abs. pressure;

$$T_{sat} = 201.4°C$$
$$= 201.4 + 273 = 474.4 \text{ K}$$
$$v_g = 0.124 \text{ m}^3/\text{kg}$$
$$h_g = 0.124 \text{ m}^3/\text{kg}$$
$$h_g - 2794.0 \text{ KJ/kg}$$
$$T_{sup} = 201.4 + 130 = 331.4°C$$
$$= 331.4 + 273 = 604.4 \text{ K}$$
$$h_s = 3104 \text{ KJ}$$

So, heat required to convert 10 kg of water at 35°C into,

Superheated steam = 10 (3104 – 35 × 4.187)

= 29574.5 KJ

Also, we know that;

$$h_s = h_g + C_p (T_{sup} - T_s)$$

or $$3104 = 2794 + C_p \times 130$$

or $$C_p = \frac{3104 - 2794}{130} = 2385 \quad \text{Ans.}$$

Now, let 'v' be the volume of one kg of superheated steam in m^3.

So, at constant pressure process, applying Charles' Law;

$$v = \frac{T_{sup}}{T_s} \times v_g = \frac{604.4}{474.4} \times 0.124 = 0.158$$

Hence,volume of superheated steam, $v = 0.158 \text{ m}^3$ **Ans.**

Example 6.14. Steam at a pressure of 7 bar abs. and dryness fraction 0.8 is throttled to a pressure of 4 bar absolute. Determine the final condition of the steam.

Solution. Given data :

Abs. pressure of steam = 7 bar

Dryness fraction, x = 0.8

Abs. pressure of steam after throttling = 4 bar

Now, using the Mollier diagram, locate the point corresponding to 7 bar abs. pressure and 0.8 dryness fraction. From this point, draw a horizontal line till it cuts the pressure line of 4 bar abs. It will be seen that steam after throttling is still wet having dryness-fraction, as read from this diagram equal to 0.82.

So, $x_2 = 0.82$ **Ans.**

To check the result analytically ;

As, $h_{f1} + x_1 . h_{fg_1} = h_{f2} + x_2 . h_{fg_2}$

or $697.2 + 0.8 \times 2066.3 = 604.7 + 2133.8 \times x_2$

or $x_2 = 0.818$ **Ans.**

Hence, the dryness fraction after throttling, $x_2 = 0.82$ **Ans.**

Example 6.15. Steam at a pressure of 8.5 bar absolute and dryness fraction of 0.98 is subjected to throttling expansion to 1 bar abs. Determine, the final condition of steam. Assume 'C_p' for superheated steam as 2.2 KJ/kg K.

Solution. Given data :

Abs. pressure of steam = 8.5 bar

Dryness-fraction = 0.98

Abs. pressure, after throttling process = 1 bar

Superheated constant pressure, C_p = 2.2 KJ/kg K

Now, using Mollier diagram, locate the point corresponding to 8.5 bar abs. pressure and 0.98 dryness-fraction. From this point draw a horizontal line till it cuts the pressure line of 1 bar abs. It will be seen that after expansion is superheated.

And, from this diagram, temp. of superheated steam,

$$T_{sup} = 125°\text{C} \quad \textbf{Ans.}$$

Now, to check the result analytically; using steam tables;

$$h_{f1} = 732.2 \text{ KJ}$$

$$h_{fg1} = 2039.4 \text{ KJ}$$

$$x_1 = 0.98$$

$$h_{f2} = 417.5 \text{ KJ}$$

$$h_{fg2} = 2258.0 \text{ KJ}$$

$$T_s = 99.6°\text{C}$$

But $h_{f1} + x_1 . h_{fg1} = h_{f2} + h_{fg2} + C_p (T_{sup2} - T_{s2})$

or $732.2 + 0.98 \times 2039.4 = 417.5 + 2258 + 2.2 (T_{sup2} - 996)$

or $T_{sup2} = 124.7°\text{C}$

Hence, superheating temp. of the steam after throttling

$$T_{sup} = 124.7°C \textbf{ Ans.}$$

ASSIGNMENT

Objective Type Questions :

(A) Fill in the Blanks :

1. Steam is a ______.
2. Saturation temperature is ______ at different pressures.
3. Total heat of steam is reckoned from ______
4. Latent heat ______ with decrease in pressure.
5. Dryness fraction of dry steam is ______.
6. Latent heat is zero at ______ pressure.
7. The value of critical pressure is ______
8. Sensible heat of one kg of water at 65°C is ______ approx.
9. Superheated vapour starts behaving like ______.
10. Entropy of superheated steam is equal to ______.
11. Saturation temperature ______ with the increase in pressure.
12. One kg of wet steam having dryness fraction 0.85 will have ______ kg of water vapour and ______ kg of dry steam.
13. The difference between the temperature of superheated steam and the saturated steam for a given pressure is known as ______.
14. The units of specific volume are ______.
15. Entropy remains constant during ______ process.
16. Enthalpy remains constant during ______ process.
17. At critical pressure latent heat will become ______.
18. A mixed phase of liquid and gas is called a ______.
19. Triple point is that point where ______.
20. The dryness fraction of steam, when expressed in percentage is called the ______ of steam.

(B) State True or False :

1. The value of dryness fraction is greater than unity.
2. The temperature remains constant during the formation of wet steam.
3. During an adiabatic expansion, the dryness fraction of steam decreases.
4. Throttling process is represented by horizontal line on Mollier diagram.
5. The value of critical temperature is 374°C.
6. Temperature of superheated steam is known as degree of superheat.
7. The volume of wet steam is more than that of dry steam at the same pressure.
8. Latent heat is equal to the sum of internal latent heat and the external work done.
9. During an adiabatic process, enthalpy remains constant.
10. The unit of enthalpy is same as that of heat.
11. When change of state takes place, temperature remains constant.
12. Specific volume increases with the increase in pressure.
13. Dryness fraction will increase if steam is throttled from a higher pressure to a lower pressure.

14. In case of wet steam, constant pressure and the isothermal process are identical.
15. For hyperbolic expansion of wet steam, the change in the internal energy is equal to the change in enthalpy.
16. The entropy remains constant during an irreverisible adiabatic process.
17. The process of superheating approximately follows the Charle's law.
18. The latent heat of steam at atmospheric pressure approximately equals 539 kcal/kg in MKS system of units.
19. The steam produced at pressures greater than critical pressure is called super pressure steam.
20. Triple point is the point where the three phases of water-solid, liquid and gas exist together.

(C) Multiple Choice Questions :

1. Vapour is a :
   (*a*) pure substance (*b*) perfect gas
   (*c*) mixed phase of liquid and gas
   (*d*) substance, homogeneous and invariable in chemical composition.
2. Which form of the vapour has a behaviour close to that of a gas ?
   (*a*) Wet vapour only (*b*) Dry and saturated vapour
   (*c*) Wet and dry vapoured (*d*) Superheated vapour
3. Critical temperature for steam is :
   (*a*) 313 °C (*b*) 347 °C
   (*c*) 375 °C (*d*) 409 °C
4. Critical pressure for steam is :
   (*a*) 252 bar (*b*) 225 bar
   (*c*) 184 bar (*d*) 163 bar
5. At critical pressure :
   (*a*) Entropy is zero
   (*b*) Latent heat of vapourization is zero
   (*c*) Saturated vapour becomes a gas
   (*d*) Sublimation occurs from solid to vapour phase.
6. The heat absorbed by water at its saturation temperature to get converted into dry steam at the same temperature is called :
   (*a*) Sensible heat (*b*) Specific heat
   (*c*) Latent heat (*d*) Total heat
7. The enthalpy of vaporization is minimum at :
   (*a*) Absolute zero (*b*) 317 °K
   (*c*) Critical temperature (*d*) Triple point
8. Dryness fraction of a steam is given by :
   (*a*) $\dfrac{m_s}{m_s + m_w}$ (*b*) $\dfrac{m_w}{m_s + m_w}$
   (*c*) $\dfrac{m_w}{m_s}$ (*d*) $\dfrac{m_s}{m_w}$
   where m_s is the mass of dry steam in a certain mass of wet steam and 'm_w is the mass of water vapour in suspension in the same mass of wet steam.
9. At any point under the dome of temperature-entropy plot the quality of steam is :
   (*a*) Less than 50% (*b*) Between 0 and 100%
   (*c*) Steam is superheated (*d*) Unpredictable

10. When an open vessel containing water is taken to hill top and heated, the water starts boiling at :
(*a*) Less than 100°C (*b*) More than 100°C temp.
(*c*) At 100°C temperature (*d*) Unpredictable

11. A wet vapour can be completely specified by its :
(*a*) Pressure (*b*) Temperature
(*c*) Pressure and dryness fraction
(*d*) Pressure and temperature

12. Which of the following parameter remains constant during superheating of steam ?
(*a*) Temperature (*b*) Pressure
(*c*) Enthalpy (*d*) Internal energy

13. For complete specification of superheated vapour, one needs
(*a*) Pressure (*b*) Temperature
(*c*) Pressure as well as temperature
(*d*) Specific volume.

14. Which of the following parameter decreases as the pressure of steam is raised ?
(*a*) Saturation temperature (*b*) Specific volume
(*c*) Sensible heat (*d*) Latent heat of vapourization

15. Mollier diagram is a plot of :
(*a*) Temperature and entropy
(*b*) Temperature and enthalpy
(*c*) Pressure and enthalpy
(*d*) Enthalpy and entropy.

16. The ordinate of a Mollier diagram represents :
(*a*) Temperature (*b*) Pressure
(*c*) Entropy (*d*) Specific enthalpy.

17. The triple point of water is :
(*a*) 54.35 k (*b*) 13.84 k
(*c*) 216.55 k (*d*) 273.16 k

18. During which process, a solid changes directly to the gaseous form without ever being a liquid.
(*a*) Condensation (*b*) Evaporation
(*c*) Sublimation (*d*) Crystallisation

19. Which of the following process is represented by a horizontal line parallel to the abscissa of a Mollier-diagram ?
(*a*) Constant press (*b*) Constant temperature
(*c*) Free expansion or throttling
(*d*) Adiabatic expansion

20. Steam coming out of the whistle of a pressure cooker is mostly likely to be :
(*a*) Wet vapour (*b*) Dry and saturated vapour
(*c*) Superheated vapour
(*d*) Very wet vapour with dryness fraction less than 0.5.

ANSWERS

(A) Fill in the Blanks :

1. vapour 2. different 3. zero degree
4. increases 5. unity 6. critical
7. 220.65 bar 8. 272.06 KJ 9. perfect gas
10. entropy of dry saturated steam + change of entropy during superheating.
11. increases 12. 0.15, 0.85 13. degree of superheat
14. m^3/kg 15. isentropic/adiabatic 16. throttling
17. zero 18. vapour
19. three phases of solid, liquid and gas exist together
20. quality

(B) State True or False :

1. False	2. True	3. True	4. True	5. True
6. False	7. False	8. True	9. False	10. True
11. True	12. True	13. True	14. True	15. True
16. False	17. True	18. True	19. True	20. True

(C) Multiple Choice Questions :

1. (*c*)	2. (*d*)	3. (*c*)	4. (*b*)	5. (*b*)
6. (*c*)	7. (*c*)	8. (*a*)	9. (*b*)	10. (*a*)
11. (*c*)	12. (*b*)	13. (*c*)	14. (*d*)	15. (*d*)
16. (*d*)	17. (*d*)	18. (*c*)	19. (*c*)	20. (*c*)

(D) Short Answer Type Questions :

Q. 1. What is 'Wet-steam' ?

Ans. If the saturated steam, contains suspended water particles in it, it is known as 'Wet-steam'.

Q. 2. What is 'Dry-steam' ?

Ans. If saturated steam does not contain any water particle suspended in it, it is known as 'Dry-steam' or 'Dry-saturated steam'.

Q. 3. What is 'Saturated steam' ?

Ans. Saturated steam is a vapour at the temperature corresponding to the boiling point of water at the imposed pressure.

Q. 4. What is 'Superheated steam' ?

Ans. The steam heated beyond the saturation temperature for a given pressure, is termed as 'superheated steam.

Q. 5. Define the term 'Dryness fraction' (x).

Ans. The ratio of the weight of actual dry-steam in a given sample of wet-steam, to the weight of wet-steam containing it is known as 'dryness-fraction'. It is generally denoted by 'x'.

If, m_s = Mass of dry-steam in kg

and m_w = Mass of water vapour in suspension

Then, dryness-fraction, $x = \dfrac{m_s}{m_s + m_w}$

Q. 6. What is 'Degree of superheat' ?

Ans. It is the difference between the temperature of the superheated steam and that of the saturated steam at the imposed pressure.

Q. 7. What is 'Latent heat of vaporization' (h_{fg}) ?

Ans. The quantity of heat required at a constant pressure to convert one kg of water at saturation temperature into dry and Saturated Steam is known as 'Latent-Heat of vapourization' or simply 'Latent-heat of steam'. It is generally denoted by (h_{fg}).

Q. 8. What is 'Enthalpy of steam' (h) ?

Ans. The quantity of heat required to convert, at constant pressure one kg of water at freezing point, into wet, dry-saturated or superheated steam, as the case may be, is known as 'Enthalpy' of that steam accordingly. It is also called 'Total-heat of Steam' and is generally denoted by 'h'.

Q. 9. What is the 'Total-heat of Wet-Steam' ?

Ans. The quantity of heat required to convert, at constant pressure one kg of water at freezing point into wet steam is known as 'Total-heat of Wet-Steam'. It is also known as 'Enthalpy of Wet-Steam' and is generally denoted by 'h'.

Q. 10. What is the 'Total-heat of Dry-Saturated Steam' ?

Ans. The quantity of heat required to convert, at constant pressure, one kg of water at freezing point into dry and saturated steam is known as 'Total-Heat of dry-saturated steam'. It is also called 'Enthalpy of Dry-saturated steam and is generally denoted by 'h'.

i.e., h = Sensible heat + Latent heat

or $h = h_f + h_{fg}$

Q. 11. What is the "Total-heat of Superheated Steam' ?

Ans. The quantity of heat required, to convert at constant pressure one kg of water, at freezing point into superheated steam, is known as 'Total-heat of superheated Steam'. It is also called 'Enthalpy of Superheated Steam and generally denoted 'h_s'.

h_s = Total heat of dry steam + Heat required to superheat the steam

or $h_s = h_f + h_{fg}$ + Heat of Superheat

or $h_s = h_f + h_{fg} = C_p\ (t_s - t)$

Q. 12. What do you know about 'Steam-Tables' ?

Ans. The various properties of steam *i.e.*, saturation temp. sensible heat, latent heat, total heat, specific volume and entropy etc. corresponding to various pressures are determined by experiments and plotted in the form of a table which is known as 'Steam-table'.

Q. 13. Define the term 'Specific volume of dry-steam' and the 'specific volume of wet-steam'.

Ans. The volume of one kg of dry-steam at a given pressure, is known as 'Specific volume of dry-steam'. It is generally denoted by 'v_g'. The volume of one kg of wet-steam which will be equal to the volume of dry portion of steam plus the volume of water in suspension is known as 'Specific volume of wet-steam and is denoted by 'v_w'.

$\therefore \quad v_w = x \,.\, v_g + (1 - x)\ v_f$

The term $(1 - x)\, v_f$ can be neglected as the value of 'v_f' is very small as compared to 'v_g'.

Q. 14. How the volume of superheated steam is calculated ?

Ans. The volume of the superheated steam can be calculated on the assumption that the dry-steam starts behaving (approximately) like a perfect gas, when it is heated further.

So, applying Charle's law;

Assuming that steam is superheated at pressure 'P'

$$\therefore \quad \frac{v_s}{t_s} = \frac{v_g}{t}$$

or

$$v_s = v_g \times \frac{t_s}{t}$$

where

v_s = Volume of one kg of superheated steam at pressure 'P'

v_g = Specific volume of dry steam at pressure 'P' (from steam table)

t_s = Final temperature of superheated steam.

t = Saturation temperature at pressure 'P'

Q. 15. What is the 'External Work done during evaporation' ?

Ans. The amount of work done due to large increase in volume when changing from water into steam, is known as 'External Work done during evaporation'.

Corresponding to the evaporation of one kg of water into dry saturated steam at pressure 'P' bar, we get :

External work done during evaporation = $10^5 p\ (v_g - v_f)$ Joule.

where; v_g = Sp. volume of dry saturated steam at pressure 'p' in m^3

and v_f = Volume of one kg of water in m^3

Q. 16. What is 'Internal Latent-heat' ?

Ans. The difference between the 'Latent heat' and the 'External work' of evaporation is known as 'Internal-heat'. It is also called 'True Latent-heat'.

Internal latent heat = $h_{fg} - 10^5\ pv_g$ Joule.

Q. 17. What is 'Internal Energy of steam' ?

Ans. The actual heat energy stored in the steam is known as 'Internal Energy of steam'. It is generally abbreviated as 'U'. Internal Energy of steam,

$$U = h_g + \text{Internal latent heat} \quad \text{... Joule}$$

or

$$U = h_g + (h_{fg} - 10^5\ pv_g)\text{J}$$

or

$$U = h - 10^5\ pv_g\ \text{J} \qquad (\because h = h_g + h_{fg})$$

If steam is superheated;

Then,

$$U = h_s - 10^5\ pv_g\ \text{Joule}$$

Q. 18. What is the 'Entropy of water' (S_f) ?

Ans. If water is heated from abs. temperature 't_0' to 't', then the increase in entropy per kg of water, at constant pressure will be given by;

$$\text{Increase in Entropy} = \text{Log}_e \frac{t}{t_0}$$

But, the entropy of steam is always reckoned above the freezing point of water;

$$\therefore \quad t_0 = 273\ \text{K}$$

$\therefore$ Entropy of water above its freezing point,

$$S_f = \text{Log}_e \frac{t}{273}$$

Q. 19. What is 'Temperature-Entropy diagram' for steam ?

Ans. In 'Temperature-Entropy Diagram' for steam the 'Absolute temperature' is plotted vertically and 'Entropy' is plotted horizontally. The diagram has three regions. These are separated out by two curves *i.e.*, water-curve and the saturation curve.

Q. 20. What is 'Mollier-Diagram' or Enthalpy-entropy (*H–S*) Diagram ?

Ans. It is the diagram representing the relation between 'Specific Enthalpy' on vertical line and 'Specific Entropy' on the horizontal line for the steam.

REVIEW QUESTIONS

1. Explain the various phases of formation of steam.
2. (*a*) What do you mean by
   (*i*) Wet steam,
   (*ii*) Dry steam,
   (*iii*) Superheated steam.
   (*b*) What is degree of superheat ?
3. Explain the terms :
   (*i*) Latent Heat
   (*ii*) Sensible Heat
   (*iii*) Total Heat
   (*iv*) Dryness Fraction
   (*v*) Saturation temperature
   (*vi*) Specific volume as applied to steam. **(SBTE Pb., 1981)**
4. Explain the terms enthalpy and 'entropy' of steam. **(Delhi 1981)**
5. Discuss how the entropy of dry and superheated steam is calculate ?
6. Discuss the H–S and T–S diagrams with neat sketches. **(SBTE Pb., Dec. 99)**
7. (*i*) What is 'entropy of water'.
   (*ii*) Write in short about 'entropy increase during evaporation'.
8. Discuss the advantages of superheated steam.
9. What do you understand by external work of evaporation ? How it is calculated.
10. What is superheated steam ? How the volume of superheated steam is calculated ?
11. (*i*) What is internal latent heat ?
    (*ii*) Explain how the internal energy of steam is determined ?
    (*iii*) What is the differences between the total heat and internal energy of steam ? **(W. Bengal, 1979)**

PROBLEMS FOR PRACTICE

1. Find the enthalpy, internal energy and entropy of one kg of steam at a pressure of 10 bar.
   (*i*) When steam is dry and saturated.
   [**Ans.** 2778 KJ/kg, 2584 KJ/kg, 6.587 KJ/kg k]
   (*ii*) When steam is 0.75 dry.
   [**Ans.** 2274.3 KJ/kg, 2128.7 KJ/kg, 5.475 KJ/Kg k]
   (*iii*) When steam is superheated to 250 °C
   [**Ans.** 2943 KJ/kg, 2710 KJ/kg, 6.92 KJ/kg k]
2. Calculate the total heat of 5 kg of steam at a pressure of 8 bar (abs.) having dryness fraction of 0.8. Also calculate heat in KJ required to convert this steam into dry and saturated steam. [**Ans.** 2039.5 KJ]

3. A pressure cooker contains 1.25 kg of steam at 5.0 bar pressure and 0.9 dry. Find the quantity of heat that must be rejected so as the quantity of steam becomes 0.5 dry. **[Ans.** Qty. = –1032.5 KJ]
4. 5000 m^3/hr wet steam 0.95 dry at 10 bar pressure is to be supplied by a boiler for a processing plant, calculate;
   (*i*) The mass of steam supplied per hour.
   (*ii*) The quantity of coal of clorific value 16000KJ to be burnt in boiler if overall efficiency of boiler is 29%.
   [Ans. (*i*) m = 27086 kg (*ii*) Qty. = 1562 t/hr]
5. Steam at 20 bar and 300°C passes through a pipe at the velocity of 120 m/s. If steam flows at the rate of 500 kg/hr. Find the diameter of the pipe. **[Ans.** d = 13 mm]
6. Steam initially dry saturated, expands isentropically from a pressure of 16 bar to 0.16 bar. Find the index of isentropic expansion. **[Ans.** n = 1.126]
7. 0.2 kg of steam at a temperature of 165°C expands reversibly at a constant pressure of 7 bar until the volume is doubled. Find the final temperature work done and heat transferred if the initial dryness fraction of steam is 0.7.
   [Ans. t_2 = 315°C, W = 26.7 KJ, Q = 189 KJ]
8. Steam originally dry is allowed to fall in pressure from 10 bar to 4.0 bar under the following conditions.
   (*i*) In a closed vessel, which loses heat by radiation and conduction.
   (*ii*) By passing through a throttle value.
   Estimate the final condition of the steam (superheat or dryness fraction) in both cases. Take 'C_p' for superheated steam as 0.65.
   [Ans. (*i*) Q = 0.42, (*ii*) T-S = 218.4°C]
9. Steam is supplied from a boiler at a pressure of 20 bar and 99% dry to steam engine. It is found that the steam loses 20 KJ/kg as it flows through the pipe line pressure remaining constant. Determine the dryness fraction and temp. of steam at the engine end of pipe line. **[Ans.** 0.979, 212.4°C]
10. 10 kg of wet saturated steam at 15 bar pressure is superheated to the temp. of 290°C at const. pressure. Find the heat required and the total heat of steam. Take, x = 0.85. **[Ans.** 4846.2 KJ, 29841.7 KJ]

❑❑❑

7

CHAPTER

Steam Generators

INSIDE THIS CHAPTER

7.1. INTRODUCTION

'Steam generator' may be defined as a closed vessel strongly constructed of steel sheet, in which the steam is produced, at the pressure much higher than the atmospheric pressure by transfer of heat, produced by burning of fuel, to the water.

The construction and appearance of a steam generator depends upon the arrangement made for burning the fuel and for transfer of heat to the water. The steam is produced in steam-generators for utilisation in steam-engines, steam turbines, industrial process work and for heating installations.

7.2. USES OF STEAM

The steam generated in steam-generators is used in the following industrial applications.

(*i*) It is used in steam-turbines of power plants for power-generation.
(*ii*) It is used in steam-engines as in locomotive steam-engines.
(*iii*) It is used in industrial process work, such as in textile industry.
(*iv*) It is also used in heating installations, in the situations, having lack of electric power.

7.3. CLASSIFICATION OF BOILERS/STEAM GENERATORS

The boilers/steam generators can be classified as follows :

(*a*) On the basis of circulation of hot gases :

(*i*) Water tube boiler.

(*ii*) Fire tube boiler.

In a water tube boiler, water circulates through tubes surrounded by hot gases.

In a fire tube boiler, got gases circulate through tubes surrounded by water.

(*b*) On the basis of disposition of the axis of boiler shell :

(*i*) Horizontal boiler.

(*ii*) Vertical boiler.

(*iii*) Inclined boiler.

In the case of a horizontal boiler, the axis of its shell is horizontal when in working, whereas in the case of a vertical boiler it is vertical. But in an inclined boiler, the axis of the shell is inclined to the horizontal in its working position.

(*c*) On the basis of system of firing.

(*i*) Externally fired boiler.

(*ii*) Internally fired boiler.

Externally fired boiler has its furnace arranged outside the boiler shell.

Internally fired boiler has its furnace arranged inside the boiler shell.

(*d*) On the basis of type of fuel used :

(*i*) Coal fired boiler.

(*ii*) Oil fired boiler.

(*iii*) Gas fired boiler.

(*iv*) Electricity heated boiler.

(*v*) Nuclear fuel boiler.

(*vi*) Waste heat boiler.

A boiler may be named according to the source of heat for raising steam, which may be coal, oil, gas, electricity, nuclear energy or waste heat of gases of a chemical plant.

(*e*) On the basis of mobility :

(*i*) Stationary boiler.

(*ii*) Mobile boiler.

Stationary boiler is fixed at one place while in working condition. Whereas, the mobile boiler is able to change its position with respect to one place while in working, such as locomotive boiler and marine boiler.

(*f*) On the basis of number of tubes :

(*i*) Single tube boiler.

(*ii*) Multi-tube boiler.

A boiler having a single tube (fire or water) is known as a single tube boiler. Whereas a boiler having more than one tube (fire or water) is known as a multi-tube boiler.

(*g*) On the basis of draught :

(*i*) Natural draught boiler.

(*ii*) Forced draught boiler.

In a natural draught boiler, the circulation of hot gases in the boiler and out to the chimney is due to natural convection. Whereas a forced draught boiler possesses circulation of hot gases forced by a fan.

(*h*) On the basis of utility for power generation of steam :

(*i*) Thermal power station boiler.

(*ii*) Locomotive boiler

(*iii*) Marine boiler

7.4. ESSENTIALS OF A GOOD BOILER

(*i*) It should be light in weight.

(*ii*) It should occupy a small space.

(*iii*) It should be capable of quick starting.

(*iv*) It should be capable of meeting large variations in load.

(*v*) It should allow easy maintenance.

(*vi*) The various joins should be accessible for inspection and as far as possible, the flames should not be made to impinge on the joints.

(*vii*) The velocity of water and flue gases should be minimum after bringing heavy frictional losses.

(*viii*) The mud and other deposits should not collect on the heated plates. The tubes should not accumulate soot or water deposits and must be sufficiently strong to allow for wear and corrosion.

(*ix*) The boiler should be capable of producing maximum weight of steam for a minimum fuel consumption, attention and initial cost etc.

(*x*) The boiler should be as per the safety regulations laid down in the boiler act.

7.5. FACTORS AFFECTING BOILER SELECTION

The following factors must be taken into view while deciding the type and size of the boiler required for specific purpose;

(*i*) The power required to be generated.

(*ii*) The working pressure.

(*iii*) Initial cost and facilities available for eraction.

(*iv*) Availability of the floor area.

(*v*) Operating and maintenance cost.

(*vi*) Location of the power house.

(*vii*) Fuel and water available.

(*viii*) The probable permanency of the station.

(*ix*) The probable load factor.

(*x*) Accessibility for cleaning and inspection.

7.6. COMPARISON OF FIRE TUBE AND WATER TUBE BOILER

The detailed comparison between the fire tube boiler and the water tube boiler is given below in a tabular form :

Water Tube Boiler	Fire Tube Boiler
1. It can produce steam upto a max. temp. of 550°C.	1. It can produce steam upto a max. temp. of 350°C.
2. It can produce steam at high pressure.	2. It can produce steam only upto 25 times the atm. pressure.
3. In this boiler the rate of steam production is more (nearly 500,000 kg/hr).	3. In this boiler the rate of steam production is less (nearly 10,000 kg/hr).
4. Area required for this type of boiler is less, (About 5 m^2 per 100 kg/hr).	4. Area required by this type of boiler is more (About 8 m^2 per 1000 kg/hr).
5. Its overall efficiency with economizer may be upto 90%.	5. Its overall efficiency with economizer may be upto 70%.
6. It can't produce steam without priming, so some antipriming arrangement is required.	6. It can produce steam without priming.
7. It can be easily transported and errected as its various parties can be separated.	7. Its transportation and errection is difficult.
8. The direction of circulation of water is well defined.	8. The direction of circulation of water is not defined.
9. It is preferred for widely fluctuating loads.	9. It can also cope readily with sudden increase of load but for a short time.
10. It is used in large power plants and ships.	10. It is not suitable for large power plants.
11. Its operating cost is high.	11. Its operating cost is less.

7.7. INTRODUCTION TO FUELS FOR BOILER

7.7.1. Fuel

The 'fuel' is a substance containing combustible elements which on combining with oxygen generate heat. The amount of heat produced during combustion is known as 'Clorific value' of the fuel. The principle constituents of any fuel are Carbon, Hydrogen, and sometimes a small amount of sulphur.

7.7.2. Classification of Fuels

Fuels may be classified according to their physical character as :

(*i*) Solid fuel

(*ii*) Liquid fuel

(*iii*) Gaseous fuel.

(*i*) Solid fuel. Of all the solid fuel coal is the most important one, which is used for power purposes. It is found in many varieties depending upon its physical and chemical properties resulting from its geological formation most important of which are given below.

(*a*) *Peat.* It is spongy and humidified substance found in boggy land and is considered to be the first stage in the formation of coal. It is not used as a fuel for power generation because of its low heat value, high moisture content and difficulty of handling and storing.

(*b*) *Lignite.* It is an intermediate product between peat and bituminous coal and has a brown fibrous structure. Because of its high moisture content (30 to 50%) and low heat value it is not especially suited for use in efficient generation of steam. Lignite when exposed to the air disintegrates into the small flakes, having the texture of crumbled clay. Due to this characteristic it can be stored for only a short time and is shipped with some difficulty.

(*c*) *Bituminous Coal.* After lignite the next stage of coal formation is bituminous coal. It contains very little moisture content 4 to 6% and a high carbon content (75 to 90%). This is the most commonly used fuel for power generation. Bituminous covers a variety of coal which possess but few distinguishing characteristics making a classification difficult. They are either 'caking' or 'non-caking' the former having a property of forming a fuse mass of coke when burned while the latter burns freely, that its without fusing action. The caking coal is however very useful for gas manufacturing purposes. While the non-caking variety is mostly used for steam raising purposes hence the term steam coal.

(*d*) *Anthracite.* It represents the final stage in coal formation and is exceptionally good for domestic purposes because of its smoothless combustion, low ash content and cleanliness in handling. It possesses a high calorific value (3500 kcal/kg gross value) and is therefore very valuable for steam raising purposes in power plants.

(*ii*) Liquid fuels. Liquid fuels are chiefly derived from natural petroleum (or crude oil) which is obtained from oil wells under the earth's crest in certain parts of the world such as Iraq, America, Russia and small quantities in India. Some of the important liquid fuels used for engineering purpose are Gasoline or Petrol, Paraffins, Diesel oil and Heavy fuel oil or Crude oil.

A brief information about their properties is given in the following Table 7.1.

Table 7.1

Fuel	Composition by Weight			Specific Gravity	Gross calorific value in kcal per kg
	C	H	S		
Gasoline or Petrol	85.4	14.6		0.74	11,200
Paraffine	86.3	13.6	0.1	0.79	11,100
Diesel oil	86.3	12.8	0.9	0.87	11,000
Heavy fuel oil	96.1	11.8	2.1	0.95	10,500

A few advantages of liquid fuels are given below :

- Higher calorific value.
- Require less space.
- Do not injure the grate bars like solid fuels.
- Cleanliness.
- Consumption is easily controllable.

***(iii)* Gaseous fuels.** Brief description of some gaseous fuels which are commonly used is given as follows.

(*a*) *Natural Gas.* It is nearly odourless and colourless which accumulates in the upper parts of oil gas wells. It consists of methane with varying quantities of heavier hydrocarbons and some non-combustible constituents such as carbon dioxide, Nitrogen and water vapour. It has a high thermal value varying from 4500 to 4950 kcal/m^3 and is used for both domestic and industrial heating purposes.

(*b*) *Liquified Petroleum Gas.* During the processing of petroleum a good amount of propane and Butane are produced. These gases are easily liquified by applying moderate pressure. Their flexibility of handling and storage makes them suitable for their use as fuel for stoves, trucks, buses and tractors etc.

(*c*) *Refinery Oil Gas.* During refinery operations of crude oil some light gases are produced which cannot economically be converted into gasoline or some other liquid product. These gases are called as refinery gases and consists of light hydrocarbons with small percentage of Hydrogen and Carbon Monoxide. They are generally used within the plant to help supply fuel for the many furnaces and heaters required in refinery operation.

(*d*) *Coal Gas.* It is obtained by the carbonization of coal and chiefly consists of hydrogen, Carbon Monoxide and various hydrocarbons. The quantity of the gas depends upon the coal used, the temp. of carbonization and the type of plant. It is generally used in gas fired boilers and for commercial purposes.

(*e*) *Producer Gas.* It is produced in a gas producer from the incomplete combustion of coke or charcoal in a current of air. Because of the large percentage of Nitrogen appearing in 'Producer Gas' resulting from the air fed to producer, the heating value of the gas is law. The average heating value is 1100 kcal/m^3. It is used extensively for firing open hearth furnaces in steel industry.

(*f*) *Blast Furnace Gas.* It is one of the by products of the blast furnace operation. It contains a large percentage of Nitrogen and the lowest heating value of any of the commercial gases.

(*g*) *Blue Water Gas.* It is produced in a generator by passing steam through red hot coke. It consists of approximately 50% Hydrogen 40% Carbon Monoxide and 10% Carbon Dioxide. The heating value of blue gas is approximately 2700 kcal/m^3.

(*h*) *Carburatted Water Gas.* The manufactured gas produced most widely for commercial distributes is Carburatted Water gas. It is obtained by enriching the water gas with oil gas (or fuel oil) by spraying the liquid oil on to a checker work of heated fire brick, called a carburettor. Its Clorific value is usually 400 kcal/m^3.

Gaseous fuels posses a number of **advantages** over solid fuels, a few of which are given below :

- The gas may be easily piped into the furnace so that no physical handling of fuel is necessary.
- Gases are generally free from ash and other foreign matter and they burn completely.
- Complete combustion with no smoke can be obtained with a low percentage of excess air.

- The control of the length of flame and nature of flame (reducing or oxidising) is relatively easy.
- Gaseous fuels afford greater economy than coal, or oil in high temperature work.

7.8. SIMPLE VERTICAL BOILER

In a vertical boiler the shell, which is usually cylindrical is placed in the vertical position. A cross tube simple vertical boiler is illustrated in Fig. 7.1.

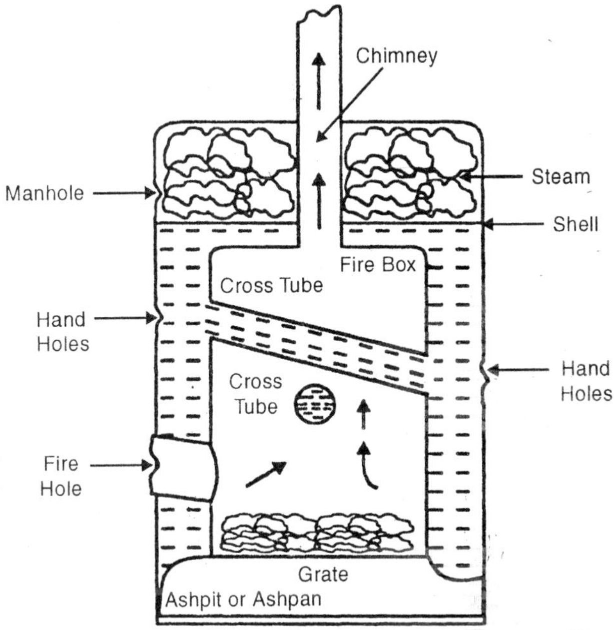

Fig. 7.1.

In this boiler the combustion of fuel takes place on the grate *G*, provided at the bottom made up of a number of bars of cast iron having air space between each pair. The firehole FH provides the passage through which coal may be introduced on to the grate. The ash left behind after the combustion of coal gets collected in the ashpit or ashpan. A from where it can be removed from time to time. The fire box FB is surrounded by the cylinderical shell *S* containing a substantial quantity of water that encloses the fire box. In order to improve the circulation of water and heating surface one or more cross-tubes *T* (also called gallway tubes) are fitted inside the fire box. As the hot products of combustion come in contact with the water, the latter is transferred into steam which gets collected in the upper portion of the shell. The hot gases then escape to the atmosphere through the chimney or stack *C* which passes from the crown of the fire box through the top of the shell.

The cross-tubes *T* are slightly inclined to facilitate the circulation of water by convection currents. The boiler may be cleaned at frequent intervals by a man who enters the boiler through the manhole M in the upper portion of the shell. The hand holes H are provided for cleaning the cross-tubes.

The working pressure in this boiler, which issued for small powers generally does not exceed 10 times that of atmospheric pressure. It has the advantages of being easily transported. It occupies less floor area and a heavy foundation is not required for this boiler. The ratio of heating surface to the grate area for a simple vertical boiler varies from 8 to 10.

7.9. COCHRAN BOILER

The cochran boiler is an improvement over the simple vertical boiler in which the heating surface is greatly increased by providing a large number of parallel tubes. Fig. 7.2 shows the brief outline of such a boiler. The combustion of coal takes

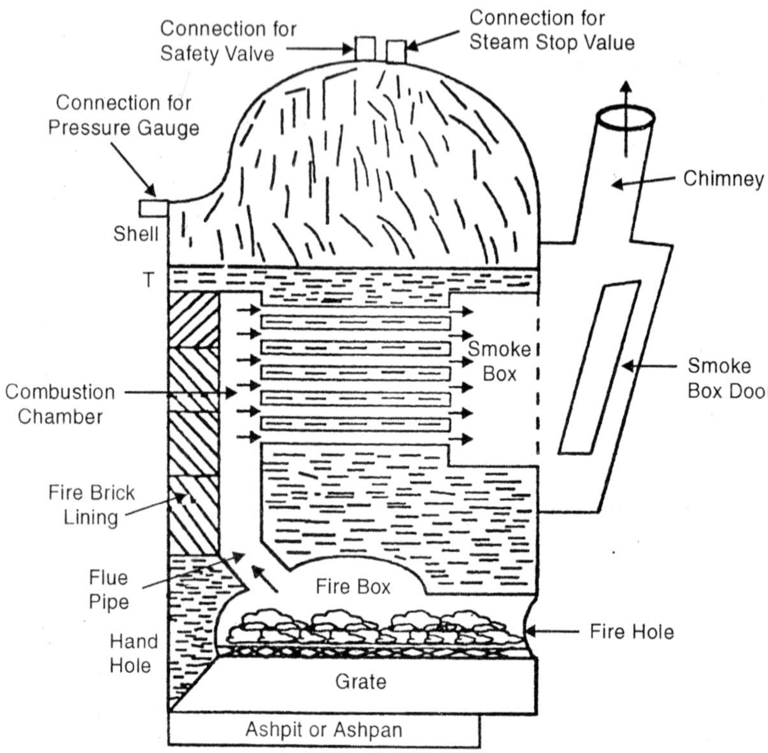

Fig. 7.2. Cochron boiler.

place on the grate (G) inside the fire box (FB) which is hemispherical in shape so that the unburnt fuel, if any, is deflected back to the grate. In order to ensure complete combustion the flue gases from the fire box are not allowed to pass directly into the fire tubes (T) but via the combustion chamber (CC) which is provided with a fire brick lining (FBL) to prevent the shell (S) from being damaged due to overheating by the flue gases. The latter then pass through a large number of horizontal fire tubes and in doing so convert water into steam which collects over the water space of the boiler from where it can be taken out through the connection for steam stop valve (CSSV). Finally, the flue gases escape to the atmosphere via the smoke box (B) and the chimney (C). The smoke box door (SBD) enables the cleaning and inspection of the smoke box and the fire tubes. The crown of the shell (S) is also hemispherical in shape. Hence, for a given weight of the plate material of the shell it has the maximum space and strength. At the top of the crown of the shell there is a man hole (M) through which a man can enter the boiler and clean it.

The coal can be introduced on the grate through the fire hole (FH). The boiler may also be oil fired in which case an oil burner will be fitted at the fire hole, there will be no grate and the connections of safety valve and pressure gauge are also shown in the diagram. This boiler is provided with usual mountings and is a self-contained unit. It is a made in sizes upto about 2.743 m diameter and 5.791 m height. Its maximum evaporative capacity from and at 100 °C is 4735 kg per hour.

The ratio of heating surface to grate area varies from 10 to 25 in this case. The steam may be generated upto a pressure of 10 kgf/cm^2. It also requires a smaller floor area and is the most popular form amongst the vertical boilers. The design is quite compact possessing a great external and internal accessibility.

7.10. CONSTRUCTIONAL FEATURES OF LANCASHIRE LOCOMOTIVE NESTLER AND BABCOCK AND WILCOX BOILERS

7.10.1. Lancashire Boiler

It is a fire tube type horizontal boiler. As shown in Fig. 7.3, it consists of a horizontal cylindrical shell usually 2 to 3 metres in dia. and 8 to 9 metres in length. The cylindrical shell contains two tubes *A* and *B*, called flue tubes, which are symmetrical in dimension and which run parallel to the axis of the shell. The flue tube are tapered at their back end, just to provide sufficient space for fitting work at this end. Each flue tube contains a furnace at its front end. There is grate '*G*' over which the fuel is burnt and a vertical fire brick wall, called bridge wall, is built at the back of each grate. The purpose of providing this bridge wall at the back of each grate is to prevent the flow of ashes alongwith the flue gases and to deflect the flue gases, so that the heat of fuel is more uniformly transmitted to the water, surrounding the flue tubes. The boiler is set in fire brick work.

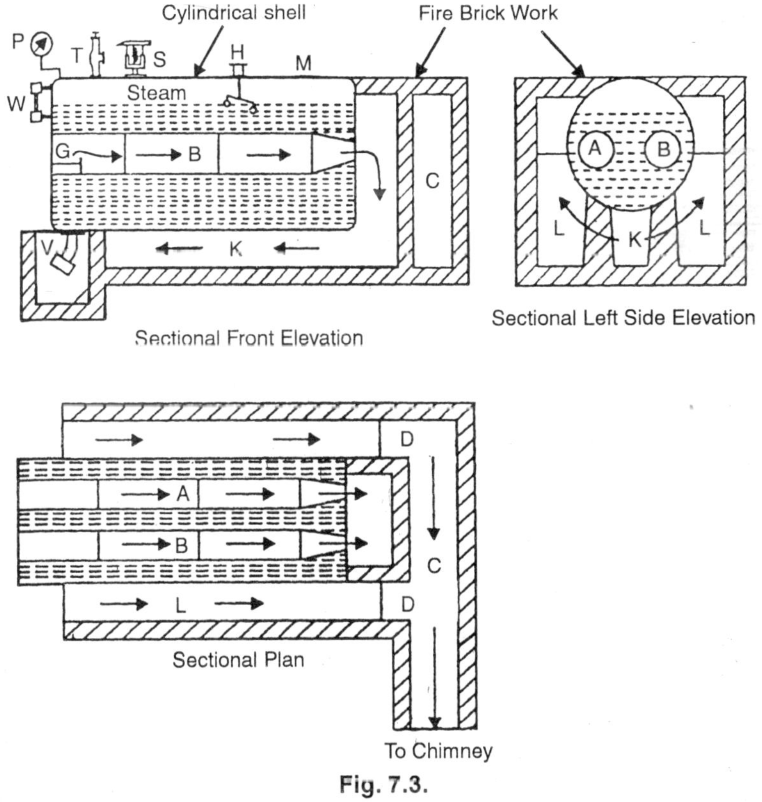

Fig. 7.3.

The flue gases pass from the furnaces over the bridge walls and then along the flue tubes. *A* and *B* to the back end of the boiler, where they dip into the bottom, external flue path *K*. Flowing along the bottom flue path, the flue gases reach the front end of the boiler, where they become divided, half of these going along each of the side flue path *L*. Thus, flowing along the side flue paths, the flue gases again reach the back end of the boiler from where they are led into a common passage C, which leads them to the chimney. The path of flow of the flue gases has been shown by arrows in Fig. 7.3.

A damper D is provided in each side flue path L to control the draught. The objective of draught is to ensure complete combustion of fuel by providing sufficient quantity of air through the fuel. Due to the continuous flow of flue gases, the water surrounding the flue tubes becomes more and more heated and ultimately it reaches its boiling point temperature and steam begins to be produced. The steam thus generated, collects in the steam space within the boiler.

In Fig. 7.3, P is the pressure gauge, W the water level indicator, S the safety valve, H the high steam and low water safety valve. T is the steam stop-valve, v the blow-off cock and M the manhole. It should be noted that Lancashire boiler has a very good steaming capacity and also the coal of inferior quality can be used in it, without any operational constrain.

7.10.2. Locomotive Boiler

The constructional features of locomotive boiler, which is a horizontal, fire tube boiler are shown in the Fig. 7.4. It consists of a large cylindrical shell, which has a fire box at its one end and a smoke box at the other end as shown in the figure. The fire box and smoke box are connected by a number of horizontal small diameter tubes, which are surrounded by water to be evaporated.

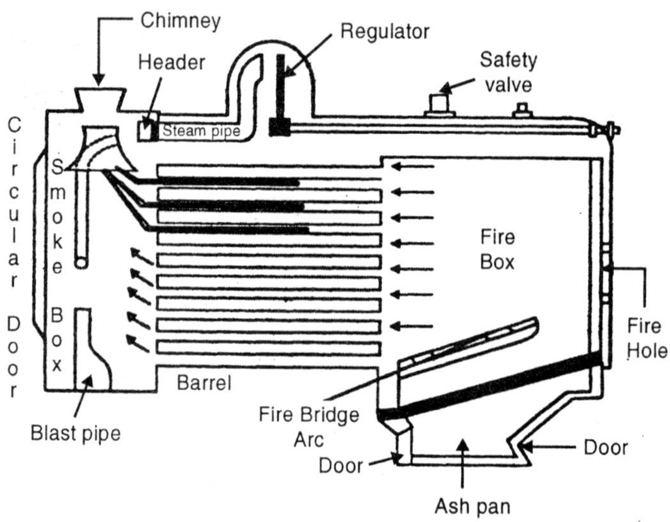

Fig. 7.4.

The hot products of combustion from the fire box pass through the horizontal tubes to the smoke box from where these are led to the chimney. The grate is made inclined towards the front and a fire brick arch is provided over here. It helps in deflecting the products of combustion so that they come in contact with the whole heating area of the fire box. Dampers are also provided below the grate which controls the supply of air for combustion.

The required draught is produced by the exhaust steam from the cylinders of the engine. The exhaust steam is discharged through the blast pipe, provided at the base of the smoke box. This jet of discharged steam from the blast pipe creates a partial vacuum in the smoke box as it drags hot gases alongwith it to the chimney. In case the engine is not operating, the required draught is obtained by means of a blower (not shown).

A regulator, which is a stop valve, is provided in a dome shaped chamber, mounted on the shell of the boiler. It is controlled by means of a rod. It regulates the supply of steam through a pipe to the header. This header has two portions which are saturated steam chambers and superheated steam chamber. The steam from saturated chamber, which is collected from the regulator, is passed through the superheated chamber through superheated pipes, from where it is given to the engine cylinder through piping.

For efficient and safe operation of the boiler, the various mountings and accessories are also provided on it.

7.10.3. Nestler Boiler

The constructional features of Nestler boiler, which is a fire tube boiler are shown in Fig. 7.5. It consists of a horizontal main shell, and in its lower part, it contains a circular furnace tube. Outer end of the furnace tube is provided with a burner, whereas its other end opens into a lower back chamber. This chamber is connected to front chamber through a number of horizontal tubes.

The front chamber is further connected to the upper back chamber through another set of horizontal tubes, as shown in the figure. The burner is supplied with oil through suitable pipe line because it is an oil fired boiler.

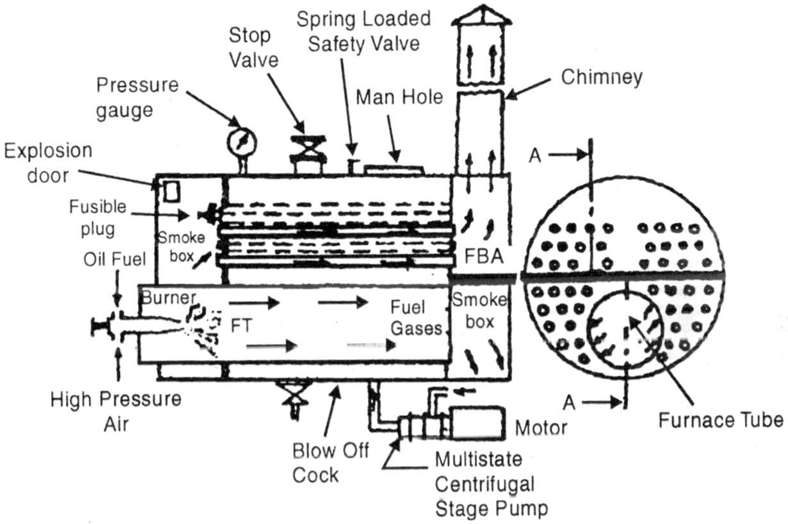

Fig. 7.5.

7.10.4. Babcock and Wilcox Boiler

The constructional features of this boiler, which is a water tube boiler are shown in Fig. 7.6. It is the boiler which is externally fired and suitable where large quantity of steam is the main requirement.

It consists of a cylindrical steam and water drum which is connected to two headers at the front and back ends. These headers are connected by a large number of water tubes, having about 10 cm bore and about one cm thickness. These tubes are inclined at an angle of about 15° with the horizontal. A mud box is provided at the bottom of the back and header, which may be cleaned whenever required. Hand hole is provided in the header, in front of each tube.

The hot products of combustion are made to flow up and down over the inclined nest of tubes, before they are allowed to pass to the chimney. It is achieved by providing baffles across the tubes. The draught is regulated by means of a damper.

When the water in the tubes is heated up, it starts rising into the upper header and enters into the boiler. Cold water from boiler flows to the lower header to take the place of water already entered into the boiler. This way the circulation of water is achieved. Steam formed is separated out in the drum.

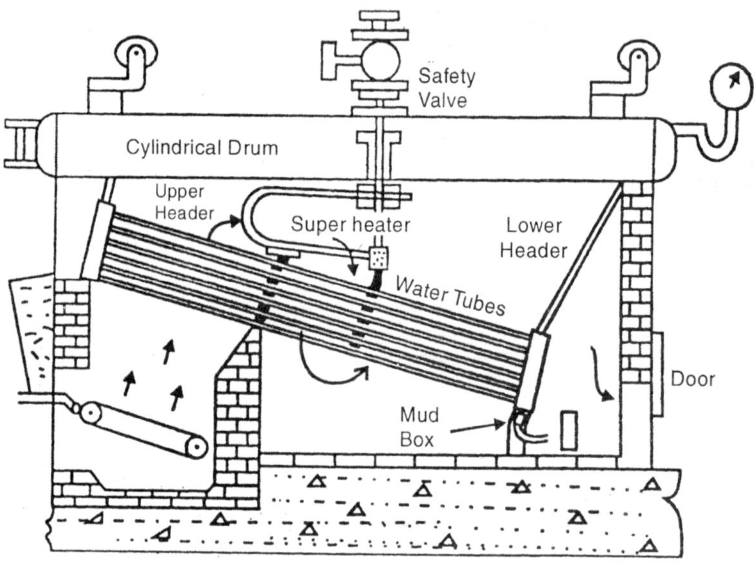

Fig. 7.6.

A superheated consisting of a number of U-tubes is also provided in the combustion chamber, as its integral part.

Various types of mountings and accessories are provided for the efficient and safe working of the boiler. It may be noted that dimensions of the drum vary from 1 m to 2 m in diameter and length from 6 m to 9 m.

7.11. INTRODUCTION TO MODERN BOILER

7.11.1. Forced Circulation Boilers

The boiler in which the circulation of water is made under pressure, is known as *'forced circulation boiler'*.

It has the following advantages over conventional boilers :

(*i*) Increased rate of heat transfer.

(*ii*) Smaller diameter tubes can be used thus minimizing overall weight.

(*iii*) Number of boiler drums required may be minimised.

(*iv*) In this, the scaling problem is also minimised.

(*v*) If external source of power is available, it can be rapidly started.

(*vi*) Fluctuation of load can be easily accommodated, without the use of any complicated control devices.

(*vii*) In this system the danger of overheating is minimised.

(*viii*) It is comparatively light in weight for the given output of boiler.

Some of the commonly used forced circulation boilers are explained below :

7.11.2. La Mont Boiler

This type of boiler is shown in Fig. 7.7. It is a water tube, forced circulation boiler.

The feed water is supplied to the boiler through on economizer. From the boiler drum, the water is circulated through the evaporator tubes, by means of a circulating pump, which develops a pressure of about 2.5 times that of the atmospheric pressure. Also, the wet steam is superheated by passing it through the superheater coils. One air preheater is also provided with this type of boiler to preheat the air to be supplied to the boiler furnace.

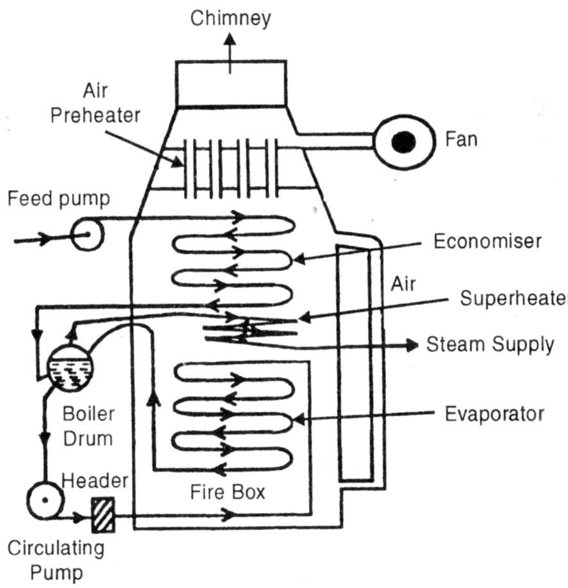

Fig. 7.7.

7.11.3. Benson Boiler

This type of boiler is shown in Fig. 7.8. It is also a water tube, forced circulation boiler.

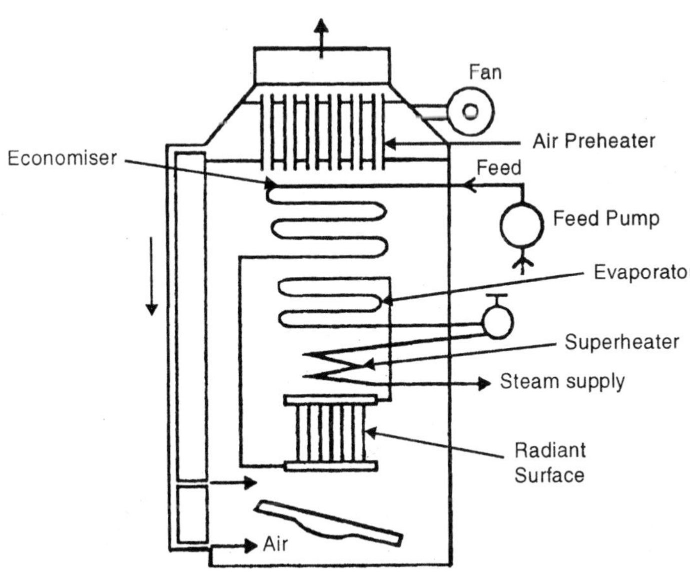

Fig. 7.8.

In 1992, Mark Benson suggested, that if the boiler pressure is raised to the critical pressure, then there is no formation of steam bubbles, because the steam and water at this pressure will have same density.

Mark Benson made this suggestion on the basis of this attempts on how to prevent steam bubble formation, as these weaken the rate of heat-transfer. In this boiler the feed water is supplied to the boiler at critical pressure, because at this pressure water will be directly converted into superheated steam, as the latent heat at critical pressure is zero.

It has been observed that this boiler gives max. efficiency if operated at a pressure slightly below the critical pressure.

7.11.4. Loeffler Boiler

This type of boiler is shown in Fig. 7.9. It is also a water tube forced circulation boiler.

In this boiler the feed water is evaporated by the heat of superheated steam. The steam produced in the evaporator drum is circulated through the superheater coils, where it is superheated by the heat produced by the combustion of fuel in the furnace. About two third portion of the superheated steam is made to pass through the evaporator drum and the remaining portion is supplied to the turbine etc.

The high pressure helps in improving the rate of heat transfer. To improve the boiler efficiency, still further economizer and air preheaters are also provided with this type of boiler.

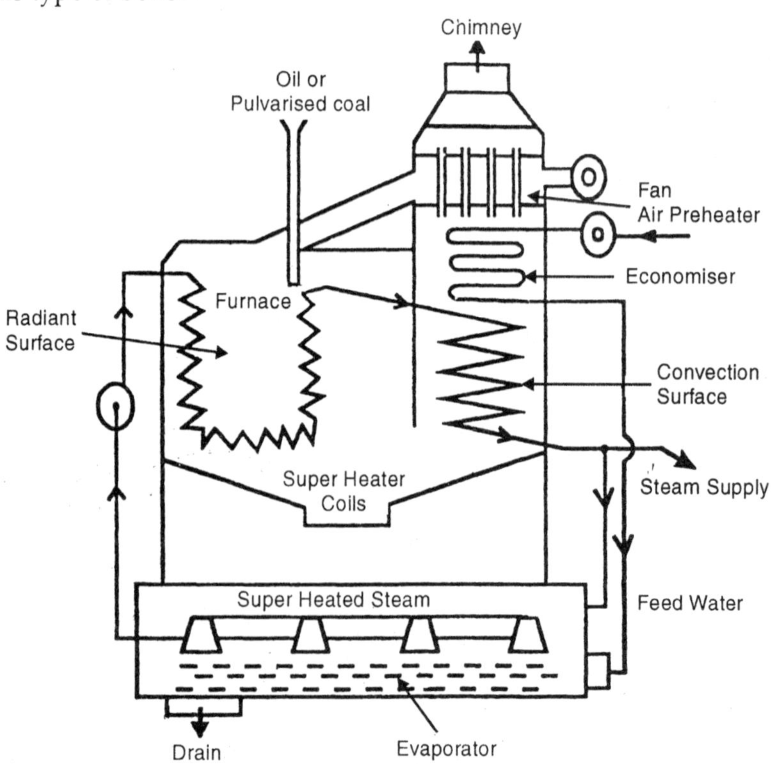

Fig. 7.9.

7.11.5. Fluidised Bed Combustion Boiler

The constructional features of this type of boiler are shown in Fig. 7.10. It is a water tube boiler. The speciality of this type of boiler is that in this boiler cheaper and easily available fuels like rice husk, saw dust or ground nut shell etc. can be used inspite of coal or coke that are being used in almost all type of conventional boilers.

The fluidised bed consists of a rectangular box, bottom of which is the distributor plate. Air nozzles are arranged in a specific order on this distributor plate. This plate also has provisions for the fuel nozzles through which the fuel is admitted pneumatically from below the fluidised bed in such a way, so as to cover the length and breadth of the furnace resulting in optimum combustion.

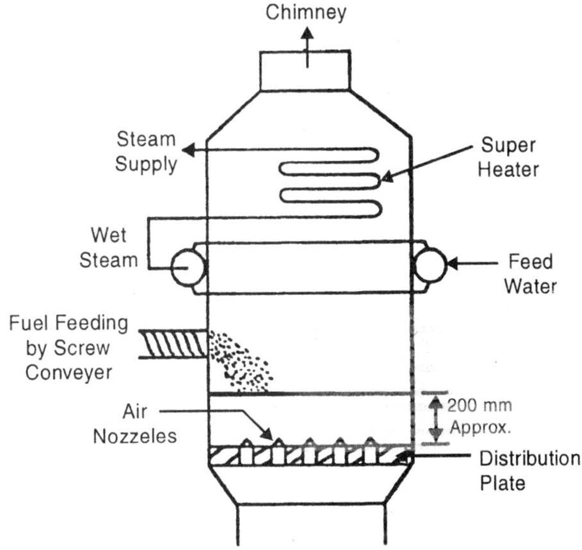

Fig. 7.10.

In some cases the fuel may be admitted over the bed with the help of screw conveyer as shown also inspite of at the bottom of bed. The bottom of the distributor plate consists of air box through which high pressure air from forced-draft fan is admitted into the fluidised bed via the air nozzles.

The temperature of the fluidised bed which is normally 800 to 860°C is controlled by immersing the water bed coils, which also helps in steam generation. The flue gas emerging from the bed then passes through the convective zone of the boiler, the heat recovery system, the dust collection system and finally the induced draft fan before finally passing into the atmosphere through the chimney.

7.12. BOILER MOUNTINGS

Boiler mountings are the components generally mounted on the surface of the boiler to accomplish safety during operation and the complete control of steam generation process.

These are the essential parts of the boiler, without which the boiler operation is not possible. The following are the various important mountings of a boiler :

(*i*) Water level indicator

(*ii*) Pressure gauge

(*iii*) Safety valve

(*iv*) Junction or stop valve

(*v*) Feed check valve

(*vi*) Blow-off-cock

(*vii*) Fusible plug

7.12.1. Water Level Indicator

It is the device which indicates exact level of water in the boiler. A glass tube is fitted on it, to make the water level inside the boiler visible from the boiler room floor. A common type of glass tube water level indicator is shown in Fig. 7.11.

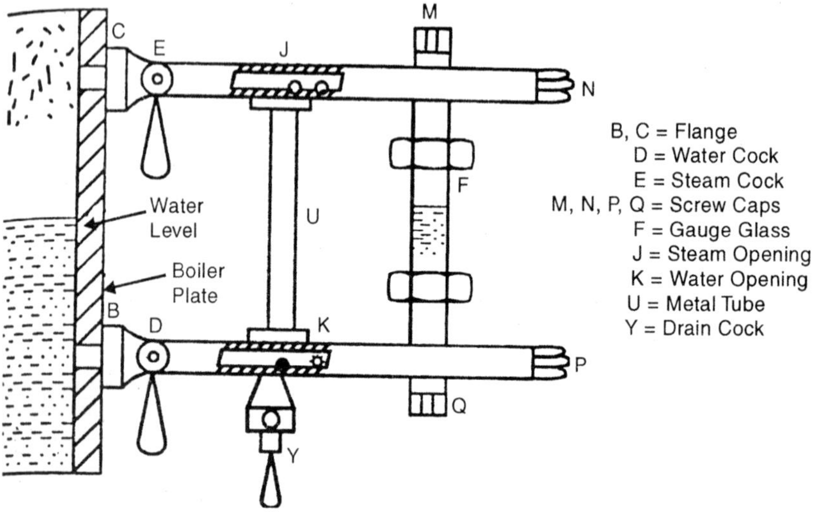

Fig. 7.11.

It is connected to the boiler plate through stuffing boxes in hollow gun metal casting having flanges for bolting to the plate. There are two cocks for controlling the passages of water and steam from the boiler. When these cocks are opened the water stands in the glass tube at the same level as inside the boiler. In case the glass tube gets broken the rush of water from bottom passage and steam from top passage carries the balls in the position shown by dotted circles and shuts off the water and steam. Then the attendant can safely close the cocks and replace the broken glass tube.

7.12.2. Pressure Gauge

This gauge is usually mounted on the front top of the boiler shell. It is mounted on the boiler to show the pressure of steam inside the boiler. Its dial is graduated to read the pressure in different units of measurement. Bourdon's pressure gauge as shown in Fig. 7.12. is commonly used.

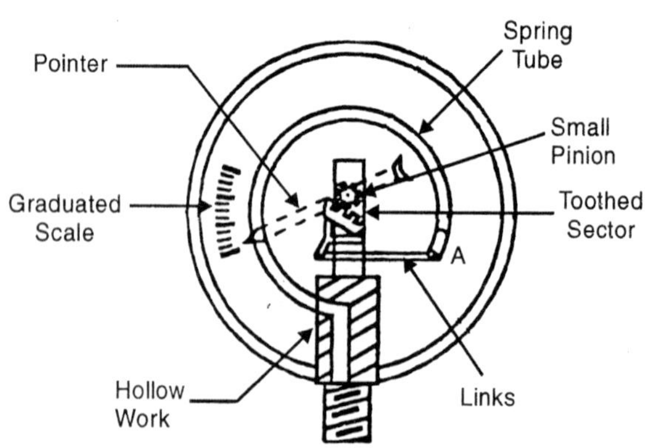

Fig. 7.12.

The essential element of this gauge is the elliptical spring tube, which is generally made of bronze and is solid drawn. One end of this tube is attached by links to a toothed quadrant which in turn meshes with a small pinion on the central spindle. The other end is connected to steam space. When pressure is applied to the interior of the elliptical tube it tries to straighten up and causes the pointer to move along the scale and indicates the pressure on graduated dial.

7.12.3. Safety Valves

All the boilers except the small boilers should have at least two safety valves which will open, when the boiler pressure exceeds the working pressure. Safely valves are classified into two types according to the method of loading the valve *i.e.*

(*i*) Weight loaded

(*ii*) Spring loaded

The weight loaded safety valves are further of following two types :

(*a*) Dead weight safety valves.

(*b*) Lever and weight types of safety valves.

7.12.3.1. Dead Weight Safety Valve

In this type of safety valve, the valve is loaded by the direct application of dead weight as shown in the Fig. 7.13.

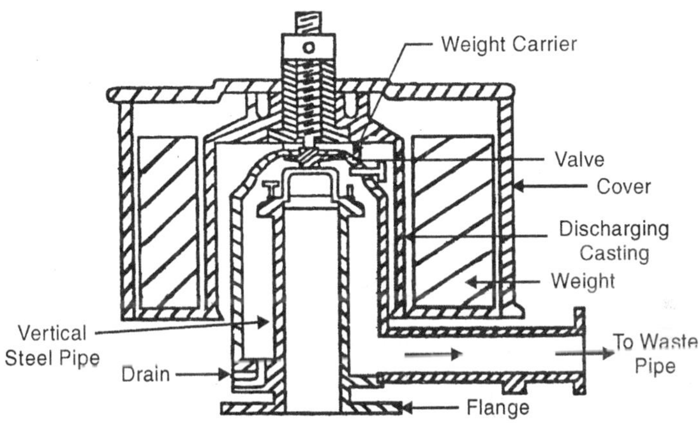

Fig. 7.13. Dead weight safety valve.

In this safety valve the value rests on the valve seat which is fixed on the top of the vertical steel pipe by a ring and screws. The dead weight and the weight of the valve itself balances the pressure on the valve. When the pressure becomes too high, the valve and weight will lift and the surplus steam will escape and it is carried to the waste pipe. Thereafter, the valve again comes down on its seat.

7.12.3.2. Lever Safety Valve

This type of safety valve is suitable for stationary boilers. Lever safety valve is shown in the Fig. 7.14.

The valve is loaded by means of weighted lever which is pivoted about a fulcrum situated close to the valve. The valve load depends

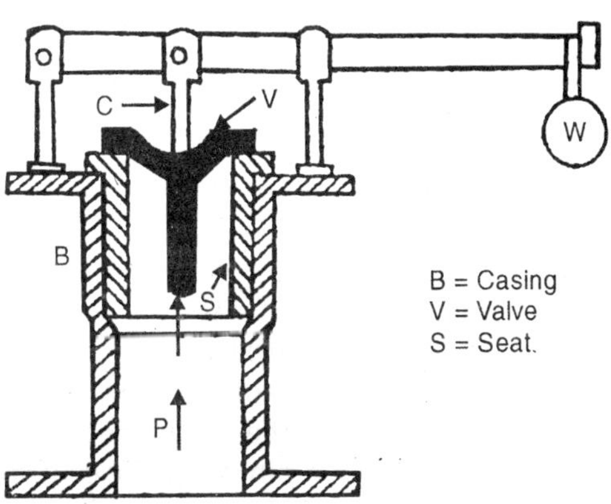

Fig. 7.14. Lever safety valve.

upon the lever ratio. Its working and function is the same as discussed in case of dead weight safety valve.

7.12.3.3. Spring Loaded Safety Valve

Spring loaded safety valves are commonly used on locomotive and marine boilers because this type of safety valve is not affected by movement and vibrations. It is shown in the Fig. 7.15.

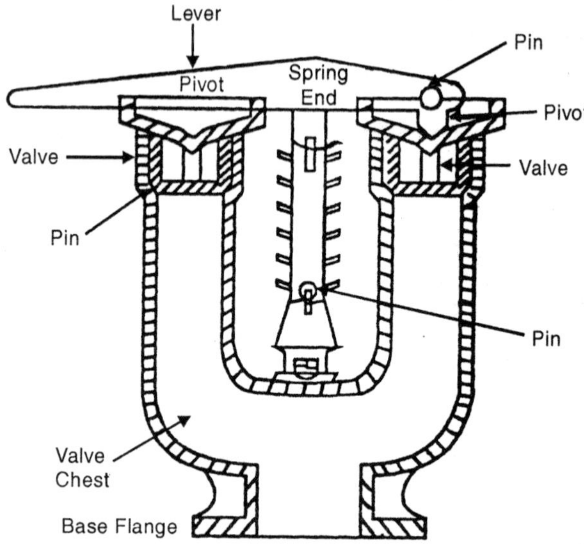

Fig. 7.15. Spring loaded safety valve.

Spring loaded safety valve consists of two separate valves and seatings. The valves are held down on their seats by the helical spring and lever. The spring is hooked to an arm of the lever midway between the valves. The lower end of the spring is hooked to the shackle which is secured to the valve chest by studs and nuts.

Its disadvantage is that the load on the valves increases as the valve lifts where the steam pressure exceeds the limit. So the pressure required to just lift the valve is less than required to open it fully.

7.12.3.4. Combined High Steam and Low Water Safety Valve

This type of safety valve is also called compound safety valve. It is suitable for stationary boilers.

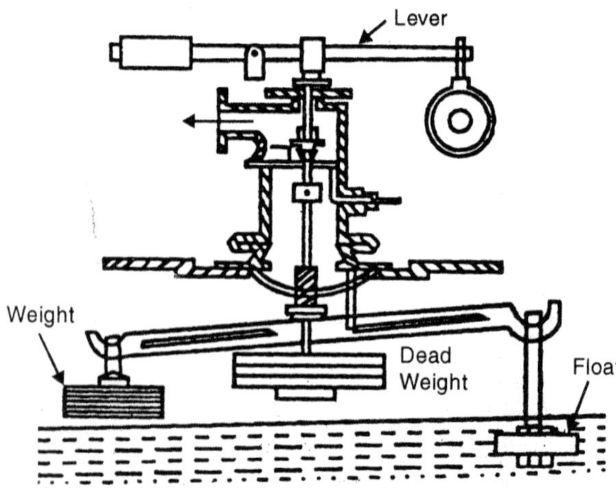

Fig. 7.16. Combined high steam and low water safety valve.

The principle of working of this safety valve is also the same as lever safety elements are present in this type of safety valve. Constructional view of this type of safety valve is shown in Fig. 7.16.

The first safety element consists of high pressure steam valve held down on its seat partly by the dead weights and partly by lever and weight. The second safety element consists of a float device which gives warning of the near approach of dangerously lower water level and then prevents the collapse of the furnace tubes.

7.12.4. Junction Valve or Stop Valve

A valve placed directly on a boiler and connected to the steam pipe which carries steam to the engine or turbine is called junction valve. This valve is necessary for shunting off steam when not required. It is also called as stop valve. A common type of junction valve is shown in Fig. 7.17.

The valve spindle passes through a gland and stuffing box fixed in the cover of the valve body. The upper portion of the spindle is threaded. By turning the hand wheel fitted on the spindle, the spindle is raised or lowered and thus the valve disc is raised or lowered to allow the flow of steam or to stop the flow of steam.

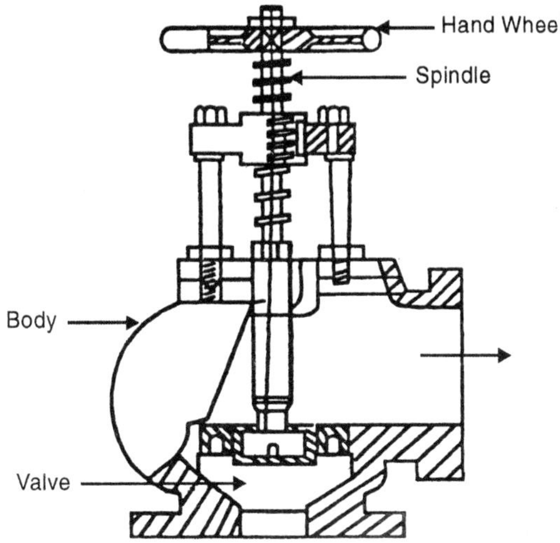

Fig. 7.17. Steam stop valve or junction valve.

7.12.5. Feed Check Valve

A valve placed at the boiler end of the feed pipe to control or regulate the flow of water is called the feed check valve. It is a non-return valve which permits flow of water in one direction only and automatically prevents the back flow of water from the boiler when the feed pump is not working. The constructional detail of the feed check valve is shown in Fig. 7.18.

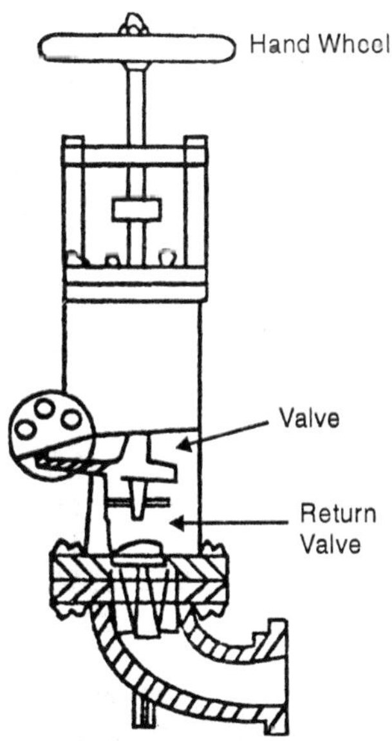

Fig. 7.18. Feed check valve.

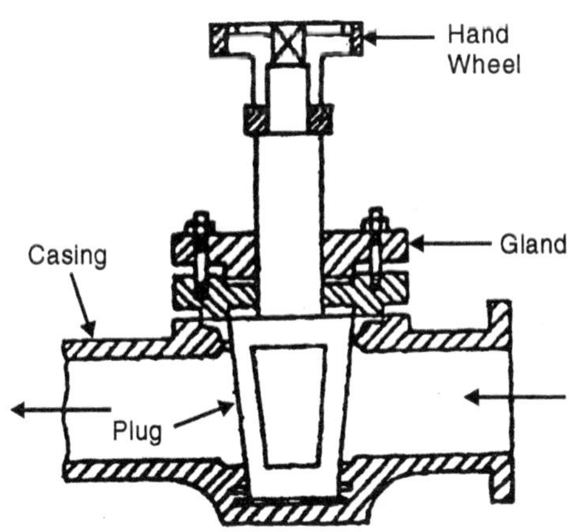

Fig. 7.19. Blow-off-cock

7.12.6. Blow-off-cock

The constructional detail and self-explanatory working diagram is shown in Fig. 7.19.

The blow-off-cock is attached at the lowest joint of every boiler for the purpose of emptying the boiler when necessary and for discharging mud etc. deposited from the feed water at the lowest point.

7.12.7. Fusible Plug

It is the mounting fitted at the crown of the boiler as shown in the Fig. 7.20.

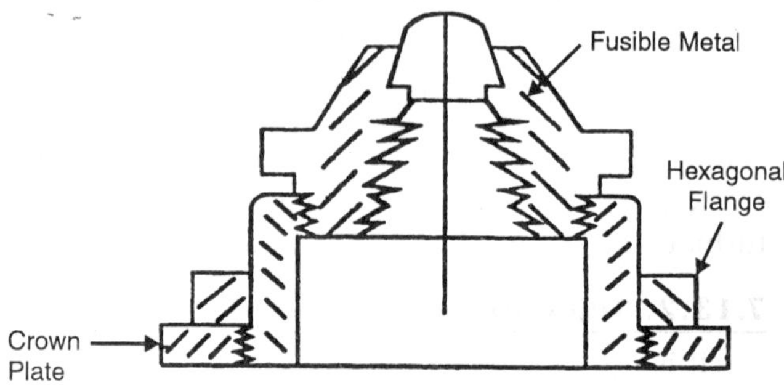

Fig. 7.20. Fusible plug.

The fusible plug under normal working of a boiler is covered with water which keeps the temperature of the fusible metal below its melting point. But when the water level in the boiler falls too low to uncover the top of the plug the fusible metal quickly melts, the plug drop out and the opening so made allows the steam to rush into the furnace. The steam in this way puts-off the fire or gives warning that the crown of the boiler furnace is in danger of being overheated.

7.13. BOILER ACCESSORIES

Boiler accessories are those appliances which are installed with a boiler in its neighbouring area to increase the efficiency of the boiler.

These are not the essential parts of the boiler, so without these the boiler operation can be accomplished although at a lower efficiency. The following are the important boiler accessories :

(*i*) Feed water pump
(*ii*) Injector
(*iii*) Economiser
(*iv*) Super heater
(*v*) Air-preheater
(*vi*) Pressure-reducing valve
(*vii*) Steam drier or separator
(*viii*) Steam trap

7.13.1. Feed Water Pump

It is used to deliver feed-water to the boiler. It may be either reciprocating or rotary pump. One type of reciprocating feed water pump has been shown in Fig. 7.21.

In this type of pump there are simple steam engine cylinder and pump cylinder placed side by side. The working of engine cylinder is similar to the working of steam engine. The pump delivers water on each forward and backward stroke, being double acting. There are suction and discharge valves for each side of the pump plunger. The two pumps work alternately, so a continuous flow of water is kept practically.

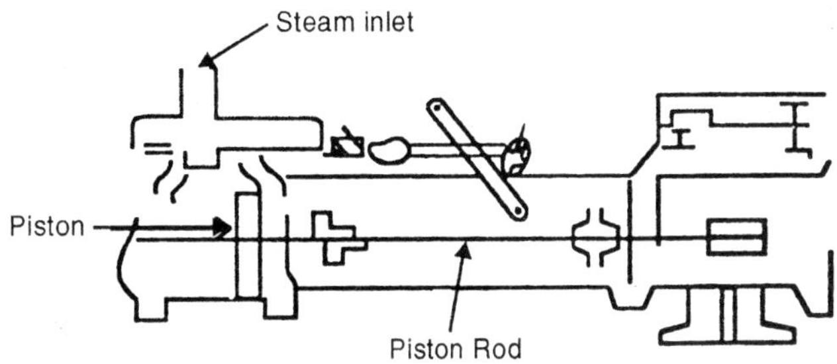

Fig. 7.21. Feed water pump.

Rotary feed pumps are of centrifugal type, driven directly by a small steam turbine or by an electric motor. These pumps are also known as turbine pumps.

7.13.2. Injector

It is used in small boilers and locomotive boilers. It can also be fitted as a stand by or as a reserve feed water pump owing to its cheapness and simplicity. It is very simple in working because there are no moving parts or plungers in it. The water is forced into the boiler by the action of steam flowing through a tapered nozzle as shown in the Fig. 7.22.

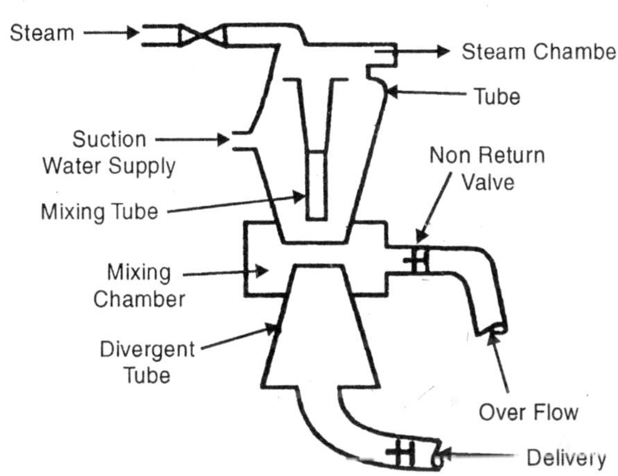

Fig. 7.22. Steam injector.

The injector is more economical because whole of the heat of the steam utilized in injector is returned to the boiler. The injector, therefore, works as a feed water heater as well as feed water pump. The only disadvantage of this is that, these are not very reliable. Therefore, these are not used in large boilers or marine boilers where large quantity of feed water is required to be pumped. In the working of the injector the steam is admitted through the steam pipe and discharged from the nozzle, and it mixes with the water contained in mixing tube, where condensation of steam takes place. The jet then passes through the mixing chamber and enters the divergent tube which reduces the velocity and increases the pressure of water. Due to this increase of pressure the non-return value fitted in the delivery pipe is opened and the water enters into the boiler.

7.13.3. Economizer

A considerable heat of the plant is carried away by the flue-gases. So, to utilize this heat, an economizer is placed in the path of the flue gases to pre-heat the feed water. It consists of a number of horizontal or vertical tubes through which the feed water from the pump is passed. The hot gases pass over the external surface of water tubes. The economiser shown in Fig. 7.23 is of vertical type.

It is fitted with scraper which moves up and down the water tubes by the mechanical drive and in this way the exterior surface of the tube are kept free from root deposits. The advantages of economisers are :

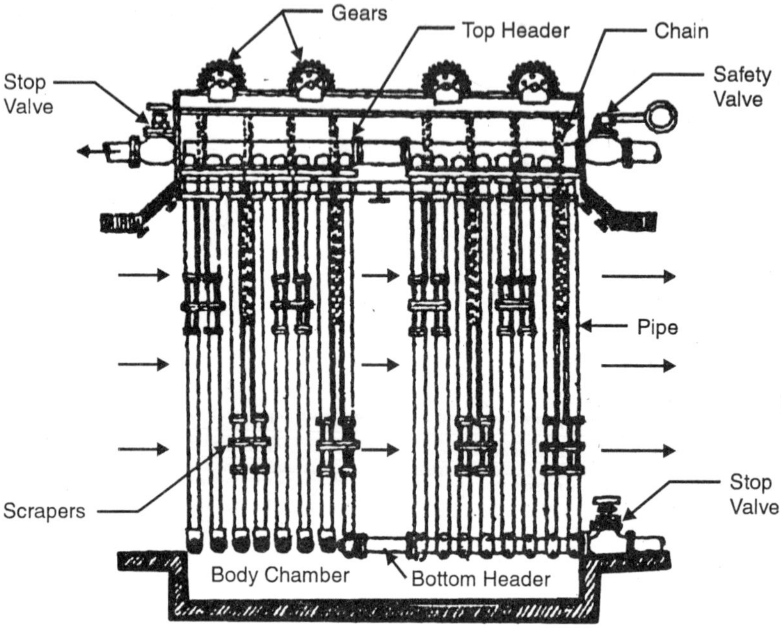

Fig. 7.23. Economiser.

(*i*) It increases the heating surface of the boiler and therefore evaporative capacity of the boiler is increased.

(*ii*) The temperature difference between different parts of the boiler is reduced and, therefore, stresses due to unequal expansion are reduced.

(*iii*) The heat carried away by the flue gases is taken up by the economiser which results in saving of heat energy. This saving of heat energy results in an increase of the overall efficiently of the plant.

7.13.4. Superheater

Super heater used for stationary boiler is shown in Fig. 7.24.

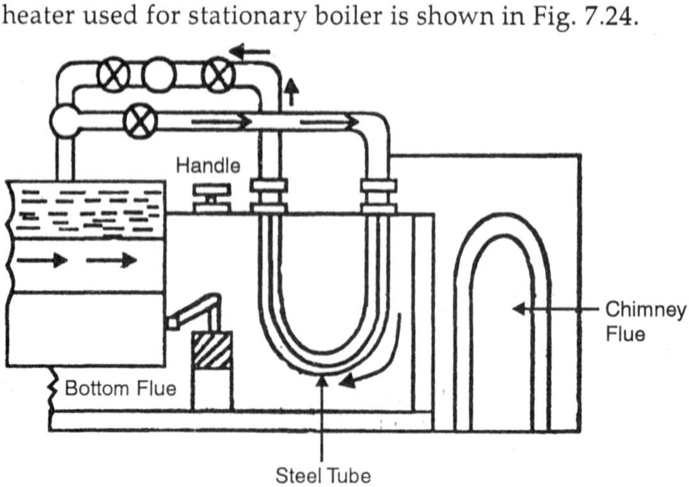

Fig. 7.24. Superheater.

The use of superheaters enables the wet steam to be completely dried and temperature is raised at constant pressure. The super heater is a nest of steel tubes bent to U-form and expanded into mild steel boxes called headers. These are generally located in the path of furnace gases so that heat is utilized by the super heater from the hot gases.

7.13.5. Air-preheater

The air entering the boiler furnace is at a lower temperature as compared to the temperature of exhaust flue gases. Instead of heating this air after it enters the furnace and take part in combustion, it will be desirable to heat this air to some extent. The air is heated by the heat carried away by the flue gases and going as a waste through the chimney.

It is situated between the economiser and the chimney. The plant efficiency is increased by the use of air pre-heater. Air pre-heaters are generally of three types :

1. Tubular type
2. Plate type
3. Regenerative type

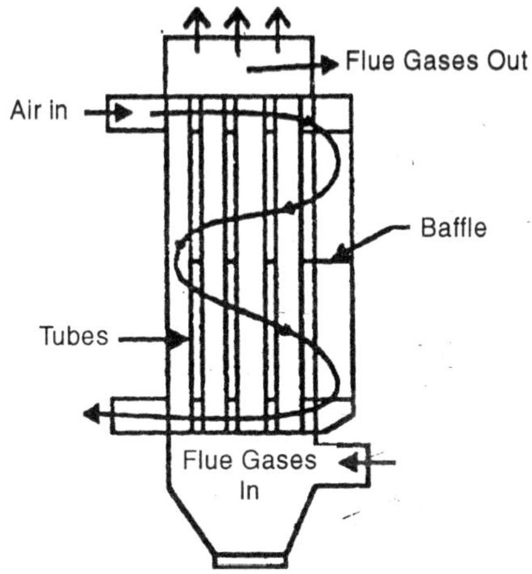

Fig. 7.25. Tubular air pre-heater.

A tubular type air pre-heater is shown in Fig. 7.25.

Air preheaters are effective and economically justified only in the large steam power plants where high capacity and high pressure are met with. They are not useful for small plants operating at low pressures.

7.13.6. Pressure-Reducing Valve

Sometimes steam is required at a lower pressure than that of the boiler for processing work in an industry or heating purpose. The high pressure steam may be passed through a pressure reducing valve which will maintain a constant reduced pressure on the delivery side.

Figure 7.26 shows one form of pressure reducing valve in which high pressure steam entering through left hand opening is throttled by the valve '*V*' to desired low pressure through the right hand opening.

The valve is connected to a loose fitting position (LFP) through the valve rod (VR). The flexible rubber diaphragm (FRD) prevents leakage past the piston. Since, the diameter of the piston is equal to the diameter of the valve, so upward pressure of steam on the valve is balanced by its downward pressure on the piston. When steam is flowing through the reducing valve, the pressure of reduced pressure steam on the top of the valve balanced by the force of the spring. The force exerted by the spring may be adjusted to the desired value with the help of adjusting screw.

Any rise or fall of steam pressure at inlet will be followed by a momentary rise or fall of pressure on the top of the valve, which being forced upward by the constant force of the spring, will fall or rise till the steam pressure above the valve becomes the pressure to which the spring is adjusted.

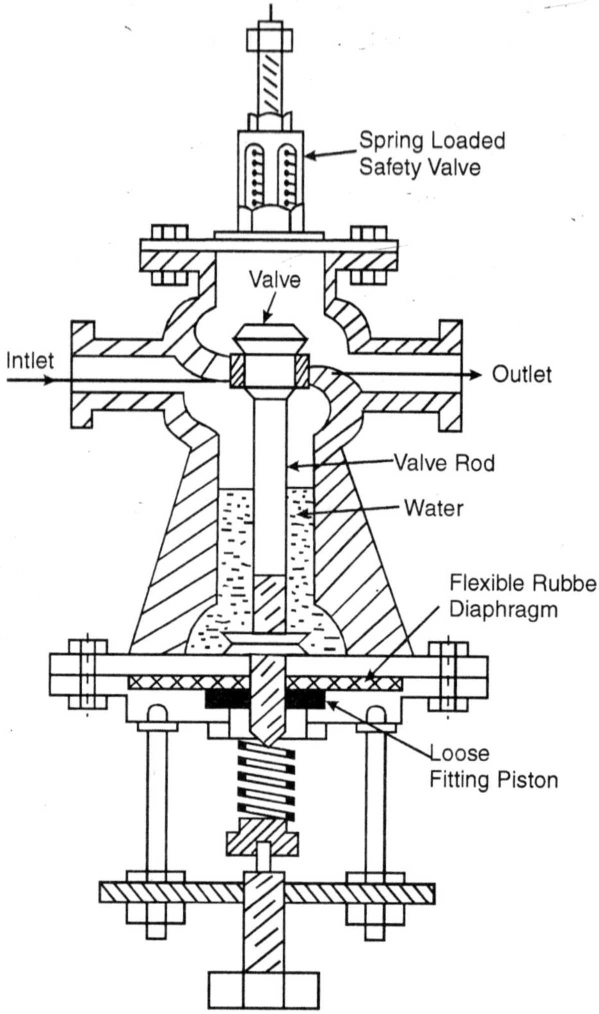

Fig. 7.26. Pressure reducing valve.

7.13.7. Steam Drier or Separator

The function of a drier or steam separator is to separate water particles from steam before it is supplied to a steam engine or turbine. It is provided as close as possible to the engine in the main steam pipe line leading from the boiler to the engine.

A simple form of steam separator is shown in Fig. 7.27.

It consists of a cylindrical vessel (CV) fitted with baffle plates (BP). The steam striking the battle plate is suddenly deflected so that its direction of flow is changed. Consequently the water particles, due to their greater weight and interia after stricking with baffle plates fall to the bottom of the separator. The water gauge indicates the level of water collected at the bottom of separator. The drain cock (DC) fitted at the bottom of the separator helps to drain away the water separator from incoming steam. The steam free from water particles goes to the engine through the outlet pipe as shown in the Fig. 7.27.

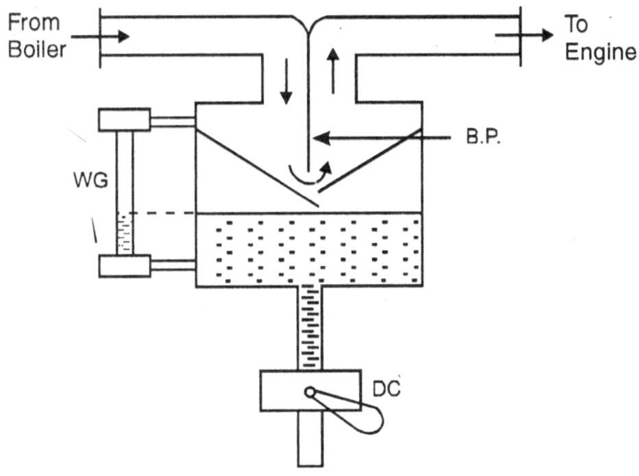

Fig. 7.27. Steam drier or separator.

7.13.8. Steam Trap

Steam traps are devices used to automatically drain away or return to the boiler, the water collected in the steam pipe or steam jacket as a result of partial, condensation of steam, without allowing the steam to escape.

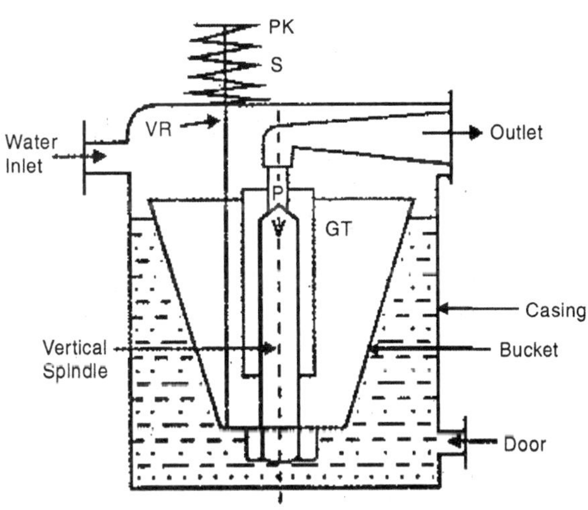

Fig. 7.28. Steam Trap.

Figure 7.28 shows a common type of steam trap known as bucket steam trap. It consists of a casing (C) which contains sufficient water, to float the bucket (B) which carried a vertical spindle (VS). The valve (V) fitted to the top of the spindle fits in a pipe (P). The guide tube (GT) guides the vertical spindle (VS).

The steam trap is so located that the water formed due to condensation of steam in the steam pipe, flows into it by gravity. When the water rises high enough in the casing (C), it overflows into the bucket (B) which ultimately sinks down and the valve (V) opens. The water is then forced up through the guide tube (GT).

The door (D) provided at the lower end of the trap is used to remove periodically the sediments or dust deposited at the bottom of the casing. The trap can be tested at any time by blowing-off by pressing the knob by hand.

7.14. PERFORMANCE OF BOILERS

The evaporative capacity of a boiler may be expressed in several ways such as kg/f steam/hour, or kg of steam/hr/m^3 of heating surface or kg of steam/

hr/m^3 of furnace volume or kg of steam/kg of fuel fired. But different boilers work under different conditions of temperature and pressure and have different evaporative capacities. The feed water temperature and the condition of steam raised also vary from one boiler to another boiler. So a mere idea of the amount of water evaporated in a boiler is not sufficient to know its capacity in comparison to another boiler.

In order to compare the performance of different boilers it is therefore necessary that the water be supposed to be evaporated under standard conditions. The standard conditions adopted are temperature of the feed water and the temperature of steam, 100°C and steam dry and saturated. Under standard conditions therefore the evaporation of one kg of water means the utilization of L units of heat. Where L is the latent heat of steam at 100°C (at atmospheric pressure) and has a value 2256.9 KJ, which is generally taken as 2257 KJ for convenience.

7.14.1. Equivalent Evaporation

Equivalent evaporation of a boiler is defined as the weight of water evaporated from water at 100°C to dry and saturated steam at 100°C by utilising the same amount of heat as would have been used under the actual working conditions.

It may be expressed in kg/kg of fuel, or per hour, or per square meter of heating surface per hour and is generally denoted by 'W_e'.

Let, h = Total heat of steam at the working pressure

h_f = Sensible heat of feed water

h_{fg} = Latent heat of steam at atmospheric pressure and is equal to 2257 KJ

W_a = Weight of water actually evaporated, into steam per kg of fuel at the working pressure.

W_e = Equivalent evaporation in kg per kg of fuel.

Then, heat absorbed by water,

$$W_e \times h_{fg} = W_a (h - h_f)$$

or

$$W_e = \frac{W_a (h - h_f)}{h_{fg}} = \frac{W_a (h - h_f)}{2257} \qquad (\because h_{fg} = 2257 \text{ J})$$

where $h = h_f + xh_{fg}$ If steam is wet with dryness fraction 'x'

$= h_f + h_{fg}$ If steam is dry saturated

$= h_f + h_{fg} + C_p (t_s - t)$ If steam is superheated

7.14.2. Factor of Evaporation

It is a quantity which when multiplied by the amount of steam generated at a given pressure from water at given temperature gives the equivalent evaporation from and at 100 °C.

If F = Factor of evaporation

Then, equivalent evaporation from and at 100°C

$$= \text{Actual evaporation} \times F$$

$$= \text{Actual evaporation} \times \frac{(h - h_f)}{2257}$$

$$F = \frac{(h - h_f)}{2257}$$

7.14.3. Boiler Efficiency

It is the ratio of the heat actually utilized in generation of steam to the heat supplied by the fuel in the same period.

$$\text{Boiler Efficiency} = \frac{W_a\left(h-h_f\right)}{C}$$

where W_a = Weight of water actually evaporated into steam per kg of fuel at the working pressure.

C = Calorific value of fuel in KJ/Kg.

7.14.4. Efficiency of Economiser

It is defined as the ratio of the heat absorbed by the feed water in the economiser to the heat in flue gases entering the economiser recknoled above the temperature of air supplied to the boiler,

If, t_1 = Temp. of feed water entering the economiser

t_2 = Temp. of feed water learning the economiser

t_f = Temp. of flue water gases entering the economiser

t_a = Temp. of air supplied to the boiler

W_f = Weight of flue gases per kg of fuel

C_P = Specific heat of flue gases

W_a = Weight of water evaporated per kg of fuel.

Then, Efficiency of Economiser $= \dfrac{W_a\,(t_2 - t_1)}{W_f\,.\,C_p\,(t_f - t_a)}$

The efficiency of boiler will include the efficiency of economiser of the sensible heat of feed water entering the boiler is reckoned at the temperature of the feed water entering the economiser.

7.14.5. Boiler Horse Power

In order to give 'horse power' a definite meaning when applied to the performance of boilers the American society of Mechanical Engineers adopted the following standard of measurement.

One 'boiler horse power' means evaporation of 15.653 kg (34.5 lbs) of water per hour from and at 100°C into dry saturated steam or the equivalent in heating effect.

∴ Boiler Horse Power

$$= \frac{\text{Equivalent evaporation from and at 100°C per Hr.}}{15.653}$$

7.15. BOILER DRAUGHT

Draught may be defined as the small difference between the pressure of air outsideand that of gases within a furnace or chimney which causes flow of gas to take place. In a boiler, it performs the following functions :

(*i*) To force air through the fuel bed to aid in the proper combustion of fuel.

(*ii*) To draw the resulting hot gases through the system.

(*iii*) To remove the products of combustion to the atmospheric after they have given their heat to the water being evaporated.

7.15.1. Classification of Draught

The draught may be classified as below :

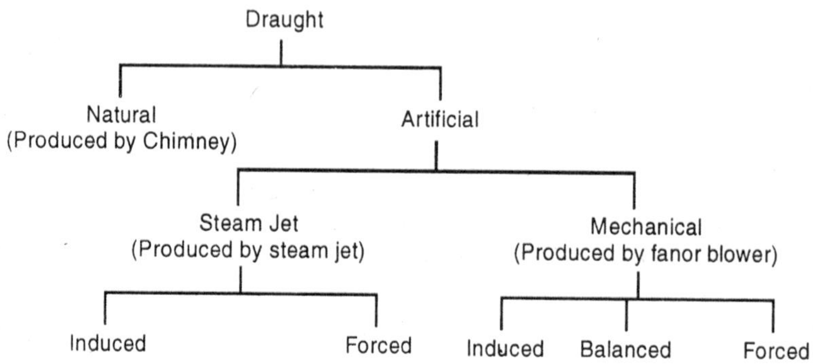

7.15.2. Natural Draught

The draught produced by chimney is called the 'Natural Draught'.

Let us consider the case before firing the boiler. The atmospheric pressure at all points on the grate level will be the same. Let us assume this pressure is equal to 'p_1'.

Referring the Fig. 7.29. Similarly the pressure at all points in a horizontal plane passing through point 2, the top of the chimney will also be the same and suppose its value is 'P_2'. The value of P_2 will be less than P_1 because atmospheric pressure decreases with the increase in altitude. The difference between the two levels considered is H. *i.e.*, the height of the chimney.

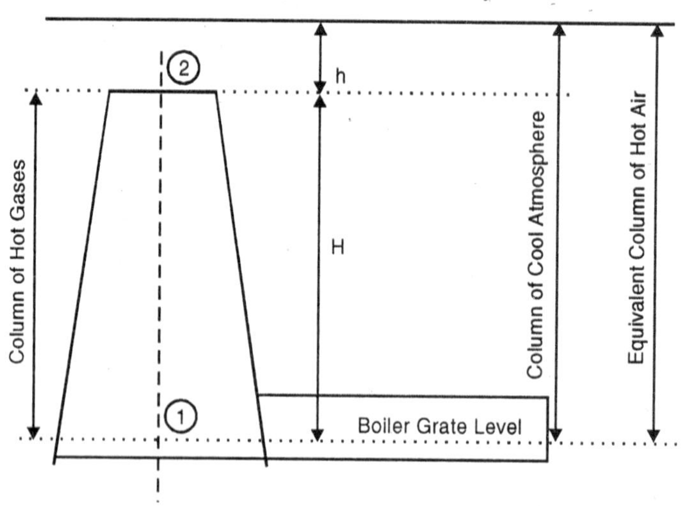

Fig. 7.29.

Now, when the boiler is fired, the hot gases will fill up the chimney. The situation will not effect any change on the atmospheric pressure at the top of the chimney which will remain the same as P_2. The pressure p_1 at grate level outside the boiler in the ashpit etc. shall also remain unaffected. But the pressure at the base of the chimney will be equal to the sum of pressure P_2 at the top and the pressure due to the hot gas column of height H. This pressure will definitely be less than the pressure P_1 which is equal to the sum of pressure P_2 and the pressure due to the cold air column of height H.

The pressure difference between the grate and the chimney base is thus equal to the pressure difference in the weight of column of height H of cold air and that of hot gases. The difference of pressure is known as theoretical maximum static draught. It is equivalent to a height '*h*' of the hot gas column expressed in mm of water.

7.15.3. Artificial Draught

In steam power plants the total static draught required varies from 25 to 350 mm of water column. In practice it will neither be economical nor convenient to build a chimney as could produce a draught of such a large magnitude. Moreover, natural draught is dependent upon the elimatic conditions. To meet the desired high draught requirements an artificial draught is produced. The artificial draught may either be produced by fans and blowers known as 'mechanical draught' or by steam jet called 'steam jet draught'.

The artificial draught has the following advantages :

(*i*) It does not depend upon climatic conditions.

(*ii*) Efficiency is much higher than the natural or chimney draught.

(*iii*) Interior quality of fuel can be economically burnt.

(*iv*) It is possible to regulate it according to the requirement.

(*v*) High temperature of flue gases as necessary for a natural draught is not required.

7.15.4. Steam Jet Draught

The simple and easy way or producing artificial draught is by steam jet. It is of two types :

7.15.4.1. Forced Steam Jet Draught

In forced draught, steam from the boiler after having been throttled to a pressure of about 1.5 to 2.0 bar gauge is supplied to the nozzles installed in the ashpit. The steam emerging out of the nozzles with a great velocity drags the air along the fuel bed, furnace, flues and then to the chimney. Forced draught produced by steam is also known as 'turbine draught'.

7.15.4.2. Induced Steam Jet Draught

In induced draught as shown in the Fig. 7.30. The jet of exhaust steam is directed into the smoke box, near the stack. The partial vacuum thus created draws the air to the smoke box through the flues, the grate and ashpit. This type of arrangement is employed on the locomotive boilers.

(*a*) Advantages of steam jet draught :

(*i*) It is quite simple and cheap in cost.

(*ii*) It has the capacity of using low grade fuels.

(*iii*) It occupies very little useful space.

(*b*) Disadvantages :

(*i*) It can operate only when steam is raised.

(*ii*) The draught produced is quite low *i.e.*, 20–30 mm of H_2O column.

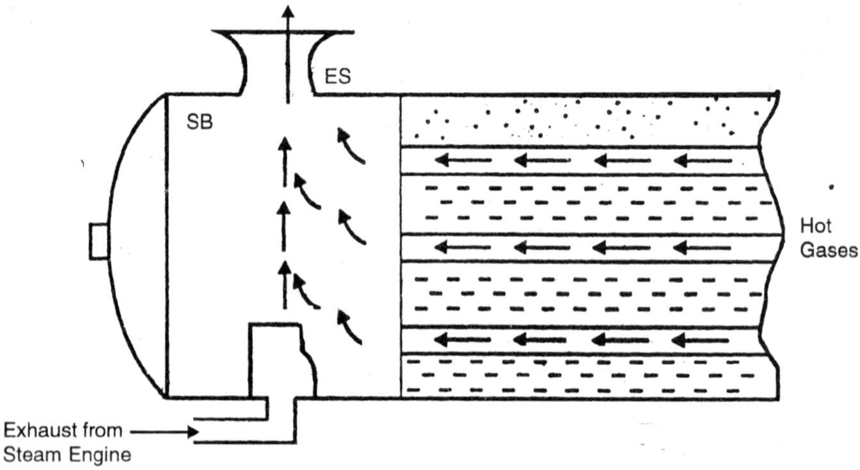

Fig. 7.30.

7.15.5. Mechanical Draught

The draught produced by fans and blowers is known as 'Mechanical Draught'.

It is of the following three types :

7.15.5.1. Induced Draught

In this type of draught steam the fan is placed near or at the base of the chimney. The partial vacuum created by the fan in the furnace and flues, draws the products of combustion from the main flue and alloys than to pass up the chimney. This type of draught is usually used when economisers and air preheaters are incorporated.

7.15.5.2. Forced Draught

In a forced draught system a fan or blower is installed near or at the base of the boiler grate to deliver air to the furnace under pressure varying from 2.5 mm to about 7.5 cm of water. Since, in this case the pressure in the furnace is higher than atmospheric there is possibility of leakage of nonios gases into the boiler room unless the setting is built air tight, which costs generally more.

7.15.5.3. Balanced Draught

The combination of forced and induced draught system is known as 'balanced draught'.

The forced draught fan supplies air at a sufficient pressure through the fuel bed on the top of the grate and also through the fuel bed on the top of the grate and also through the boiler flues, economiser and air preheater etc. and then sends then to the chimney. The pressure on the fire bed is slightly negative so that the leakage is inwards and there is no danger of blow out of flame and ash particles.

7.16. INDIAN BOILER ACT

The boiler being high pressure vessels, are required to be operated within the safe pressure limits with regular upkeep and maintenance because the failure of boiler may prove extremely disastrous. To ensure fully safety some standard rules and regulations have been enforced through a legislation all over the country which is commonly known as Indian Boiler act 1923. All boiler owners are required to follow this legislation in a strict sense.

The main provisions of this act are given below :

(*i*) A boiler cannot be put to use unless it has been registered with the chief inspector of Boilers.

(*ii*) The maximum working pressure of the boiler has to be determined by the Boiler Inspector who will issue a certificate to this effect. The owner is not allowed to run his boiler at a pressure higher than this, under any circumstances whatsoever.

(*iii*) In case of any accident, if at all it should occur, the boiler owner has to report giving full details of the nature and cause of accident, with in twenty four hours of the occurrence of the accident.

(*iv*) The rules, regulations and by-laws governing the upkeep and maintenance of boilers, the procedure for their registration inspection and determination of maximum pressure and inside safety conditions etc. are subject to revision by a Central Board under the control of Government of India.

(*v*) Violation of Law is liable to prosecution and punishment with fine.

7.17. BOILER INSPECTION

The boiler inspector carries out periodic inspection of all the boilers installed in his area. The inspection is generally carried out annually but in case of complaint or accident etc; the boiler is subject to inspection at any time. The inspection includes the following :

7.17.1. Inspection for Registration

It is the first inspection which is carried out when a new boiler is got to be registered. This inspection is made after the boiler is completely stripped off all its mountings covering and lagging etc. All parts are carefully examined for defects which might have taken place either in transit or would have already been there due to some defective design or workmanship. All these defects are noted in the registration book meant for this purpose. The working pressure of the boiler is fixed after testing its various parts. A certificate is also issued to this effect. If the boiler undergoes a satisfactory test, it is registered with the government and registration is assigned to it.

7.17.2. Hydraulic Test

This test is carried out to check the tightness of the boiler joints, fillings and the leakage arising from the repairs of defective spots. Under this test the boiler is completely filled with water and the pressure raised in accordance with the following specifications.

(*i*) For an old boiler, the hydraulic pressure applied will be one and a half times the prescribed pressure for operation the boiler should not leak at all under this pressure. For a new boiler the specifications are :

(*a*) If the working pressure is less than 100 lbs/m^2 the boiler will be subjected to a pressure double than the working pressure and it should not leak under this pressure.

(*b*) If the working pressure is more than 100 lbs/m^2 the pressure applied for test will be one and a half times the working pressure plus 50 lbs/m^2 and the boiler should again be free from any leakage under this pressure.

7.17.3. Steam Test

This test is carried out to check the setting of safety valves at the working pressure. After performing this test the safety valves are properly sealed so that these could not be tempered with afterwards, by the owner.

7.17.4. Annual Inspection

This type of inspections carried out annually after removing all parts, fittings, mountings and accessories. The boiler is inspected both internally and externally for any defect that might have crept in during the course of its working. The certificate is only renewed if the boiler is found to be completely free from any defect.

7.17.5. Inspection Under Steam

This inspection is carried out when the boiler is generating the steam. Such an inspection is very rare and is only done in cases where the boiler cannot be stopped due to some urgent public or private work.

In this case prior approval of the government is obtained to run the boiler beyond the expiry date of its current certificate.

7.17.6. Internal Inspection

This type of inspections carried out in case of the small boiler such as cochran and Locomotive where it is not possible for the inspector to enter inside and make close inspection of the internal parts. The inspection is done when internal parts like the tubes and furnaces etc. are taken out for repair or renewals etc.

7.17.7. Accident Inspection

When some accident occurs the boiler inspector holds an inquiry and looks into the cause of accident, the damage to the boiler and the men, by inspection at site.

7.17.8. Casual Inspection

This type of inspection is carried out by surprise visits from time to time to ensure that the boiler owners do not run their boilers in contradiction to the laws of the act.

NUMERICAL EXAMPLES

Example 7.1. In a throttling calorimeter, the pressure of steam before throttling is 10 bar. If the pressure of steam after throttling is 1 bar, what is the limiting value of dryness fraction of steam on high side.

Solution. One of the essential requirements for the satisfactory operation of a throttling calorimeter is that steam must be superheated after throttling. In order to achieve this condition, there is certain minimum value of dryness fraction, which the steam must have (before throttling) so that superheated steam may be obtained after throttling.

As throttling is a constant enthalpy process :

$\therefore$ Enthalpy of steam before throttling = Enthalpy of steam after throttling

Let. 'x' be the limiting dryness fraction of steam before throttling, then,

Enthalpy of saturated steam at 1 bar = $h_f + xh_{fg}$ ...(*i*)

But at 10 bar pressure (from steam tables)

$$h_f = 762.6 \text{ KJ/kg.}$$
$$h_{fg} = 2013.6 \text{ KJ/kg.}$$

also at 1.0 bar pressure;

$$h_s = 2675.4 \text{ KJ/kg.}$$

So, substituting the values in eqn. (*i*)

$$2675.4 = 762.6 + x \times 2013.6$$

or $x = 0.95$ **Ans.**

$\therefore$ Dryness fraction of steam at 10 bar pressure must not be less than 0.95 in order that throttling calorimeter be used to determine the dryness fraction.

Example 7.2. In a test to find the dryness fraction of steam with a combined separating and throttling calorimeter the following observations were recorded.

(*i*) Water collected in separating calorimeter = 4.5 kg.

(*ii*) Steam condensed after the throttling calorimeter = 45.5 kg.

(*iii*) Initial pressure of steam = 11.5 bar (abs.).

(*iv*) Temperature of steam after throttling = 140°C.

(*v*) Monometer reading = 98 mm Hg.

(*vi*) Barometer reading = 752 mm Hg.

Estimate the dryness fraction of steam as it enters the throttling calorimeter. Also, determine the dryness fraction of steam before separately calorimeter. Take 'C_p' for the superheated steam as 2.0 KJ/kg°k.

Solution. Steam from separating calorimeter flows to the throating calorimeter where it is ultimately condensed.

Weight of steam leaving the separating calorimeter, w_1 = 45.5 kg.

Weight of the suspended moisture collected in the separating calorimeter, w = 4.5 kg

then, $$x_1 = \frac{w_1}{w + w_1} = \frac{45.5}{45.5 + 4.5} = 0.91$$

Throttling calorimeter : Absolute pressure of the steam after throttling

= Barometer reading + Monometer reading

= 752 + 98 = 850 mm Hg.

Also, 1.0 bar = 750.06 mm Hg.

Hence, pressure of steam after throttling $= \frac{850}{750.06} = 1.1232$ bar

For steam at 11.5 bar pressure (before throttling)

$$h_f = 789.75 \text{ KJ/kg (By interpolation)}$$
$$h_{fg} = 1991.45 \text{ KJ/kg}$$

And for steam at 1.1332 bar pressure (after throttling)

We have from the steam tables; (by interpolation)

Saturation temperature = 102.3 + (104.8 – 102.3) × 332 = 103.13°C

Enthalpy of saturated steam

$$h_s = 2679.6 + (2683.4 - 2679.6) \times 0.332 = 2680.86 \text{ KJ/kg.}$$

Let 'x_2' be the dryness fraction of steam as it enters the throttling calorimeter. Then,

Enthalpy of steam before throttling at 11.5 bar pressure = Enthalpy of steam after throttling at 1.1332 bar pressure

Hence, $789.75 + x_1' \times 1991.45 = 2680.86 + 2.0\,(140 - 1{,}3.13)$

or $x_1' = 0.987$ (where 'x_1'' is the apparent dryness fraction)

Now, dryness fraction of steam before separately calorimeter

$$x = x_1' \times x_1$$

or
$$x = 0.987 \times 0.91 = 0.898 \textbf{ Ans.}$$

Example 7.3. The following observation were made in a boiler trial;

Coal used = 255 kg

Calorific value of coal= 30400 KJ/kg

Water evaporated = 2100 kg

Steam pressure = 11.5 bar

Dryness fraction of steam = 0.94

Feed water temperature = 25°C

Calculate the equivalent evaporation from and at 100°C per kg of coal and the efficiency of the boiler.

Solution. Enthalpy of 0.94 dry steam at 11.5 bar pressure

$$h = 789.8 + 0.94 \times 1991.5$$
$$= 2662 \text{ KJ/kg.}$$

Sensible heat of feed water at 25°C

$$h_f = 104.8 \text{ KJ/kg.}$$

Weight of water evaporated per kg of fuel

$$W_a = \frac{W_s}{W_f} = \frac{2100}{255} = 8.235 \text{ KJ/kg of coal.}$$

Equivalent evaporation from and at 100°C

$$W_e = \frac{W_a\,(h - h_f)}{2256.9} = \frac{8.235\,(2662 - 104.8)}{2256.9}$$
$$= 9.33 \text{ kg/kg of coal } \textbf{Ans.}$$

Thermal efficiency of boiler

$$\eta = \frac{W_a\,(h - h_f)}{C}$$

$$= \frac{8.235(2662 - 104.8)}{30400}$$

$$= 0.693 \text{ i.e., } 69.3\%. \textbf{ Ans.}$$

Example 7.4. The following data is provided for two boilers :

	Steam Pressure (bar)	Quantity of Steam	Rate of Evaporation
Boiler *A*	10 bar	0.91 superheated	8.5 kg/kg. of coal
Boiler *B*	16 bar	to 275°	6.95 kg/kg. of coal

Both the boilers are supplied with water of 30°C. Select the boiler which has higher rate of heat utilization per kg of coal fired. Take C_p = 2.184 for superheated steam.

Solution. Enthalpy of 0.91 dry steam at 10 bar

$$h_A = 762.6 + 0.91 \times 2013.6$$
$$= 2595 \text{ KJ/kg.}$$

Enthalpy of superheated steam at 16 bar and 275°C

$$h_B = 2791.7 - 2.184\ (275 - 201.4)$$
$$= 2952.4 \text{ KJ/kg.}$$

Sensible heat of water at 30°C = 125.7 KJ/kg.

Heat utilization per kg of coal for boiler A

$$= \frac{h_A - h_{fA}}{2256.9} \times 8.5 = \frac{(2595 - 125.7)}{2256.9} \times 8.5$$
$$= 9.3 \text{ kg/kg of coal}$$

Heat utilization per kg of coal per boiler A.

$$= \frac{h_A - h_{fA}}{2256.9} \times 8.5 = \frac{(2952.4 - 125.7)}{2256.9} \times 6.95$$
$$= 8.7 \text{ kg/kg of coal}$$

Thus, boiler A has better heat utilization rate per kg of coal fired. **Ans.**

Example 7.5. The observations recorded during a boiler trial are given below :

Duration	:	**60 minutes**
Steam generated	:	**35500 kg**
Steam pressure	:	**12 bar**
Steam temperature	:	**250°C**
Temp. of water entering economiser	:	**17°C**
Temp. of water leaving economiser	:	**77°C**
Oil burnt	:	**3460 kg**
Calorific value of oil	:	**39500 KJ/kg.**

Calculate :

(*a*) The equivalent evaporation per kg of fuel.

(*b*) The thermal efficiency of plant.

(*c*) The percentage heat energy of the fuel energy utilized by the economiser.

Solution. (*a*) Saturation temperature of steam at 12 bar = 188°C

Hence steam supplied by the boils is superheated.

Enthalpy of superheated steam at 12 bar and 250°C

$$h = 2935.4 \text{ KJ/kg.}$$

Sensible heat of water entering the economiser

$$h_f = 71.3 \text{ KJ/kg (from steam tables)}$$

Weight of water evaporated per kg of fuel

$$W_a = \frac{35500}{3460} = 10.26 \text{ kg/kg of fuel}$$

Equivalent Evaporation,

$$W_e = \frac{W_a (h - h_f)}{2256.9}$$

$$\therefore \quad W_e = \frac{10.26\,(2935.4 - 71.3)}{2256.9} = 13.02 \text{ kg } \textbf{Ans.}$$

(*b*) Thermal efficiency of plant,

$$\eta = \frac{W_a (h - h_f)}{C} = \frac{10.26\,(2935.4 - 71.3)}{39500}$$

$$= 0.744 \text{ i.e., } 74.4\% \textbf{ Ans.}$$

(*c*) Heat energy utilized by the economiser per kg of fuel

$$= W_e (h_{f2} - h_f)$$

Where h_2 is the sensible heat of water leaving the economiser which is 322.3 KJ/kg (from steam tables)

Hence heat energy utilized by economiser

$$= 10.26\,(322.3 - 71.3)$$

$$= 2575.3 \text{ KJ/kg of fuel}$$

∴ Percentage of heat utilized by economiser

$$= \frac{2575.3}{39500} \times 100 = 6.52\% \textbf{ Ans.}$$

ASSIGNMENT

Objective Type Questions :

(A) Fill in the Blanks :

1. In water tube boiler ______ flows through the tubes.
2. High pressure boilers are ______ tube boilers.
3. Steam injector is used for ______.
4. The function of pressure reducing valve is to ______.
5. Feed check valve is used ______.
6. Fusible plug is mounted on the ______ of the ______.
7. Safety valve is provided at the ______ of the ______.
8. Steam boiler cannot be operated without______.
9. In economiser the heat of ______ is utilized.
10. Water tubes for a Babcock and Wilcox steam generators are ______.
11. ______ water is used for the boilers.
12. Steam injector used for boiler is a ______.

13. Economiser is used to ______ the efficiency of the boiler.
14. Efficiency of steam power plant ______ when feed water is heated.
15. Latent heat of steam is ______.
16. Water tube boilers produce steam at a ______ pressure than that of fire tube boilers.
17. A ______ in a boiler is a place in the combustion chamber upon which coal is burnt.
18. For the same diameter and thickness of tube, a water tube boiler has ______ heating surface as compared to a fire tube boiler.
19. The locomotive boiler has ______ fire tubes and ______ superheater tubes.
20. The draught in locomotive boiler is produced by ______.

(B) State True or False :

1. Babcock and Wilcox boiler is a water tube boiler.
2. Benson boiler is a low pressure boiler.
3. Locomotive boiler is a water tube boiler.
4. Safety valve is a mounting.
5. There are generally two water level indicators mounted on a boiler.
6. Superheater is a mounting.
7. Generating pressure is more in the case of fire tube boilers.
8. Accessories increase the efficiency of the boiler.
9. Pressure gauges record the absolute pressure of the steam.
10. Steam power plants use low pressure steam.
11. Water tube boilers start giving steam more quickly as compared to fire tube boiler.
12. Fusible plug is situated on the top of the boiler.
13. Hot water is discharged when the blow off cock is opened while the boiler is working.
14. Steam generators cannot be operated without accessories.
15. Superheating of steam is done by using the heat of the flue gases.
16. Steam generator can be operated without mountings.
17. A safety valve mainly used with locomotive and marine boiler is spring loaded safety valve.
18. A fusible plug in a boiler is used to put off fire in the furnace when the level of water falls to an unsafe limit.
19. The chimney draught varies with height of chimney, temp. of furnace gases and climatic conditions.
20. The draught in locomotive boiler is produced by steam jet.

(C) Multiple Choice Questions :

1. The fuel mostly used in boilers is :
   (*a*) anthracite (*b*) peat
   (*c*) lignite (*d*) bituminous
2. Clorific value of steam coal lies in the range:
   (*a*) 1000-2500 kcal/kg (*b*) 3000-45000 kcal/kg
   (*c*) 5000-8000 kcal/kg (*d*) 9000-125000 kcal/kg
3. Which is not the desirable property of a good fuel?
   (*a*) economical in cost (*b*) minimum sulphur content
   (*c*) high ignition point (*d*) high heating value

4. The presence of ...... in boiler exhaust indicates incomplete combustion.
   (*a*) Oxygen (*b*) Water vapours
   (*c*) CO (*d*) CO_2
5. Scale is formed on the boiler tube surface when the boiler feed water contains.
   (*a*) dissolved gases
   (*b*) oils and suspended solids
   (*c*) calcium and magnesium salts
   (*d*) insoluble compounds
6. What is the object of providing baffles in between water tubes in a Babcock-Wilcox boiler?
   (*a*) to regulate water flow
   (*b*) to regulate steam flow
   (*c*) to have better contact of flue gases with water tubes
   (*d*) to prevent buckling of water tubes.
7. Which aspect is not valid for a fire tube locomotive boiler?
   (*a*) externally fired (*b*) horizontal placement
   (*c*) multi-tubular (*d*) mobile unit
8. Maximum steam pressure in a locomotive boiler is limited to
   (*a*) 5 kgf/cm^2 (*b*) 10 kgf/cm^2
   (*c*) 18 kgf/cm^2 (*d*) 25 kgf/cm^2
9. Forced circulation of water takes place in :
   (*a*) Locomotive boiler (*b*) Lamont boiler
   (*c*) Babcock-Wilcox boiler (*d*) Stirling boiler
10. Circulation of water is by natural convection currents in:
    (*a*) Lamont boiler (*b*) Benson boiler
    (*c*) Velcox boiler (*d*) Babcock-Wilcox boiler
11. Which of the following is a once through boiler?
    (*a*) Lancashire boiler (*b*) Velcox boiler
    (*c*) Benson boiler (*d*) Babcock-Wilcox boiler
12. Which is the following equipment is a boiler accessory?
    (*a*) Stop valve (*b*) Feed water pump
    (*c*) Fusible plug (*d*) Blow of cock
13. Which of the following filling is a boiler mounting?
    (*a*) superheater (*b*) economiser
    (*c*) feed check valve (*d*) air pre heater
14. Which of the following is fitted on a boiler to improve the boiler efficiency?
    (*a*) safety valve (*b*) steam trap
    (*c*) economiser (*d*) blow-down cock
15. An economiser is fitted with a boiler to :
    (*a*) increase the steam pressure
    (*b*) heat the fuel of low calorific value
    (*c*) heat the air entering the fire grate
    (*d*) heat the feed water by exhaust flue gases
16. The economiser unit of a boiler is located between :
    (*a*) forced draught fan and furnace
    (*b*) furnace and superheater
    (*c*) superheater and air preheater
    (*d*) air preheater and chimney
17. By the use of an economiser, saving in fuel consumption is about:
    (*a*) 5 – 10% (*b*) 10 – 15%
    (*c*) 15 – 20% (*d*) 20 – 25%

18. Which of the following is fitted in the lowest part of the boiler for periodic cleaning and inspection?
    (*a*) feed check valve (*b*) sliding valve
    (*c*) blow off-cock (*d*) manhole
19. Which of the following is not a safety valve for the boiler?
    (*a*) lever type (*b*) sliding type
    (*c*) dead weight type (*d*) spring type
20. The amount of water evaporated from and at 100° C into dry saturated steam at atmospheric pressure is called:
    (*a*) evaporative capacity (*b*) equivalent evaporation
    (*c*) generation factor (*d*) boiler horse power

ANSWERS

(A) Fill in the Blanks :

1. Water **2.** Fire **3.** Feeding water into the boiler
4. Control pressure on the delivery side of boiler
5. To control flow of water from feed pump to the boiler
6. Top, boiler, **7.** Top, boiler
8. Water **9.** Waste flue gases
10. Inclined **11.** Soft
12. Mounting **13.** Increase
14. Increase **15.** 334.96 KJ/kg
16. Lower **17.** grate
18. Lesser **19.** Horizontal, vertical
20. The exhaust steam from the engine cylinder.

(B) State True or False :

1. True	**2.** False	**3.** False	**4.** True	**5.** True
6. False	**7.** True	**8.** True	**9.** False	**10.** False
11. True	**12.** True	**13.** True	**14.** False	**15.** True
16. False	**17.** True	**18.** True	**19.** True	**20.** True

(C) Multiple Choice Questions :

1. (*d*)	**2.** (*c*)	**3.** (*c*)	**4.** (*c*)	**5.** (*c*)
6. (*c*)	**7.** (*a*)	**8.** (*c*)	**9.** (*b*)	**10.** (*d*)
11. (*c*)	**12.** (*b*)	**13.** (*c*)	**14.** (*c*)	**15.** (*d*)
16. (*c*)	**17.** (*c*)	**18.** (*c*)	**19.** (*b*)	**20.** (*b*).

(D) Short Answer Type Questions :

Q. 1. What is a 'Steam Generator'?

Ans. 'Steam Generator' may be defined as a closed vessel strongly constructed of steel sheet, in which the steam is produced, at the pressure much higher than the atmospheric pressure by the transfer of heat, produced by the burning of fuel, to the water.

Q. 2. What do you know about 'Water tube boiler' and 'Fire tube boiler'?

Ans. In a 'Water tube boiler', water circulates through tubes, surrounded by hot gases.
In a 'Fire tube boiler', hot gases circulate through tubes, surrounded by water.

Q. 3. What is 'fuel'? Name its important types.

Ans. The 'fuel' is a substance containing combustible elements, which combining with oxygen generate heat.

Fuels may be classified according to their physical character as :

(*i*) **Solid fuels.** A few examples of the Solid fuels are, Peat, Lignite, Bituminous Coal, Anthracite etc.

(*ii*) **Liquid fuels.** Some of the important liquid fuels used for engineering purpose are Gasoline or Petrol, Paraffins, Diesel oil and Heavy fuel oil or crude oil.

(*iii*) **Gaseous fuels.** A few examples of the gaseous fuels are, Natural gas, Liquified Petroleum gas (LPG), Refinery Oil Gas, Coal Gas, Producer Gas, Blast Furnace Gas, Blue Water Gas and Carburetted Water gas etc.

Q. 4. What do you know about 'Boiler Mountings'? Name them.

Ans. 'Boiler Mountings' are the components generally mounted on the surface of the boiler to accomplish safety during operation and the complete control of steam generation process.

The following are the various important mountings of a boiler :

(*i*) Water Level Indicator (*ii*) Pressure Gauge

(*iii*) Safety Gauge (*iv*) Junction Valve or Stop Valve

(*v*) Feed Check Valve (*vi*) Blow-off-Cock

(*vii*) Fusible Plug

Q. 5. What do you know about 'Boiler Accessories'? Name them.

Ans. 'Boiler-Accessories' are these appliances, which are installed with a boiler in its neighbouring are to increase the efficiency of the boiler. These are not the essential parts of the boiler, so without these the boiler operation can be accomplished, although at a lower efficiency. The following are the important boiler accessories :

(*i*) Feed Water Pump (*ii*) Injector

(*iii*) Economiser (*iv*) Super-heater

(*v*) Air-preheater (*vi*) Pressure-Reading Value

(*vii*) Steam drier or separator (*viii*) Steam trap.

Q. 6. What is a 'Calorimeter'? Name five important types of calorimetres.

Ans. A 'Calorimeter' is a device generally used to determine the dryness fraction of steam.

The following are the important types of Calorimetres :

(*i*) Barrel Calorimeter

(*ii*) Separating Calorimeter

(*iii*) Throttling Calorimeter

(*iv*) Combined Separating and Throttling Calorimeter

(*v*) An Electric Calorimeter.

Q. 7. Define the term 'Equivalent evaporation' as applied to the boilers.

Ans. 'Equivalent Evaporation' of a boiler is defined as the weight of water evaporated from water at 100°C to dry and saturated steam at 100°C by utilising the same amount of heat as would have been used under the actual working conditions. It is generally denoted by 'W_e'.

Q. 8. What is 'Factor of Evaporation'?

Ans. It is a quantity which when multiplied by the amount of steam generated at a given pressure from water at given temperature gives the equivalent evaporation from and at 100°C.

If, F = Factor of Evaporation

Then, equivalent evaporation from and at 100°C

$$= \text{Actual evaporation} \times F$$

$$= \text{Actual evaporation} \times \frac{(h - h_f)}{2257}$$

So, $$F = \frac{(h - h_f)}{2257}$$

Q. 9. What is the 'Boiler-Efficiency'? Define it.

Ans. It is ratio of heat actually utilized in the generation of steam to the heat supplied by the fuel in the same period

$$\text{Boiler efficiency} = \frac{W_a (h - h_f)}{C}$$

where W_a = Weight of water actually evaporated into steam per kg of fuel at the working pressure.

and C = Calorific value of the fuel in KJ/kg.

Q. 10. Define the 'Efficiency of Economiser'.

Ans. It is defined as the ratio of the heat absorbed by the feed water in the economiser to the heat in flue gases entering the economiser reckoned above the temperature of air supplied to the boiler.

Q. 11. Define the term 'Boiler Horse Power'?

Ans. 'Boiler Horse Power' means evaporation of 15.653 kg (34.5 lbs) of water per hour from and at 100 °C into dry saturated steam or the equivalent in heating effect.

$$\therefore \text{Boiler Horse Power} = \frac{\text{Equivalent evaporation from at } 100^\circ \text{ C per hour}}{15.653}$$

Q. 12. What is 'Boiler Draught'?

Ans. 'Boiler Draught' may be defined as the small difference between the pressure of air outside and that of the gases within a furnace or chimney which causes flow of gas to take place.

In a boiler, it perform the following functions;

(*i*) to force air through the fuel bed to aid in the proper combustion of fuel,

(*ii*) to draw the resulting hot gases through the system

(*iii*) to remove the products of combustion to the atmosphere after they have given their heat to the water being evaporated

Q. 13. Classify the draught as applied to boilers with the help of a flow diagram.

Ans. The draught may be classified as below :

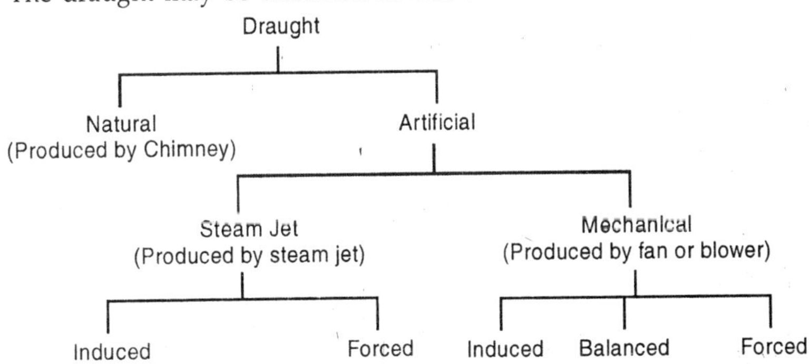

Q. 14. What is 'Natural Draught' ?

Ans. The draught produced by the Chimney is called the 'Natural-Draught'.

Q. 15. What do you know about 'Artificial Draught'?

Ans. The artificial draught may either be produced by fans or blowers known as 'mechanical draught' or by steam jet called 'steam jet draught'.

Q. 16. What is 'Mechanical Draught'? Name its various types.

Ans. The draught produced by fans and blowers is known as 'Mechanical Draught'. It is of the following three types :

(*i*) Induced draught.
(*ii*) Forced draught.
(*iii*) Balanced draught.

Q. 17. What do you know about the term 'Boiler Inspection'?

Ans. As per the Indian Boiler Act the boilers are inspected annually for their fitness physically and for registration etc. In case of complaint or accident etc. the boiler is subject to inspection at any time.

Q. 18. How boilers are classified, on the basis system of firing?

Ans. The boilers, on the basis of system of firing may be classified as follows :

(*i*) *Externally fired* : It has its furnace arranged outside the boiler shell.
(*ii*) *Internally fired* : It has its furnace arranged inside the boiler shell.

Q. 19. How the boilers are classified, on the basis of fuel used?

Ans. The boilers, on the basis of fuel used may be classified as follows :

(*i*) Coal fired boiler,
(*ii*) Oil fired boiler,
(*iii*) Gas fired boiler,
(*iv*) Electricity heated boiler,
(*v*) Nuclear fuel boiler,
(*vi*) Waste heat boiler.

Q. 20. Define the term 'Calorific value of fuel'.

Ans. The amount of heat produced per unit wt or volume of the fuel during combustion is known as the 'Caloric Value' if the fuel.

REVIEW QUESTIONS

1. (*a*) Define steam generator. What are the main uses of steam ?
   (*b*) How the boilers or steam generators are classified ?
   (SBTE Pb., Dec. 1999)
2. What is the different between fire tube boilers and water tube boilers?
3. Describe the construction and working of a locomotive boiler, with the help of a neat sketch.
4. Describe the construction and working of a Nestler-Boiler with a neat sketch. **(SBTE Haryana, 1976)**
5. Describe the construction and working of a babcock and Wilcox boiler.
6. Describe the construction and working of any modern boiler with the help of a neat sketch.
7. Describe in brief the various mountings and accessories of a boiler.
8. What do you understand by draught? What are the advantages of forced draught over the induced draught? **(W. Bengal, 1979)**
9. Differentiate between Nestler boiler and Babcock and Wilcox boiler. **(SBTE Pb., Dec. 1999)**

❑❑❑

8

CHAPTER

Fuels and Combustion

INSIDE THIS CHAPTER

8.1. INTRODUCTION

■ Fuel may be *chemical* or *nuclear*. Here we shall consider briefly *chemical fuels* only.

A chemical fuel is a substance which releases heat energy on combustion. The principal combustible elements of each fuel are *carbon* and *hydrogen. Though sulphur* is a combustible element too but its presence in the fuel is considered to be *undesirable*.

■ In **chemical thermodynamics** the study of systems involving chemical reactions is an important topic. *A chemical reaction may be defined as the rearrangement of atoms due to redistribution of electrons.* In a chemical reaction the terms, **reactants** and the **products** are frequently used. *'Reactants'* comprise of initial constituents which start the reaction while *'products'* comprise of *final constituents* which are formed by the chemical reaction. Although the basic principles which will be discussed in this chapter apply to any chemical reaction, here main attention will be focused on an important type of chemical reaction—**combustion**.

8.2. CLASSIFICATION OF FUELS

Fuels can be classified according to whether :

1. They occur in nature called **primary fuels** or are prepared called **secondary fuels;**
2. They are in solid, liquid or gaseous state. The detailed classification of fuels can be given in a summary form as follows :

Type of fuel	Natural (Primary)	Prepared (Secondary)
Solid	Wood Peat Lignite coal	Coke Charcoal Briquettes
Liquid	Petroleum	Gasoline Kerosene Fuel oil Benzol Alcohol Shale oil
Gaseous	Natural gas	Petroleum gas Producer gas Coal gas Coke-oven gas Blast furnace gas Carburetted gas Sewer gas

8.3. SOLID FUELS

(*i*) **Coal.** Its main constituents are carbon, hydrogen, oxygen, nitrogen, sulphur, moisture and ash. Coal passes through different stages during its formation from vegetation. These stages are enumerated and discussed below :

(*ii*) Plant debris – Peat – Lignite – Brown coal – Sub-bituminous coal – Bituminous coal – Semi-bituminous coal – Semi-anthracite coal – Anthracite coal – Graphite.

(*iii*) **Peat.** It is the first stage in the formation of coal from wood. It contains huge amount of moisture and therefore it is dried for about 1 to 2 months before it is put to use. It is used as a domestic fuel in Europe and for power generation in Russia. In India it does not come in the categories of good fuels.

(*iv*) **Lignites and brown coals.** These are intermediate stages between peat and coal. They have a woody or often a clay like appearance associated with high moisture, high ash and low heat contents. Lignites are usually amorphous in character and impose transport difficulties as they break easily. They burn with a smoky flame. Some of this type are suitable for local use only.

(*v*) **Bituminous coal.** It burns with long yellow and smoky flames and has high percentages of volatile matter. The average calorific value of bituminous coal is about 31350 KJ/kg. It may be of two types, namely *caking* or *noncaking*.

(*vi*) **Semi-bituminous coal.** It is softer than the anthracite. It burns with a very small amount of smoke. It contains 15 to 20 percent volatile matter and has a tendency to break into small sizes during storage or transportation.

(*vii*) **Semi-anthracite.** It has less fixed carbon and less lustre as compared to true anthracite and gives out longer and more luminous flames when burnt.

(*viii*) **Anthracite.** It is very hard coal and has a shining black lustre. It ignites slowly unless the furnace temperature is high. It is non-caking and has high percentage of fixed carbon. It burns either with very short blue flames or without flames. The calorific value of this fuel is high to the tune of 35500 kJ/kg and as such is *very suitable for steam generation.*

(*ix*) **Wood charcoal.** It is obtained by destructive distillation of wood. During the process the volatile matter and water are expelled. The physical properties of the residue (charcoal), how depends upon the rate of heating and temperature.

(*x*) **Coke.** It consists of carbon, mineral matter with about 2% sulphur and small quantities of hydrogen, nitrogen and phosphorus. It is solid residue left after the destructive distillation of certain kinds of coals. It is smokeles and clear fuel and can be produced by several processes. It is *mainly used in blast furnace* to produce heat and at the same time to reduce the iron ore.

(*xi*) **Briquettes.** These are prepared from fine coal or coke by compressing the material under high pressure.

8.4. LIQUID FUELS

The chief source of liquid fuels is *petroleum* which is obtained from wells under the earth's crust. These fuels have proved *more advantageous in comparison to solid fuels* in the following aspects.

Advantages :

1. Require less space for storage.
2. Higher calorific value.
3. Easy control of consumption.
4. Staff economy.
5. Absence of danger from spontaneous combustion.
6. Easy handling and transportation.
7. Cleanliness.
8. No ash problem.
9. Non-deterioration of the oil in storage.

Petroleum. There are different opinions regarding the origin of petroleum. However, now it is accepted that petroleum has originated probably from organic matter like fish and plant life etc. by bacterial action or by their distillation under pressure and heat. It consists of mixture of gaseas, liquids and solid hydrocarbons with small amounts of nitrogen and sulphur compounds. In India, the main sources of Petroleum are Assam and Gujarat.

Heavy fuel oil or crude oil is imported and then refined at different refineries. The refining of crude oil supplies the most important product called *petrol*. Petrol can also be made by polymerization of refinery gases.

Other liquid fuels are kerosene, fuels oils, colloidal fuels and alcohol.

8.5. GASEOUS FUELS

(*i*) **Natural gas.** The main constituents of natural gas are *methane* (CH_4) and *ethane* (C_2H_6). It has calorific value nearly 21000 kJ/m^3. Natural gas is used alternately or simultaneously with oil for internal combustion engines.

(*ii*) **Coal gas.** Mainly consists of *hydrogen, carbon monoxide* and *hydrocarbons.* It is prepared by carbonization of coal. It finds its use in boilers and sometimes used for commercial purposes.

(*iii*) **Coke-oven gas.** It is obtained during the production of coke by heating the bituminous coal. The volatile content of coal is driven off by heating and major portion of this gas is utilized in heating the ovens. This gas *must be thoroughly filtered before using in gas engines.*

(*iv*) **Blast furnace gas.** It is obtained from smelting operation in which air is forced through layers of coke and iron ore, the example being that of pig iron manufacture where this gas is produced as by product and contains about 20% carbon monoxide (CO). After filtering it may be blended with richer gas or used in gas engines directly. The heating value of this gas is very low.

(*v*) **Producer gas.** It results from the partial oxidization of coal, coke or peat when they are burnt with an insufficient quantity of air. It is produced in specially designed retorts. It has low heating value and in general is suitable for large installations. It is also used in steel industry for firing open hearth furnaces.

(*vi*) **Water or illuminating gas.** It is produced by blowing steam into white hot coke or coal. The decomposition of steam takes place liberating free hydrogen, and oxygen in the steam combines with carbon to form carbon monoxide according to the reaction :

$$C + H_2O \longrightarrow CO + H_2$$

The gas composition varies as the hydrogen content if the coal is used.

(*vii*) **Sewer gas.** It is obtained from sewage disposal vats in which fermentation and decay occur. It consists of mainly marsh gas (CH_4) and is collected at large disposal plants. It works as a fuel for gas engines which in turn drive the plant pumps and agitators.

Gaseous fuels are becoming popular because of following *advantages* they possess.

Advantages :

1. Better control of combustion.
2. Much less excess air is needed for complete combustion.
3. Economy in fuel and more efficiency of furnace operation.
4. Easy maintenance of oxidizing or reducing atmosphere.
5. Cleanliness.
6. No problem of storage if the supply is available from public supply line.
7. The distribution of gaseous fuels even over a wide area is easy through the pipe lines and as such handling of the fuel is altogether eliminated.
8. Gaseous fuels give economy of heat and produce higher temperatures (as they can be preheated in regenerative furnaces and thus heat from hot flue gaseous can be recovered).

8.6. BASIC CHEMISTRY

Before considering combustion problems it is necessary to understand the construction and use of chemical formulae. This involves elementary concepts which are discussed below briefly.

Atoms. It is not possible to divide the chemical elements *indefinitely*, and the *smallest particle which can take part in a chemical change* is called an '**atom**'. If an atom is split as in nuclear reaction, the *divided atom does not retain the original chemical properties*.

Molecules. It is rare to find elements to exist naturally as single atom. Some elements have atoms which exist in pairs, each pair forming a molecule (*e.g.*, oxygen), and the atoms of each molecule are held together by stronger *inter-atomic forces*. The isolation of a molecule of oxygen would be a tedious, but possible; the isolation of an atom of oxygen would be different prospect. The molecules of some substances are formed by the mating up of atoms of different elements. For example, water has a molecule which consists of two atoms of hydrogen and one atom of oxygen. The atoms of different elements have different masses and these values are important when a quantitative analysis is required. The actual masses are infinitesimally small, and the ratios of the masses of atoms are used. These ratios are indicated by **atomic weight** quoted on a scale which defines the atomic weight of oxygen as 16.

The symbols and molecular weights of some important elements, compounds and gases are given in Table 8.1.

Table 8.1. Symbols and Molecular Weights.

Elements/Compounds/Gases	Molecule		Atom	
	Symbol	Molecular weight	Symbol	Molecular weight
Hydrogen	H_2	2	H	1
Oxygen	O_2	32	O	16
Nitrogen	N_2	28	N	14
Carbon	C	12	C	12
Sulphur	S	32	S	32
Water	H_2O	18	–	–
Carbon monoxide	CO	28	–	–
Carbon dioxide	CO_2	44	–	–
Sulphur dioxide	SO_2	64	–	–
Marsh gas (Methane)	CH_4	16	–	–
Ethylene	C_2H_4	28	–	–
Ethane	C_2H_6	30	–	–

8.7. COMBUSTION EQUATIONS

■ In a combustion chamber proportionate masses of air and fuel enter where the chemical reaction takes place, and then the combustion products pass to the exhaust. By the conservation of mass the mass flow remains constant (*i.e.*, total mass of *products* = total mass of *reactants*), but the reactants are chemically different from the products, and the products leave at a higher temperature. The *total number of atoms of each element concerned in the combustion remains constant, but the atoms are rearranged into groups having different chemical properties*. This information is expressed in the chemical equation which shows (*i*) the reactants and the products of combustion, (*ii*) the relative quantities of the reactants and products. The two sides of the equation must be **consistent**, each having the same number of atoms of each element involved.

■ The oxygen supplied for combustion is *usually* provided by *atmospheric air*, and it is necessary to use accurate and consistent analysis of air by *mass* and by *volume*. It is usual in combustion calculations to take air as 23.3% O_2, 76.7% N_2 by mass, and 21% O_2, 79% N_2 by volume. The small traces of other gases in dry air are included in nitrogen, which is sometimes called '*atmospheric nitrogen*'.

Some important *combustion equations* are given below :

1. Combustion of hydrogen

$$2H_2 + O_2 \longrightarrow 2H_2O \quad \text{...(8.1)}$$

The above equation of combustion of hydrogen tell us that :

(*i*) Hydrogen reacts with water to form steam or water,

(*ii*) Two molecules of hydrogen react with one molecule of oxygen to give two molecules of steam or water,

i.e., 2 volumes H_2 + 1 volume O_2 $\longrightarrow$ 2 volumes H_2O

The H_2O may be liquid or a vapour depending on whether the product has been cooled sufficiently to cause condensation.

The *proportions by mass* are obtained by using *atomic weights* as follows :

$$2H_2 + O_2 \longrightarrow 2H_2O$$

$$\therefore \quad 2\,(2 \times 1) + 2 \times 16 \longrightarrow 2\,(2 \times 1 + 16)$$

i.e., $$4 \text{ kg } H_2 + 32 \text{ kg } O_2 \longrightarrow 36 \text{ kg } H_2O$$

or $$\mathbf{1 \text{ kg } H_2 + 8 \text{ kg } O_2 \longrightarrow 9 \text{ kg } H_2O}$$

The same proportions are obtained by writing the eqn. (8.1.) as :

$$H_2 + \frac{1}{2}O_2 \longrightarrow H_2O, \text{ and this is sometimes done.}$$

It will be noted from eqn. (8.1) that the total volume of the *reactants* is

2 volumes H_2 + 1 volume O_2 = 3 volumes

The total volume of the *product* is only 2 volumes. There is therefore a *volumetric contraction on combustion.*

Since the oxygen is accompanied by nitrogen if air is supplied for the combustion, then this nitrogen should be included in the equation. As nitrogen is *inert* as far as chemical reaction is concerned, it will appear on both sides of the equation.

With one mole of oxygen there are 79/21 moles of nitrogen, hence Eqn. (8.1) becomes,

$$2H_2 + O_2 + \frac{79}{21}N_2 \longrightarrow 2H_2O + \frac{79}{21}N_2 \quad \text{...(8.2)}$$

2. Combustion of carbon

(*i*) Complete combustion of carbon to carbon dioxide.

$$C + O_2 \longrightarrow CO_2 \quad \text{...(8.3)}$$

and including the nitrogen

$$C + O_2 + \frac{79}{21}N_2 \longrightarrow CO_2 + \frac{79}{21}N_2 \quad \text{...(8.4)}$$

By volume :

$$0 \text{ volume C} + 1 \text{ volume } O_2 + \frac{79}{21} N_2 \text{ volumes} \longrightarrow$$

$$1 \text{ volume } CO_2 + \frac{79}{21} \text{ volumes } N_2 \quad ...[8.4(a)]$$

The volume of carbon is written as zero since the volume of solid is negligible in comparison with that of a gas.

By mass :

$$12 \text{ kg C} + (2 \times 16) \text{ kg } O_2 + \frac{79}{21} + (2 \times 14) \text{ kg } N_2$$

$$\longrightarrow (12 + 2 \times 16) \text{ kg } CO_2 + \frac{79}{21} + (2 \times 14) N_2$$

i.e., $12 \text{ kg C} + 32 \text{ kg } O_2 + 105.3 \text{ kg } N_2 \longrightarrow 44 \text{ kg } CO_2 + 105.3 \text{ kg } N_2$

or $1 \text{ kg C} + \frac{8}{3} \text{ kg } O_2 + \frac{105.3}{12} \text{ kg } N_2 \longrightarrow \frac{11}{3} \text{ kg } CO_2 + 105.3 \text{ kg } N_2$ ...[8.4 (*b*)]

(*ii*) **The incomplete combustion of carbon.** The incomplete combustion of carbon occurs when there is an insufficient supply of oxygen to burn the carbon completely to carbon dioxide.

$$2C + O_2 \longrightarrow 2CO \quad ...(8.5)$$

and including the nitrogen,

$$2C + O_2 + \frac{79}{21} N_2 \longrightarrow 2CO + \frac{79}{21} N_2 \quad ...(8.6)$$

By mass :

$$(2 \times 12) \text{ kg C} + (2 \times 16) \text{ kg } O_2 + \frac{79}{21} (2 \times 14) \text{ kg } N_2$$

$$\longrightarrow 2\,(12 + 16) \text{ kg CO} + \frac{79}{21} (2 \times 14) \text{ kg } N_2$$

i.e., $24 \text{ kg C} + 32 \text{ kg } O_2 + 105.3 \text{ kg } N_2 \longrightarrow 56 \text{ kg CO} + 105.3 \text{ kg } N_2$

or $1 \text{ kg C} + \frac{4}{3} \text{ kg } O_2 + \frac{10.53}{24} \text{ kg } N_2 \longrightarrow \frac{7}{3} \text{ kg CO} + \frac{105.3}{24} \text{ kg } N_2$ ...[8.6(*a*)]

If a further supply of oxygen is available then the combustion can continue to completion,

$$2CO + O_2 + \frac{79}{21} N_2 \longrightarrow 2CO_2 + \frac{79}{21} N_2 \quad ...(8.7)$$

By mass :

$$56 \text{ kg CO} + 32 \text{ kg } O_2 + \frac{79 \times 28}{21} \text{ kg } N_2 \longrightarrow 88 \text{ kg } CO_2 + \frac{79 \times 28}{21} \text{ kg } N_2$$

or $1 \text{ kg CO} + \frac{4}{7} \text{ kg } O_2 + \frac{105.3}{56} \text{ kg } N_2 \longrightarrow \frac{11}{7} \text{ kg } CO_2 + \frac{105.3}{56} \text{ kg } N_2$ ...[8.7(*a*)]

8.8. THEORETICAL AIR AND EXCESS AIR

The *minimum amount of air* that supplies sufficient oxygen for the complete combustion of all the carbon, hydrogen, and any other elements in the fuel that may oxidise is called the "**theoretical air**". When complete combustion is achieved with theoretical air, the products contain no oxygen.

In practice, it is found that complete combustion is not likely to be achieved unless the amount of air supplied is somewhat greater than the theoretical amount. Thus 150 percent theoretical air means that air actually supplied is 1.5 times the theoretical air.

The complete combustion of methane with minimum amount of theoretical air and 150 percent theoretical air respectively is written as :

$$CH_4 + 2O_2 + 2\left(\frac{79}{21}\right) N_2 \longrightarrow CO_2 + 2H_2O + 2\left(\frac{79}{21}\right) N_2 \quad ...(8.8)$$

$$CH_4 + 2(1.5)\,O_2 + 2\left(\frac{79}{21}\right)(1.5)\,N_2 \longrightarrow 2CO_2 + 2H_2O + O_2 + 3\left(\frac{29}{21}\right) N_2 \quad ...(8.9)$$

(with 150 percent theoretical air)

The amount of air actually supplied may also be expressed in terms of percent excess air. The excess air is the amount of air supplied over and above the theoretical air. This 150 percent theoretical air is equivalent to 50 percent excess air.

Note. For complete combustion of fuel we need air. As per theoretical basis there is a minimum amount of air which is required by the fuel to burn completely, *but always, air in excess is used because whole of air supplied for combustion purposes does not come in contact with the fuel completely and as such portion of fuel may be left unburnt.* But if a large quantity of excess air is used it exercises a *cooling effect on combustion process which however can be avoided by preheating the air. The weight of excess air supplied can be determined from the weight of oxygen which is left unused. The* amount of excess air supplied varies with the type of fuel and the firing conditions. It may approach a value of 100% but modern practice is to use 25% to 50% excess air.

8.9. STOICHIOMETRIC AIR FUEL (A/F) RATIO

Stoichiometric (or chemically correct) *mixture* of air and fuel is one that contains *just sufficient oxygen for complete combustion of the fuel.*

A *weak mixture is one which has an excess of air.*

A *rich mixture* is one which has a *deficiency* of air.

The percentage of excess air is given as under:

$$\text{\%age excess air} = \frac{\text{Actual A/F ratio - Stoichiometric A/F ratio}}{\text{Stoichiometric A/F ratio}} \quad ...(8.10)$$

(where A and F denote *air* and *fuel* respectively)

The ratios are expressed as follows :

For *gaseous fuels* — *By volume*

For solid and liquid fuels — *By mass*

For *boiler* plant the mixture is usually greater than 20% weak; for *gas turbines* it can be as much as 300% weak. *Petrol engines* have to meet various conditions of load and speed, and operate over a wide range of mixture strength. The following definition is used :

$$\text{Mixture strength} = \frac{\text{Stoichiometric A / F ratio}}{\text{Actual A / F ratio}} \quad ...(8.11)$$

The working value range between 80% (weak) and 120% (rich).

Note. The reciprocal of the air fuel ratio is called the *fuel-air (F/A) ratio.*

8.10. AIR-FUEL RATIO FROM ANALYSIS OF PRODUCTS

When analysis of combustion products is known air-fuel ratio can be calculated by the following methods :

1. **Fuel composition known**
   (*i*) Carbon balance method
   (*ii*) Hydrogen balance method
   (*iii*) Carbon-hydrogen balance method.
2. **Fuel composition unknown**
   (*i*) Carbon-hydrogen balance method.

1. Fuel composition known :

(*i*) *Carbon balance method.* When the fuel composition is known, the carbon balance method is quite accurate if *composition takes place with excess air and when free (solid) carbon is not present in the products.* It may be noted that the Orsat analysis will not determine the quantity of solid carbon in the products.

(*ii*) *Hydrogen balance method.* This method is used when solid carbon is suspected to be present.

(*iii*) *Carbon-hydrogen balance method.* This method may be employed *when there is some uncertainty about the nitrogen percentage reported by the orsat analysis.*

2. Fuel composition unknown. When the fuel composition is not known the carbon-hydrogen balance method has to be employed.

8.11. HOW TO CONVERT VOLUMETRIC ANALYSIS TO WEIGHT ANALYSIS ?

The conversion of volumetric analysis to weight analysis involves the following steps :

1. Multiply the volume of each constituent by its molecular weight.
2. Add all these weights and then divide each weight by the total of all and express it as percentage.

8.12. HOW TO CONVERT WEIGHT ANALYSIS TO VOLUMETRIC ANALYSIS ?

1. Divide the weight of each constituent by its molecular weight.
2. Add up these volumes and divide each volume by the total of all and express it as a percentage.

8.13. WEIGHT OF CARBON IN FLUE GASES

The weight of carbon contained in one kg of flue or exhaust gas can be calculated from the amounts of CO_2 and CO contained in it.

In eqn. [8.4(*b*)], it was shown that 1 kg of carbon produces 11/3 kg of CO_2 when completely burnt. Hence 1 kg of CO_2 will contain 3/11 kg of carbon.

In eqn. [8.6(*a*)], it can be seen that 1 kg of carbon produces 7/3 kg of CO, hence 1 kg CO contains 3/7 kg of carbon.

Therefore, weight of carbon per kg of fuel

$$= \left(\frac{3}{11}CO_2 + \frac{7}{3}CO\right)$$

where CO_2 and CO are the quantities of carbon dioxide and carbon monoxide present in 1 kg of flue or exhaust gas.

8.14. WEIGHT OF FLUE GASES PER KG OF FUEL BURNT

Due to supply of air, the weight of flue gas or exhaust gas is always more than that of fuel burnt. The actual weight of dry flue gases can be obtained by comparing the weight of carbon present in the flue gases with the weight of carbon in the fuel, since there is no loss of carbon during the combustion process. As the analysis of the exhaust gases is volumetric, so this *must first be reduced to weight analysis.*

Also, *total weight of carbon in one kg of flue gas is*

$$= \left(\frac{3}{11}CO_2 + \frac{3}{7}CO\right)$$

∴ The *weight of flue gas/kg of fuel burnt*

$$= \frac{\text{Weight of carbon is one kg of fuel}}{\text{Weight of carbon in one kg of fuel gas}}$$

8.15. ANALYSIS OF EXHAUST AND FLUE GAS

The combustion products are mainly gaseous. When a sample is taken for analysis it is usually cooled down to a temperature which is below the saturation temperature of the steam present. The *steam content* is therefore *not included* in the analysis, which is then quoted as the *analysis of the dry products.* Since the products are gaseous, it is usual to quote the analysis by *volume. An analysis which includes the steam in the exhaust is called a wet analysis.*

Practical analysis of combustion products :

The most common means of analysis of the combustion products is the ***Orsat apparatus*** which is described below :

Construction. An Orsat's apparatus consists of the following :

(*i*) A burette

(*ii*) A gas cleaner

(*iii*) Four absorption pipettes 1, 2, 3, 4.

The pipettes are interconnected by means of a manifold fitted with cocks S_1, S_2, S_3 and S_4 and contain different chemicals to absorb carbon dioxide (CO_2), carbonmonoxide (CO) and oxygen (O_2). Each pipette is also fitted with a number of small glass tubes which provide a greater amount of surface. These tubes are wetted by the absorbing agents and are exposed to the gas under analysis. The measuring burrette is surrounded by a *water jacket* to prevent, changes in temperature and density of the gas. The pipettes 1, 2, 3, 4 contain the following chemicals :

Pipette 1. Contains *'KOH'* (caustic soda) *to absorb* CO_2 (carbon dioxide)

Pipette 2. Contains *an alkaline solution of 'pyrogallic acid'* to absorb O_2 (oxygen)

Pipette 3, 4. Contain an *acid solution of 'cuprous chloride'; to absorb* CO (carbonmonoxide)

Furthermore the apparatus has a *levelling bottle* and a three way cock to connect the apparatus either to gases or to atmosphere.

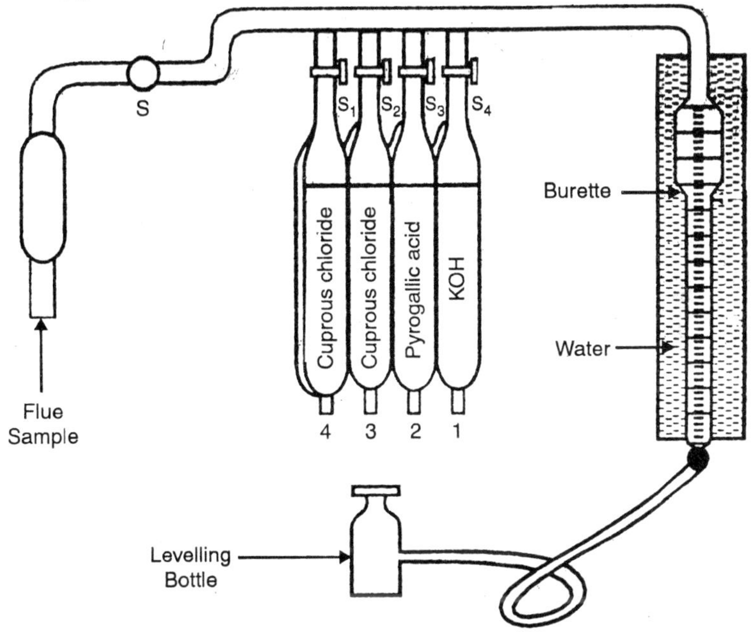

Fig. 8.1. Orsat's apparatus.

Procedure. 100 cm^3 of gas whose analysis is to be made is drawn into the bottle by lowering the levelling bottle. The stop cock S_4 is then opened and the whole flue gas is forced to pipette 1. The gas remains in this pipette for sometime and most of the carbon monoxide is absorbed. The levelling bottle is the lowered to allow the chemical to come to its carbon monoxide is absorbed. The leveling bottle is then lowered to allow the chemical to come to its original level. The volume of gas thus absorbed is read on the scale of the measuring bottle. The flue gas is then forced through the pipette 1 for a number of times to ensure that the whole of the CO_2 is absorbed. Further, the remaining flue gas is then forced to the pipette 2 which contains pyrogallic acid to absorb whole of O_2. The reading on the measuring burette will be the sum of volume of CO_2 and O_2. The oxygen content can then be found out by substraction. Finally, as before, the sample of gas is forced through the pipettes 3 and 4 absorb coarbon monoxide completely.

The amount of nitrogen in the sample can be determined by subtracting from total volume of gas the same of CO_2, CO and O_2 contents.

Orsat apparatus *gives an analysis of the dry products of combustion.* Steps may have been taken to remove the steam from the sample by condensing, but as the sample is collected over water it becomes saturated with water. The resulting analysis is nevertheless a true analysis of the dry products. This is because the volume readings are taken at a constant temperature and pressure, and the partial pressure of the vapour is constant. This means that the sum of the Typical values of ΔH_f for different substances at 25°C (298 K) in **kJ/mole** are given below:

S. No.	Substance	Formula	State	ΔH_f
1.	Oxygen	O	gas	249143
		O_2	gas	zero
2.	Water	H_2O	liquid	–285765
			vapour	–241783
3.	Carbon	G	gas	714852
			diamond	1900
			graphite	zero
4.	Carbon monoxide	CO	gas	–111508
5.	Carbon dioxide	CO_2	gas	–393443
6.	Methane	CH_4	gas	–74855
7.	Methyl alcohol	CH_3OH	vapour	–240532
8.	Ethyl alcohol	C_2H_5OH	vapour	–281102
9.	Ethane	C_2H_6	gas	–83870
10.	Ethene	C_2H_4	gas	51780
11.	Propane	C_3H_8	gas	–102900
12.	Butane	C_4H_{10}	gas	–125000
13.	Octane	C_8H_{18}	liquid	–247600

8.16. CALORIFIC OR HEATING VALUES OF FUELS

The *"calorific value or heating value" of the fuel is defined as the energy liberated by the complete oxidation of a unit mass or volume of a fuel.* It is expressed in kJ/kg for solid and liquid fuels and kJ/m^3 for gases.

If a fuel contains hydrogen water will be formed as one of the products of combustion. If this *water is condensed, a large amount of heat will be released* than if the water exists in the vapour phase. For this reason *two heating values* are defined; the *higher or gross heating value* and the *lower* or *net heating value.*

The *higher heating value, HHV, is obtained when the water formed by combustion is completely condensed.*

The *lower heating value,* LHV, is obtained when the water formed by combustion exists *completely in the vapour phase.*

Thus :

$$(HHV)_p = (LHV)_p + m\, h_{fg} \quad ...(12)$$

$$(HHV)_p = (LHV)_v + m\, (u_g - u_f) \quad ...(13)$$

where m = Mass of water formed by combustion,

h_{fg} = Enthalpy of vaporization of water, kJ/kg,

u_g = Specific internal energy of vapour, kJ/kg, and

u_f = Specific internal energy of liquid, kJ/kg.

In almost all practical cases, the water vapour in the products is vapour, the lower value is the one *which usually applies.*

8.17. DETERMINATION OF CALORIFIC OR HEATING VALUES

The calorific value of fuels can be determined either from chemical analysis or in the laboratory.

8.17.1. Solid and Liquid Fuels

Dulong's formula. Dulong suggested a formula for the calculation of the calorific value of the solid or liquid fuels from chemical composition which is as given below.

Gross calorific value

$$\text{or HHV} = \frac{1}{100}\left[33800\,C + 144000\left(H - \frac{0}{8}\right) + 9270\,S\right] \text{kJ/kg} \quad ...(8.14)$$

where C, H, O and S are carbon, hydrogen, oxygen and sulphur in percentages respectively in 100 kg of fuel. In the above formula the oxygen is assumed to be in combination with hydrogen and only extra surplus hydrogen supplies the necessary heat.

Laboratory Method (Bomb Calorimeter)

The calorific value of solid and liquid fuels is determined in the laboratory by *'Bomb calorimeter'*. It is so named because its shape resembles that of a bomb. Figure 8.2 shows the schematic sketch of a bomb calorimeter.

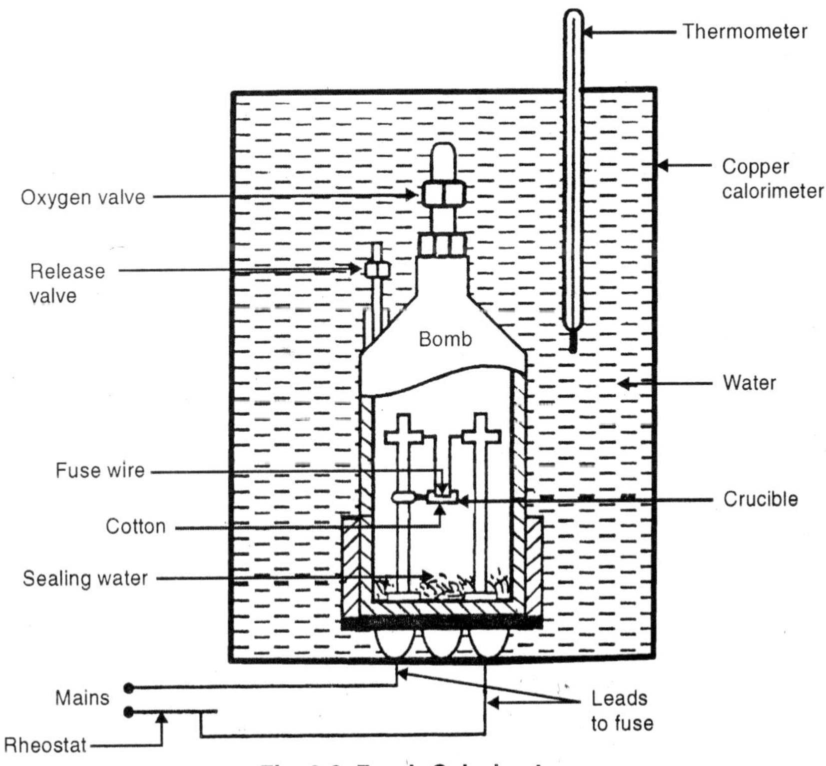

Fig. 8.2. Bomb Calorimeter.

The *calorimeter* is made of austenitic steel which provides considerable resistance to corrosion and enables it to withstand high pressure. In the calorimeter is a *strong cylindrical bomb* in which combustion occurs. The bomb has *two valves* at the top. One supplies oxygen to the bomb and other releases the exhaust gases. A *crucible* in which a weighted quantity of fuel sample is burnt is arranged between the *two electrodes* as shown in Fig. 8.2. The calorimeter is fitted with *water jacket* which surrounds the bomb. To reduce the losses due to radiation, calorimeter is further provided with a jacket of water and air. A stirrer for keeping the temperature of water uniform and a thermometer to measure the temperature up to an accuracy of 0.001°C are fitted through the lid of the calorimeter.

Procedure. To start with, about 1 gm of fuel sample is accurately weighed into be crucible and a fuse wire (whose weight is known) is stretched between the electrodes. It should be ensured that *wire is in close contact with the fuel.* To absorb the combustion products of sulphur and nitrogen 2 ml of water is poured in the bomb. Bomb is then supplied with pure oxygen through the valve to an amount of 25 atmosphere. The bomb is then placed in the weighed quantity of water, in the calorimeter. The stirring is started after making necessary electrical connections, and when the thermometer indicates a steady temperature fuel is fired and temperature readings are recorded after 1/2 minute intervals until maximum temperature is stained. The bomb is then removed; the pressure slowly released through the exhaust valve and the contents of the bomb are carefully weighted for further analysis.

The heat released by the fuel on combustion is absorbed by the surrounding water and the calorimeter.

From the above data the calorific value of the fuel can be found in the following way :

Let w_f = Weight of fuel sample (kg),

w = Weight of water (kg),

C = Calorific value (higher) of the fuel (kJ/kg),

w_e = Water equivalent of calorimeter (kg),

t_1 = Initial temperature of water and calorimeter,

t_2 = Final temperature of water and calorimeter,

t_c = Radiation corrections, and

c = Specific heat of water.

Heat released by the fuel sample = $w_f \times C$

Heat received by water and calorimeter

$$= (w_w + w_e) \times c \times [(t_2 - t_1) + t_c]$$

$$\text{Heat lost} = \text{Heat gained}$$

$$\therefore \quad w_f \times C = (w + w_e) \times c \times [(t_2 - t_1) + t_c]$$

i.e.,

$$C = \frac{(w + w_e) \times c \times [(t_2 - t_1) + t_c]}{w_f} \qquad ...(8.15)$$

[Value on c is 4.18 in SI units and unity in MKS units.]

Notes 1. Corrections pertain to the heat of oxidation of fuse wire, heat liberated as a result of formation of sulphuric and nitric acids in the bomb itself.

2. It should be noted that *bomb calorimeter measures the higher or gross calorific value because the fuel sample is burnt at a constant volume in the bomb. Further* the bomb calorimeter will measure the H.C.V. directly if the bomb contains adequate amount of water before firing to saturate the oxygen. Any water formed from combustion of hydrogen will, therefore, be condensed.

The procedure of determining calorific value of *liquid fuels* is similar to that described above. However, if the liquid fuel sample is volatile, it is weighted in a glass bulb and broken in a tray just before the bomb is closed. In this way the loss of volatile constituents of fuels during weighting operation is prevented.

8.17.2. Gaseous Fuels

The calorific value of gaseous fuels can be determined by ***Junker's gas calorimeter.***

Figure 8.3 illustrates *Junker's gas calorimeter.* Its principle is some what similar to Bomb calorimeter; in respect that heat evolved by burning the gas is taken away by the water. In its simplest construction it consists of a combustion chamber in which the gas is burnt (in a gas burner). A water jacket through which a set of tubes called flues pass surrounds this chamber. Thermometers are incorporated at different places (as shown in Fig. 8.3) to measure the temperatures.

Procedure. A metered quantity of gas whose calorific value is to be determined is supplied to the gas burner via a gas meter which records its volume and a gas pressure regulator which measures the pressure of the gas by means of a manometer. When the gas burns the hot products of combustion travel upwards in the chamber and them downwards through the flues and finally escape to the atmosphere through the outlet. The temperature of the escaping gas is recorded by the thermometer fitted at the exit and this temperature should be as close to room temperature as possible so that entire heat of combustion is absorbed by water. The cold water enters the calorimeter near the bottom and leaves near the top. Water which is formed by condensation of steam is collected in a pot.

The quantity of gas used during the experiment is accurately measured by the meter and temperature of ingoing and outgoing water are indicated by the thermometers. From the above data the calorific value of the gas can be calculated.

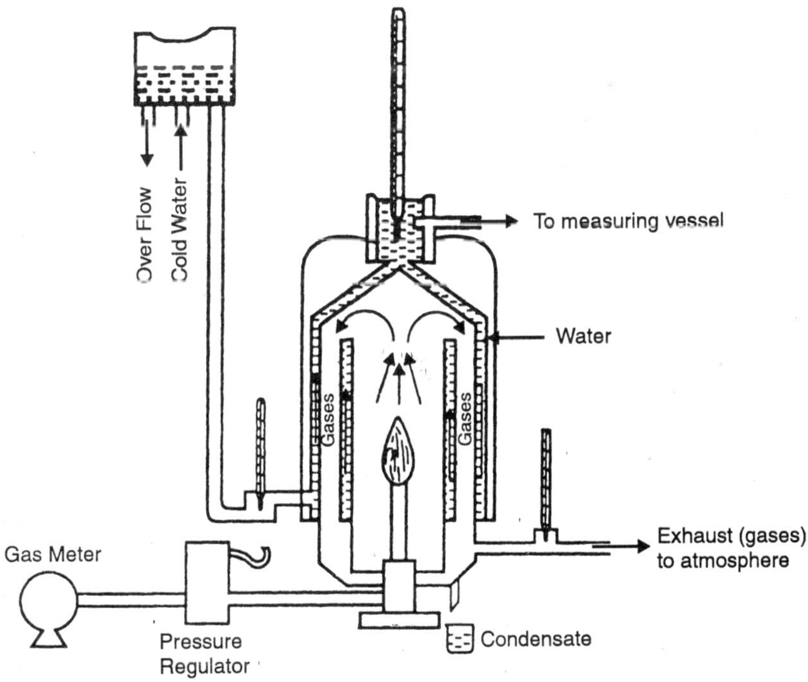

Fig. 8.3. Junkers's gas calorimeter.

Notes 1. The volume of gas used during the experiment should be converted to volume at S.T.P. (standard temperature and pressure *i.e.,* 15°C, 760 mm respectively).

2. The gross calorific value is obtained by dividing the heat given out by corrected volume of gas. The net or lower calorific value of the gas is obtained by substracting from total heat the heat associated with condensed water (which is obtained by multiplying the weighed condensate by latent heat of vapourization of water).

8.18. ADIABATIC FLAME TEMPERATURE

In a given combustion process, that takes place *adiabatically* and with no work or changes in kinetic or potential energy involved, *the temperature of the products is referred to as the 'adiabatic flame temperature'*. With the assumptions of no work and no changes in kinetic or potential energy, this is the *maximum temperature* that can be achieved for the given reactants because any heat transfer from the reacting substances and any incomplete combustion would tend to lower the temperature of the products.

The following points are worthnoting :

(*i*) *The maximum temperature achieved through adiabatic complete combustion varies with the type of reaction and percent of theoretical air supplied.*

An *increase in the air-fuel* ratio *will effect a decrease in the maximum temperature.*

(*ii*) For a given fuel and given pressure and temperature of the reactants, the *maximum adiabatic flame temperature that can be achieved is with a 'stoichiometric mixture'.*

(*iii*) The adiabatic flame temperature can be *controlled* by the *amount of excess air* that is used. This is important, for example, in *gas turbines,* where the maximum permissible temperature is determined by *metallurgical considerations* in the turbine, and *close control of the temperature of the products is essential.*

8.19. CHEMICAL EQUILIBRIUM

The calculation of the *adiabatic flame temperature* is based, in part, on the assumption that the reaction goes to *completion.* Owing to *dissociation,* complete conversion of the reactants to the products is not accomplished. As a consequence of the failure to achieve complete conversion of the reactants, the maximum reaction temperature cannot attain the level of the theoretical adiabatic flame temperature.

The combination of CO and O_2 produces CO_2 together with a release of energy. In an adiabatic system no heat is transferred to the surroundings, hence the temperature of the mixture of the products and reacting substances rises rapidly. As the mixture temperature increases to higher levels the rate of dissociation of the CO_2 becomes increasingly more pronounced. Since the dissociation of CO_2 requires absorption of energy, *a condition is reached where the rate of evolution and the rate of absorption of energy are in balance. At this point no further increase in temperature can be observed* and the reaction is in *chemical equilibrium.* For this condition

$$C + O_2 \rightleftharpoons CO_2$$

At each temperature of the equilibrium mixture the substances participating in the reaction exist in unique proportions. For the combustion of CO the right-hand side of the equation

$$CO + \frac{1}{2}O_2 = (1-x)\,CO_2 + x\,CO + \frac{1}{2}O_2 \qquad ...(8.16)$$

represents the distribution of the equilibrium products resulting from a reaction between CO and O_2. In this equation x denotes the fraction of dissociated CO_2. At low temperatures the fraction $(1 - x)$ approaches unity while at high temperatures $(1 - x)$ shows a substantial reduction in magnitude.

For the combustion of H_2 with O_2

$$H_2 + \frac{1}{2}O_2 \rightleftharpoons H_2O$$

and

$$H_2 + \frac{1}{2}O_2 = (1-x)H_2O + x\,H_2 + \frac{x}{2}\,O_2 \qquad ...(8.17)$$

It is essential to distinguish between the effects of dissociation and the losses resulting from incomplete combustion of fuel. Incomplete combustion, which may be attributed to a number of factors, results in a discharge from the system of combustible substances. *Dissociation,* on the other hand, is of *transient nature.* Usually any appreciable degree of dissociation extends over a very short time interval at the highest level of temperature attained in the reaction. The gaseous products are likely to be discharged from the system at a temperature that is indicative of a low degree of dissociation. For example, *dissociation does not influence the heating value determined in a fuel calorimeter.* Although the maximum temperature attained in the calorimeter is limited by chemical equilibrium, the combustion process moves to completion with the decrease in the temperature of the products. The reduction in temperature is a result of heat transfer to the jacket water. Dissociation of the products is negligible at room temperature, which is essentially the calorimeter reaction temperature.

The temperature of the products discharged from the combustion chamber of the gas turbine power plant is limited to approximately 870°C by introduction of a *large quantity of excess air.* Absorption of energy in the water walls of a boiler furnace limits the outlet gas temperature to approximately 1100°C. The quantity of dissociated products at temperatures ranging upward to 1100°C is not appreciable. In the cylinder of I.C. engine, considerably higher maximum temperatures – that is, in excess of 1100°C are attained, hence in the analysis of this thermal power system consideration must be given to the effects of dissociation. Of particular significance is the effect of reduced maximum temperature on the system availability. As a result of heat transfer and work performed by the gaseous medium the products are discharged from the system at a temperature below the level at which an appreciable degree of dissociation is observed.

The proportions of the dissociated products in chemical equilibrium at temperature T are established from the *equilibrium constant.* The evaluation of the equilibrium constant is achieved in accordance with the analysis presented by Van't Hoff.

8.20. ACTUAL COMBUSTION ANALYSIS

In evaluating the performance of and actual combustion process a number of different parameters can be defined depending on the nature of the process and the system considered. The combustion efficiency in a gas turbine for instance can be defined as

$$\eta_{combustion} = \frac{(F/A)_{ideal}}{(F/A)_{actual}} \quad ...(8.18)$$

where $(F/A)_{ideal}$ = Fuel-air ratio required or adiabatic and complete combustion and in which the products would attain the adiabatic flame temperature.

In case of a *steam generator (boiler)*

$$\eta_{steam\ generator} = \frac{\text{Heat transferred to steam / kg fuel}}{\text{Higher heating value of the fuel}} \quad ...(8.19)$$

In case of an *internal combustion engine,*

$$\eta_{thermal} = \frac{W_{actual}}{\text{Heating value}} \quad ...(8.20)$$

NUMERICAL EXAMPLES

Example 8.1. The following particulars refer to an experimental determination of the calorific value of a sample of coal containing 88% C and 4.2%. Weight of coal = 0.848 gm, weight of fuse wire 0.027 gm, of calorific value 6700 J/gm, weight of water in the calorimeter = 1950 gm, water equivalent of calorimeter = 380 gm, observed temperature rise = 3.06°C, cooling correction = + 0.017°C.

Find the higher and lower calorific values of the coal.

Solution. Percentage of carbon in coal = 88%

Percentage of hydrogen in coal = 4.2%

Weight of coal, w_f = 0.848 gm

Weight of fuse wire, w_{fw} = 0.027 gm

Weight of water in the calorimeter, w = 1950 gm

Water equivalent of calorimeter, w_e = 380 gm

Observed temperature rise $(t_2 - t_1)$ = 3.06°C

Cooling correction, t_c = + 0.017°C

∴ Corrected temperature rise

$$= (t_2 - t_1) + t_c$$
$$= 3.06 + 0.017 = 3.077°\text{C}$$

Calorific value of fuse wire = 6700 J/gm

Heat received by water $= (w - w_e) \times 4.18 \times [(t_2 - t_1) + t_c]$

$$= (1950 + 380) \times 4.18 \times 3.077 = 29968 \text{ J}$$

Heat given out by fuse wire

$$= w_{fw} \times \text{calorific value}$$
$$= 0.027 \times 6700 = 180.9 \text{ J}$$

Heat produced due to combustion of fuel

$$= 29968 - 180.9 = 29787 \text{ J}$$

∴ Higher calorific value of fuel,

$$\text{H.C.V.} = \frac{29787}{0.848} = 35126 \text{ J/gm} = 35126 \text{ kJ/kg}. \textbf{ Ans.}$$

Steam produced per kg of coal = 9 × 0.042 = 0.378 kg

Lower calorific value of coal,

$$\text{L.C.V.} = \text{H.C.V.} - 2465 \times 0.378 = 35126 - 931.7$$
$$= 34194.3 \text{ kJ/kg}. \quad \textbf{Ans.}$$

Example 8.2. Following results were obtained when a sample of gas was tested by Junker's gas calorimeter.

Gas burnt in the calorimeter = 0.08 m^3, Pressure of gas supply = 5.2 cm of water, Barometer = 75.5 cm of Hg. Temperature of gas = 13°C, Weight of water heated by gas = 28 kg, Temperature of water at inlet = 10°C, Temperature of water at outlet = 23.5°C, Steam condensed = 0.06 kg.

Determine the higher and lower calorific values per m^3 of the gas at a temperature of 15°C and barometric pressure of 76 cm of Hg.

Solution. The volume of the gas is measured at a temperature of 13°C and pressure of 5.2 cm of water. Let us reduce this volume to S.T.P. by using the general gas equation,

$$\frac{p_1V_1}{T_1} = \frac{p_2V_2}{T_2}$$

$p_1 = 75.5 + \left(\frac{5.2}{13.6}\right) = 75.882$ cm of Hg; $T_1 = 273 + 13 = 286$ K

$V_1 = 0.08$ m^3; $p_2 = 76$ cm of Hg

$V_2 = ?$; $T_2 = 273 + 15 = 288$ K

$$\therefore \quad \frac{75.882 \times 0.08}{286} = \frac{76 \times V_2}{288}$$

i.e.,
$$V_2 = \frac{75.882 \times 0.08 \times 288}{286 \times 76} = 0.0804 \text{ m}^3$$

Heat received by water $= 28 \times 4.18 \times (23.5 - 10) = 1580$ kJ

Higher calorific value of fuel $= \frac{1580}{0.08} = 19750$ kJ/m^3. **Ans.**

Amount of water vapour formed (*i.e.*, steam condensed) per m^3 of gas burnt

$= \frac{0.06}{0.08} = 0.75$ kg

Lower calorific value,

$$\text{L.C.V.} = \text{H.C.V.} - 2465 \times 0.75$$
$$= 19750 - 1848.7 = 17901.3 \text{ kJ/kg.} \quad \textbf{Ans.}$$

Example 8.3. A coal sample gave the following analysis by weight, Carbon 85 percent, Hydrogen 6 percent, Oxygen 6 percent, the remainder being incombustible. Determine minimum weight of air required per kg of coal for chemically correct composition.

Solution.

Element, wt. (kg)	O_2 required (kg)
C = 0.85	$0.85 \times \frac{8}{3} = 2.27$
$H_2 = 0.06$	$0.06 \times 8 = 0.48$
$O_2 = 0.06$	
	Total $O_2 = 2.75$

Weight of O_2 to be supplied

= Wt. of O_2 needed – Wt. of O_2 already present in fuel

= 2.75 0.06 = 2.69 kg

Weight of air needed $= 2.69 \times \frac{100}{23} = 11.70$ kg. **Ans.**

Example 8.4. The percentage composition of sample of liquid fuel by weight is, C = 84.8 percent, and H_2 = 15.2 percent. Calculate (*i*) the weight of air needed for the combustion of 1 kg of fuel; (*ii*) the volumetric composition of the products of combustion if 15 percent excess air is supplied.

Solution.

Element, wt. (kg)	O_2 used (kg)	Dry products (kg)
C = 0.848	$0.848 \times \frac{8}{3} = 2.261$	$\frac{0.848 \times 11}{3} = 3.109\ (CO_2)$
H_2 = 0.152	$0.152 \times 8 = 1.216$	
	Total O_2 = 3.477	

(*i*) **Minimum weight of air needed for combustion**

$$= \frac{3.477 \times 100}{23} = 15.11 \text{ kg.} \quad \textbf{Ans.}$$

$$\text{Excess air supplied} = \frac{15.11 \times 15}{100} = 2.266 \text{ kg}$$

$$\text{Wt. of oxygen in excess air} = \frac{2.266 \times 23}{100} = 0.521 \text{ kg}$$

Total air supplied for combustion = Minimum air + Excess air

$$= 15.11. + 2.266 = 17.376 \text{ kg}$$

$$\therefore \quad \text{Wt. of nitrogen } (N_2) \text{ in flue gases} = \frac{17.376 \times 77}{100} = 13.38 \text{ kg.}$$

(*ii*) **To get volumetric composition of the product of combustion let us use tabular method.**

Name of gas	Weight (x)	Molecular weight (*y*)	Proportional volume volume (*z*) = $\frac{(x)}{(y)}$	Percentage volume = $\frac{(z)}{\Sigma(z)} \times 100$
CO_2	3.109	44	0.0707	12.51 percent **Ans.**
O_2	0.521	32	0.0163	2.89 percent **Ans.**
N_2	13.38	28	0.4780	84.60 percent **Ans.**
			$\Sigma z = 0.5650$	

Example 8.5. Percentage volumetric analysis of a sample of flue gases of a coal fired boiler gave CO_2 = 10.4; CO = 0.2; O_2 = 7.8 and N_2 = 81.6 (by difference). Gravimetric percentage analysis of coal was C = 78, H_2 = 6, O_2 = 3 incombustible = 13. Estimate:

(*i*) **Weight of dry flue gases per kg of fuel.**

(*ii*) **Weight of excess air per kg of fuel.**

Solution.

Element, wt. (kg)	O_2 reqd. (kg)	Dry products (kg)
C = 0.78	$0.78 \times \frac{8}{3}$	$0.78 \times \frac{11}{3} = 2.86\ (CO_2)$
H_2 = 0.06	$0.06 \times 8 = 0.48$	
O_2 = 0.03		
	Total O_2 = 256	

$$\text{Minimum wt. of air needed for combustion} = (2.56 - 0.03) \times \frac{100}{23} = 11 \text{ kg.}$$

(*i*) **Weight of dry flue gases per kg of fuel :** To determine the wt. of flue gases per kg of fuel let us use tabular method to convert volumetric analysis to analysis by weight.

Name of gas	Volume per m^3 of flue gas (x)	Molecular weight (y)	Relative volume $z = x \times y$	Weight per kg of flue gas $\frac{z}{\Sigma z}$
CO_2	0.104	44	4.576	0.1525
CO	0.002	28	0.056	0.0019
N_2	0.816	28	22.848	0.7616
O_2	0.078	32	2.496	0.832
			$\Sigma z = 29.976$ (say 30)	

Amount of carbon present per kg of gases

= Amount of carbon in 0.1525 kg of CO_2

+ Amount of carbon present in 0.0019 kg of CO

$$= \frac{3}{11} \times 0.1525 + \frac{3}{7} \times 0.0019 = 0.0416 + 0.0008 = 0.0424 \text{ kg.}$$

Also carbon in the fuel = 0.78 kg.

∴ Weight of dry flue per kg of fuel

$$= \frac{\text{Weight of carbon in 1 kg of fuel}}{\text{Weight of carbon in 1 kg of flue gas}}$$

$$= \frac{0.78}{0.0424} = 18.4 \text{ kg. } \textbf{Ans.}$$

(*ii*) **Weight of excess air per kg of fuel :**

Weight of excess oxygen per kg of flue gas $= 0.0832 - \frac{4}{7} \times .0019$

$$\begin{bmatrix} 2CO + O_2 = 2CO_2 \\ 56 \quad 32 \quad 88 \\ 1 \text{ kg} + \frac{4}{7} \text{ kg} = \frac{11}{7} \text{ kg} \end{bmatrix}$$

= 0.0832 – 0.0011 (allowing for unburnt carbon monoxide)

= 0.0821 kg.

Weight of excess O_2 per kg of fuel = 18.4 × 0.0821 = 1.51 kg

$$\therefore \text{ Weight of excess air per kg of fuel} = \frac{1.51 \times 100}{23} = 6.56 \text{ kg. } \textbf{Ans.}$$

Example 8.6. A single cylinder was supplied with a gas having the following percentage volumetric analysis; CO = 5, CO_2 = 10, H_2 = 50, CH_4 = 25, N_2 = 10. The percentage volumetric analysis of dry gases was CO_2 = 8, O_2 = 6 and N_2 = 86. Determine the air-fuel ratio by volume.

Solution. Combustion equations are :

$$2H_2 + O_2 = 2H_2O$$
$$1 \text{ vol} + 1/2 \text{ vol} = 1 \text{ vol}$$
$$2CO + O_2 = 2CO_2$$

$$1 \text{ vol} + 1/2 \text{ vol} = 1 \text{ vol}$$
$$CH_4 + 2O_2 = CO_2 + 2H_2O$$
$$1 \text{ vol} + 2 \text{ vol} = 1 \text{ vol} + 2 \text{ vol.}$$

Gas	Vol (m^3)	O_2 needed (m^3)	Products (m^3)	
			CO_2	N_2
CO	0.05	0.025	0.05	–
CO_2	0.10	–	0.10	–
H_2	0.50	0.25	–	–
CH_4	0.25	0.50	0.25	–
N_2	0.10	–	–	0.10
Total	1.0	0.775	0.4	0.10

Volume of air required = $0.775 \times 100/21 = 3.69 \text{ m}^3$

Volume of nitrogen in the air = $3.69 \times 79/100 = 2.92 \text{ m}^3$

Dry combustion products of 1 m_3 of gases (V) contain 0.4 m^3 of CO_2 + 0.1 m^3 of N_2 (as given in the table) + 2.92 m^3 of N_2 (from air supplied for complete combustion) = 3.42 m^3.

$$\text{Excess air supplied} = \frac{O_2 \times V}{(21 - O_2)} = \frac{6.0 \times 3.42}{(21 - 6)} = \frac{20.52}{15} = 1.37 \text{m}^3$$

Total quantity of air supplied = $3.69 + 1.37 = 5.06 \text{ m}^3$.

$$\text{Air fuel ratio} = \frac{\text{Volume of air}}{\text{Volume of fuel}} = \frac{5.06}{1} = 5 \text{ (say).}$$ **Ans.**

Example 8.7. The following is the ultimate analysis of a sample of petrol by weight :

Carbon = 85 percent; Hydrogen = 15 percent.

Calculate the ratio of air to petrol consumption by weight if the volumetric analysis of the dry exhaust gas is:

CO_2 = 11.5 percent; CO = 12 percent; O_2 = 0.9 percent; N_2 = 86 percent.

Also find percentage excess air.

Solution.

Name of gas (x)	Volume per m^3 of flue gas (y)	Molecular weight $z = x \times y$	Relative weight $\frac{z}{\Sigma z}$	Weight per kg of flue gas
CO_2	0.115	44	5.06	1.1700
CO	0.012	28	0.336	0.1113
O_2	0.009	32	0.288	0.0096
N_2	0.86	28	24.08	0.8091
			$\Sigma z = 29.76$	

∴ Weight of carbon per kg of flue gas

= Weight of carbon in 0.17 kg of CO_2 + Weight of carbon in 0.0113 kg of CO

$$= \frac{3}{11} \times 0.17 + \frac{3}{7} \times 0.0113 = 0.0512 \text{ kg}$$

$\therefore$ Weight of dry flue gas per kg of fuel $= \dfrac{0.85}{0.0512} = 16.6$ kg

Vapour of combustion $= 9 \times 0.15 = 1.35$ kg

Total weight of gas $= 16.6 + 1.35 = 17.95$ kg per kg of fuel

$\therefore$ Air supplied $= (17.95 - 1) = 16.95$ kg/kg of fuel

$\therefore$ Ratio of air to petrol = 16.95 : 1. **Ans.**

$$\text{Stoichiometric air} = \left[\left(0.85 \times \frac{8}{3}\right) + (0.15 \times 8)\right] \times \frac{100}{23}$$

$$= 15.07 \text{ kg per kg of fuel}$$

$\therefore$ Excess air $= 16.95 - 15.07 = 1.88$ kg

$\therefore$ Percentage excess air $= \dfrac{1.88}{15.07} \times 100 = 12.47\%$. **Ans.**

Example 8.8. A sample of fuel has the following percentage composition : Carbon = 86 percent; Hydrogen = 8 percent; Sulphur = 3 percent; Oxygen = 2 percent; Ash = 1 percent.

For an air-fuel ratio of 12 : 1, calculate :

(*i*) Mixture strength as a percentage rich or weak.

(*ii*) Volumetric analysis of the dry products of combustion.

Solution.

Element, wt. (kg)	O_2 reqd. (kg)
C = 0.86	$0.86 \times \frac{8}{3} = 2.29$
H_2 = 0.08	$0.08 \times 8 = 0.64$
S = 0.03	$0.03 \times \frac{1}{1} = 0.03$
O_2 = 0.02	
	Total O_2 = 2.96

Weight of oxygen to be supplied per kg of fuel = 2.96 – 0.02 = 294 kg.

Weight of minimum air required for complete combustion

$$= \frac{2.94 \times 100}{23} = 12.78 \text{ kg}$$

Hence "correct" fuel-air ratio $= \dfrac{1}{12.78} : 1$

But actual ratio is $\dfrac{1}{12} : 1$

(*i*) Mixture strength $= \dfrac{12.78}{12} \times 100 = 106.5\%$

This show that mixture is 6.5% rich **Ans.**

Deficient amount of air = 12.78 – 12 = 0.78 kg

Amount of air saved by burning 1 kg of C to CO instead of CO_2

$$= \text{Oxygen saved} \times \frac{100}{23}$$

$$= \left[\frac{8}{3}CO_2 - \frac{4}{3}(CO)\right] \times \frac{100}{23} = 5.8 \text{ kg}$$

Hence $\frac{0.78}{5.8}$ = 0.134 kg of carbon burns to CO and as such 0.86 – 0.134 = 0.726 kg of carbon burns to CO_2.

$$\therefore \quad \text{CO formed} = 0.134 \times \frac{7}{3} = 0.313 \text{ kg}$$

$$CO_2 \text{ formed} = 0.726 \times \frac{11}{3} = 2.662 \text{ kg}$$

$$N_2 \text{ supplied} = 12 \times 0.77 = 9.24 \text{ kg}$$

$$SO_2 \text{ formed} = 0.03 \times 2 = 0.06 \text{ kg.}$$

(*ii*) The percentage composition of dry flue gases is given as below :

Dry products	Weight (kg) x	Molecular weight y	Proportional volume $z = \frac{x}{y}$	Percentage volume $\frac{z}{\Sigma z} \times 100$
CO	0.313	28	0.0112	2.78 percent **Ans.**
CO_2	2.662	44	0.0605	15.03 percent **Ans.**
N_2	9.24	28	0.3300	81.97 percent **Ans.**
SO_2	0.06	64	0.0009	0.22 percent **Ans.**
			$\Sigma z = 0.4026$	

ASSIGNMENT

OBJECTIVE TYPE QUESTIONS :

(A) Multiple Choice Questions :

Choose the Correct Answer :

1. The smallest particle which can take part in a chemical change is called
   (*a*) atom (*b*) molecule
   (*c*) electron (*d*) compound
2. A chemical fuel is a substance which releases ______ on combustion.
   (*a*) chemical energy (*b*) heat energy
   (*c*) sound energy (*d*) magnetic energy.
3. The most important solid fuel is :
   (*a*) wood (*b*) charcoal
   (*c*) coal (*d*) all of the above.
4. For each mole of oxygen, number of moles of nitrogen required for complete combustion of carbon are :
   (*a*) 20/21 (*b*) 2/21
   (*c*) 77/21 (*d*) 79/21.
5. Modern practice is to use ______ excess air.
   (*a*) 5 to 10 percent (*b*) 15 to 20 percent
   (*c*) 20 to 25 percent (*d*) 25 to 50 percent
6. Stoichiometric air-fuel ratio by mass for combustion of petrol is
   (*a*) 5 (*b*) 10
   (*c*) 12 (*d*) 15.05

7. An analysis which includes the steam in the exhaust is called
   (a) dry analysis (b) wet analysis
   (c) dry and wet analysis (d) none of the above.
8. The Orsat apparatus gives
   (a) volumetric analysis of the dry products of combustion
   (b) gravimetric analysis of the dry products of combustion
   (c) gravimetric analysis of products of combustion including H_2O
   (d) volumetric analysis of products of combustion including H_2O.
9. In the Orsat apparatus KOH solution is used to absorb :
   (a) carbon monoxide (b) carbon dioxide
   (c) oxygen (d) none of the above.
10. Enthalpy of formation is defined as enthalpy of compounds at
   (a) 25°C and 10 atmosphere (b) 25°C and 1 atmosphere
   (c) 0°C and 1 atmosphere (d) 100°C and 1 atmosphere.
11. Bomb calorimeter is used to find the calorimeter value of ______ fuels.
   (a) solid (b) gaseous
   (c) solid and gaseous (d) none of the above.
12. When the fuel is burned and the water appears in the vapour phase, the heating value of fuel is called
   (a) enthalpy of formation (b) lower heating value
   (c) higher heating value (d) none of the above.
13. Heat released in a reaction at constant pressure is called :
   (a) entropy change (b) enthalpy of reaction
   (c) internal energy of reaction (d) none of the above.
   (e) all of the above
14. When the fuel is burned and water is released in the liquid phase, the heating value of fuel is called :
   (a) higher heating value (b) lower heating value
   (c) enthalpy of formation (d) none of the above.
15. Choose the correct statement :
   (a) Number of atoms of each constituent are not conserved in a chemical reaction.
   (b) The mass of all the substances on one side of the equation may not ne equal to the mass of all the substances on the other side.
   (c) The number of atoms of each constituent are conserved in a chemical reaction.
   (d) The number of moles of the reactants in a chemical equation are equal to the number of moles of the products.

ANSWERS

1. (a)	2. (b)	3. (c)	4. (d)	5. (d)
6. (d)	7. (b)	8. (a)	9. (b)	10. (b)
11. (a)	12. (b)	13. (b)	14. (a)	15. (c).

(B) Short Answer Type Questions :

Q. 1. What is a 'fuel' ?

Ans. A 'Fuel' or 'Chemical fuel' is a substance which releases that energy units combustion in the presence of air.

Q. 2. How the fuels are categories as primary and secondary fuels ?

Ans. (i) Primary fuels are the fuels which occur in nature.

(*ii*) Secondary fuels are the fuels which are prepared but not available in nature.

Q. 3. Give the practical examples of primary and secondary fuels which are solid in their states.

Ans. (*a*) Primary Solid Fuels are :

(*i*) Wood (*ii*) Peat (*iii*) Lignite Coal

(*b*) Secondary Solid Fuels :

(*i*) Coke (*ii*) Charcoal (*iii*) Briquettes.

Q. 4. Give the examples of Liquid, primary and secondary fuels.

Ans. (*a*) Liquid Primary Fuel is :

(*i*) Petroleum

(*b*) Liquid Secondary fuels are :

(*i*) Gasoline (*ii*) Kerosene (*iii*) Fuel oil

(*iv*) Alcohol (*v*) Benzol (*vi*) Shale Oil

Q. 5. Give examples of the Gaseous Primary and Secondary Fuels.

Ans. (*a*) Gaseous Primary fuel :

Natural Gas

(*b*) Gaseous Secondary Fuels are :

(*i*) Petroleum Gas (*ii*) Producer gas

(*iii*) Coal Gas (*iv*) Coke-Oven Gas

(*v*) Blast-furnace Gas (*vi*) Carburatted Gas

(*vii*) Sewer Gas.

Q. 6. Enlist the advantages of 'Liquid Fuels' over 'Solid Fuel'.

Ans. (*i*) Requires less space for storage.

(*ii*) Higher calorific value.

(*iii*) Easy control of combustion.

(*iv*) Staff-economy.

(*v*) Absence of danger from spontaneous combustion.

(*vi*) Easy handling and transportation.

(*vii*) Cleanliness.

(*viii*) No ash problem.

(*ix*) Non-determination of the oil in storage.

Q. 7. Define the theoretical air as applied to combustion.

Ans. The minimum amount of air that supplies sufficient oxygen for the complete combustion of all the carbon, Hydrogen and any other elements in the fuel that may oxidise is called the 'Theoretical Air'.

Q. 8. Define 'Excess Air" as applied to combustion process.

Ans. The amount of air given in excess to the theoretical air for the complete combustion of fuel is called 'Excess Air'.

Q. 9. Define 'Stoichiometric' mixture of air and fuel.

Ans. Stoichiometric or chemically correct mixture of air and fuel is one that contains 'just sufficient' oxygen for the complete combustion of the fuel.

Q. 10. Define a 'Weak Mixture'.

Ans. A 'Weak Mixture' is one which has an excess of air.

Q. 11. Define a 'Rich Mixture'.

Ans. A 'Rich Mixture' is one which has a deficiency of air.

Q. 12. Write the formula for the percentage of excess-air in terms of Actual and Stoichiometric A/F ratio.

Ans. $$\text{Percentage of Excess Air} = \frac{\text{Actual A / F ratio} - \text{Stoichimetric} - \text{A / F ratio}}{\text{Stoichiometric A / F ratio}}$$

Q. 13. Write down the main points which explain, that how to convert Volumetric Analysis to Weight Analysis.

Ans. (*i*) Multiply the volume of each constituent by its molecular weight.

(*ii*) Add all these weights and then divide each weight by the total of all and express it as percentage.

Q. 14. How weight analysis is converted into volumetric analysis ?

Ans. (*i*) Divide the weight of each constituent by its molecular not.

(*ii*) Add up these volumes and divide each volume by the total of all and express it as a percentage.

Q. 15. Write the formula to calculate the weight of Carbon per kg of fuel in flue gases.

Ans. Weight of carbon per kg of fuel,

$$= \left(\frac{3}{11}CO_2 + \frac{3}{7}CO\right)$$

where CO_2 is the quantity of Carbon dioxide and CO the quantity of Carbon monoxide present in one kg of fuel or exhaust gas.

Q. 16. Write the formula for the weight of the flue gas per kg of the fuel burnt.

Ans. The Weight of Flue gas/kg of fuel burnt

$$= \frac{\text{Weight of Carbon in one kg of fuel}}{\text{Weight of Carbon in one kg of fluegas}}$$

Q. 17. How the flue-gas analysis is done ? Name the main parts of the apparatus.

Ans. The most common means of analysis of the combustion products is the 'Orsat Apparatus'. It generally consists of :

(*i*) A burette.

(*ii*) A gas cleaner

(*iii*) Four absorption pipettes.

Q. 18. What is contained in the four absorption pipettes in the 'Orsat Apparatus' and why ?

Ans. (*i*) Pipette '1' contains 'KOH' (Caustic Soda) to absorb CO_2 (carbon dioxide).

(*ii*) Pipette '2' contains an alkaline solution of 'Pyrogalic Acid to absorb O_2 (Oxygen).

(*iii*) Pipettes '3' and '4' contain an acid solution of 'Cuprous Chloride' to absorb CO (Carbon monoxide). Further more, the apparatus has a levelling bottle and a three way Cock to connect the apparatus either to gases or to atmosphere.

Q. 19. Define the 'Calorific value' or 'Heating Value' of fuels :

Ans. The 'Calorific value or Heating value' of the fuel is defined as the heat energy liberated by the complete oxidation of a unit mass or volume of a fuel. It is expressed in kJ/kg for solid and liquid fuels and KJ/m^3 for gaseous fuels.

Q. 20. Name the apparatus by which calorific value of fuels is determined generally.

Ans. (*i*) Bomb Calorimeter. (For solid and Liquid fuels)

(*ii*) Junker's gas Calorimeter (For gaseous Fuels)

REVIEW QUESTIONS

1. What is chemical thermodynamics ?
2. What is chemical fuel ?
3. What are primary fuels ? List some important primary fuels.
4. What are secondary fuels ? List some important secondary fuels.
5. Write a short note on 'excess air'.
6. What do you mean by stoichiometric air-fuel (A/F) ratio ?
7. Enumerate the methods by which air fuel ratio can be calculated when analysis of combustion products is known.

8. How is analysis of exhaust and flue gas carried out ?
9. Define heating value of fuel.
10. What is the difference between higher heating value (HHV) and lower heating value (LHV) of the fuel ?
11. Describe with the help of neat sketches the following calorimeters used for the determination of heating values :
    (*i*) Bomb calorimeter (*ii*) Junkers gas calorimeter
12. What is 'adiabatic flame temperature' ?
13. Write a short note of chemical equilibrium.

PROBLEMS FOR PRACTICE

1. A fuel has the following composition by weight : Carbon = 86%; hydrogen = 11.75% and Oxygen = 2.25%. Calculate the theoretical air supply per kg of fuel, and the weight of products of combustion per kg of fuel.
   [Ans. 13.98 kg; 4.21 kg]
2. The volumetric analysis of *a* fuel gas is : CO_2 = 14%; CO = 1%, O_2 = 5% and N_2 = 80%. Calculate the fuel gas composition by weight.
   [Ans. 20.24%, 0.93%; 5.25%, 73.58%]
3. The ultimate analysis of a dry coal burnt in a boiler gauge C = 84%; H_2 = 9% and incombustible 7% by weight. Determine the weight of dry flue gases per kg of coal burnt; if volumetric combustion of the flue gas is :
   CO_2 = 8.75%, CO = 2.25%; O_2 = 80% and N_2 = 81% **[Ans.** 18.92 kg]
4. During a trial in a boiler, the dry flue gas analysis by volume was obtained as CO_2 = 135, CO = 0.3%, O_2 = 6%, N_2 = 80.7%. The coal analysis by weight was reported as C = 62.4%, H_2 = 4.2%, O_2 = 4.5%, moisture = 15% and ash 13.9%. Estimate :
   (*a*) Theoretical air required to burn 1 kg of coal.
   (*b*) Weight of air actually supplied per kg of coal.
   (*c*) The amount of excess air supplied per kg of coal burnt.
   [Ans. 8.5 kg; 11.5 kg; 3 kg]
5. A steam boiler uses pulverized coal in the furnace. The ultimate analysis of coal (by weight) as received : C = 78%; H_2 = 3%, O_2 = 3%; ash 10% and moisture 5%. Excess air supplied is 30%. Calculate the weight of air to be supplied and weight of gaseous product formed per kg of coal burnt.
   [Ans. 13 kg; CO_2 = 2.86 kg; H_2 = 0.27 kg; excess O_2 = 0.69 kg and N_2 = 9.81 kg per kg of coal]
6. The percentage composition by mass of a crude oil is given as follows : C = 90%; H_2 = 3.3%, O_2 = 3%; N_2 = 0.8%; S = 0.9% and remaining incombustible. If 50% excess air is supplied find the percentage of dry exhaust gases found by volume. **[Ans.** CO_2 = 12.7%; SO_2 = 0.05%; O_2 = 7%, N_2 = 80.25%]
7. In a boiler trial, the analysis of the coal used is as follows : C = 20%, H_2 = 4.5%, O_2 = 7.5%, remainder-incombustible matter.
   The dry flue gas has the following composition by volume :
   CO_2 = 8.5%, CO = 1.2%, N_2 = 80.3%, O_2 = 10%. Determine :
   (*i*) Minimum weight of air required per kg of coal.
   (*ii*) Percentage excess air. **[Ans.** (*i*) 3.56 kg, (*ii*) 63.2%]
8. The ultimate analysis of a sample of petrol by weight is : Carbon 0.835; hydrogen 0.165. Calculate the ratio of air to petrol consumption by weight, if the volumetric analysis of the dry exhaust gas is : Carbon dioxide 12.1; Carbon monoxide 1.1; Oxygen 0.8; Nitrogen 85.4 percent. Also find the percentage excess air. **[Ans.** 16.265 : 1; 5.6%]

❑❑❑

9

CHAPTER

Air Standard Cycles

INSIDE THIS CHAPTER

9.1. INTRODUCTION

The first and second law of thermodynamics suggest the method of producing mechanical power by transferring heat from a reservoir to a working fluid, which is taken through a thermodynamic cycle. A part of the heat received by the fluid is rejected to the sink. Such cycles are termed as 'power cycles'.

Power cycles may be grouped under two heads :

(*i*) Vapour power cycles

(*ii*) Gas power cycles.

In vapour power cycles, the working fluid usually employed is steam because it is cheap and chemically stable. However any condensible vapour can be used as working fluid in vapour power cycles.

In case of gas power cycles, the working fluid is a gas. The gas is normally air or the products of combustion of the fuel and air. In gas power cycles the supply of heat is by means of the combustion of a suitable mixture of air and fuel within the cylinder of engine, which are known as Internal combustion engines.

9.2. CONCEPT OF REVERSIBILITY

Any thermodynamic process, which can be operated in the reverse direction is known as a 'Reversible process'. The reversible process must obey the second law of thermodynamics *i.e.*, during any part of a reversible process, heat cannot flow from a body at lower temperature to a body at higher temperature.

In thermodynamically reversible process, there should be thermal equilibrium between the source of heat and the working fluid because if the temperature of the heat source will not be able to flow back from the working fluid to the source of heat, the process will not be reversed. It means that the process is not reversible.

To obtain thermodynamically reversible process the following *conditions* must be satisfied;

(*i*) There should be no temperature difference between the working fluid and the heat source;

For a process to be reversible, heat, which is received by the working fluid from the heat source, is to be returned back to the heat source, during its reverse action. It is only possible if the temperature of the working fluid and the heat source is the same.

(*ii*) There should be no friction between the working fluid and the walls of the container.

It means that the process should be frictionless; which is not possible, practically. Moreover, heat is produced due to friction, which is given to the heat source. For the process to be reversible, this heat should be absorbed by the working fluid. It is not possible because the heat produced by friction cannot be converted back into the friction during its reverse action.

From the above conditions, it is clear that the concept of thermodynamic reversibility is purely hypothetical. It is because of the fact that as the condition of the reversibility is approached the transfer of heat becomes less. Moreover, the friction in the process cannot be completely eliminated.

9.2.1. Carnot Cycle

This cycle consists of two adiabatic and two isothermal processes, as shown in the P–V and T–S diagrams of the Fig. 9.1.

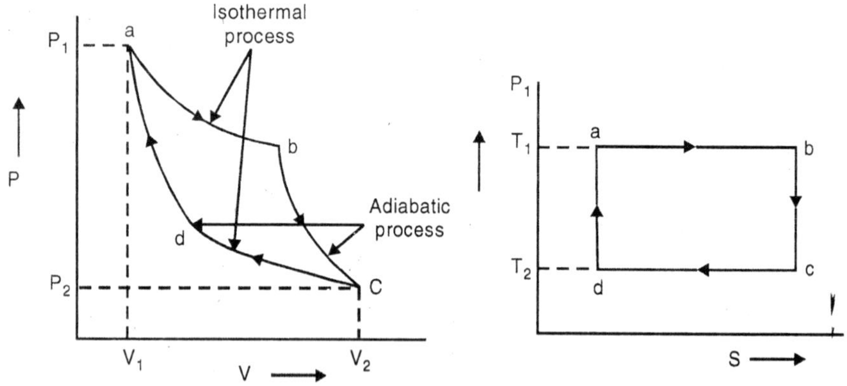

Fig. 9.1.

Air is assumed to be the working fluid, which is contained in a cylinder. The walls of this cylinder are made up of some non-conducting (insulating) materials. The cylinder cover, which is also made of non-conducting material can be removed or placed on the cylinder head as and when required.

To start with, air at point '*a*' is considered. The pressure volume and temperature of air at '*a*' be $p_1 V_1$ and T_1 respectively.

A body at temperature T_1 is brought in contact with cylinder end (cylinder head). The air expands isothermally to point '*b*' absorbing heat from the hot body. Due to expansion of the air, the piston is forced out, thus doing some useful work.

At point 'b', the hot body is removed and the end of the cylinder is covered with the non-conducting cover. The air at point 'b' expands adiabatically to point 'c'. During adiabatic expansion, the temperature of the air decreases and at the same time work is done. At point 'C' let the temperature of air be 'T_2'.

Now, at point 'c', the cold body at temperature T_2 is brought in contact with the cylinder end in place of non-conducting cover. The air compressed isothermally, to point 'd', rejecting heat to the cold body.

At point 'd' the cold body is removed and again the cylinder end is covered with the non-conducting cover. The air is compressed adiabatically to point 'a'. The temperature of the air rises from T_2 to T_1. This way the cycle is completed.

It is clear that heat is supplied during the isothermal expansion and is rejected during the isothermal compression. During the remaining two processes, no heat is transferred.

Now, let p_1, V_1 and T_1 be the pressure, volume and temperature of air at point 'a' and p_2, V_2 and T_2 their corresponding values at point 'c'.

Also let there be only one kg of air in the cylinder.

$\therefore$ Heat supplied $= p_1V_1 \cdot \log_e r$ (r = expansion ratio)

$= RT_1 \log_e r$ $(\because = p_1V_1 = mRT_1)$

and Heat rejected $= p_2V_2 \log_e r$ (r = compression ratio)

$= RT_2 \log_e r$ $(\because = p_2V_2 = mRT_2)$

The value of expansion ratio and compression ratio must be equal, otherwise the cycle will not close.

According to law of conservation of energy;

Work done = Heat supplied – Heat rejected,

$\therefore$ Work done $= RT_1 \log_e r - RT_2 \log_e r = R(T_1 - T_2) \log_e r$

$$\therefore \text{Efficiency carnot cycle} = \frac{\text{Workdone}}{\text{Heat supplied}}$$

or
$$\eta = \frac{R(T_1 - T_2) \log_e r}{RT_1 \log_e r}$$

or
$$\eta = \frac{T_1 - T_2}{T_1} \qquad ...(9.1)$$

From Eqn. (9.1), it is clear that the efficiency of the cycle depends upon the temperature range only. No other property of the working fluid (air) effects it.

From this above expression, it is clear that the efficiency increases with the decrease in the temperature T_2. It becomes 100% when T_2 is zero, which is not practicable.

In order to obtain the max. efficiency of cycle, temperature T_1 should be as high as possible and on the other hand the temperature T_2 should be as low as possible. But, the temperature T_2 is generally the atmospheric temperature. Therefore, the temp. T_1 is the aim of the design engineer.

9.2.1.1. Implications of Carnot Cycle

The efficiency of the Carnot cycle is the maximum possible, that can be achieved. So, it is the aim for an engineer to design an engine which should work on this cycle but in actual practice, it is not possible due to the following practical constraints;

(*i*) For the process to be isothermal, the speed of the piston should be very slow, so as to allow the temp. during the process to remain constant.

(*ii*) On the other hand for the process to be adiabatic, the speed of the piston should be very high, so that no heat is transferred during the process.

Practically it is not possible to construct an engine, whose piston speed is very slow during one process (isothermal) and is very high during the other subsequent process (adiabatic).

(*iii*) It is clear from the P–V diagram of Carnot cycle that the cycle is very narrow as compared to its length. It means that a very small amount of useful work is obtained correspondingly to a very long stroke length. Practically, it is not feasible.

(*iv*) In the actual working of the engine the friction is to exist in between the moving parts which cannot be eliminated completely.

9.3. OTTO, DIESEL AND DUAL COMBUSTION CYCLE

9.3.1. Otto or Constant Volume Cycle

This cycle is shown in Fig. 9.2 and presently being used in petrol or gasoline engines.

It is assumed that medium is pure air, which does not undergo any chemical change. The air is simply heated and cooled during the cycle. Also, the working fluid is assumed, to be a perfect gas having constant specific heats and the indicator diagram is strictly followed *i.e.* the heat losses are supressed and suction and exhaust pressure are equal.

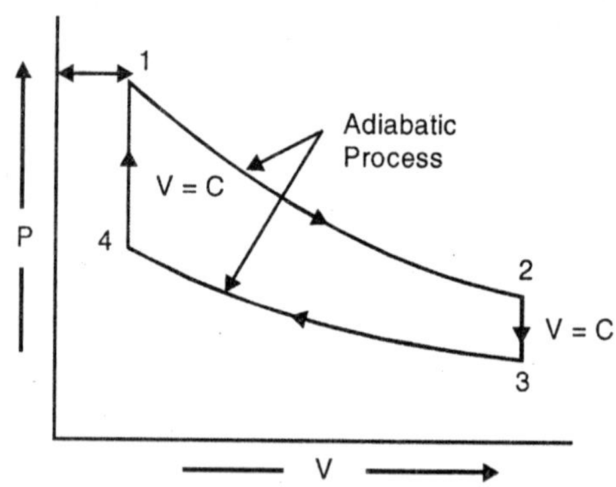

Fig. 9.2.

The cycle consists of the following expression :

(*i*) Operation 1-2, and Adiabatic expansion,

(*ii*) Operation 2 -3, heat rejection at constant volume,

(*iii*) Operation 3-4, an Adiabatic Compression,

(*iv*) Operation 4-1, heat addition at constant volume.

Let, p_3, V_3 and T_3 be the pressure, volume and absolute temperature of the air in the cylinder at point 3. The piston compresses the air adiabatically from point 3 to point 4 achieving corresponding values of pressure, volume and abs. temperature as p_4, V_4 and T_4 respectively. The air now occupies the clearance volume at the end of compression.

Now, a hot body is brought in contact with the cylinder and to supply heat at constant volume. This will increase the pressure, till point 1 is reached at which p_1 and T_1 are the corresponding pressure and absolute temperature respectively. At this stage the hot body is removed and air is allowed to expand adiabatically upto point 2, having condition p_2, V_2 and T_2.

Now, a cold body is brought in contact with the cylinder end, so that the pressure drops at constant volume till the absolute temperature, and pressure T_3 and p_3 are reached respectively and thus the air finally returns to its original condition after completing the cycle.

Let, clearance Volume $= V_1 = V_4 = 1$

Ratio of expansion during the adiabatic process 1–2

$$r = \frac{V_2}{V_1} = \frac{V_2}{1} = V_2$$

$$\therefore \quad V_2 = r$$

and ratio of compression during the adiabatic process 3–4,

$$= \frac{V_3}{V_4} = \frac{V_2}{V_1} = r \qquad (\because V_3 = V_2 \text{ and } V_4 = V_1)$$

Considering one kg of air, in the cylinder

Heat supplied, by the hot body

= Heat supplied during Constant Volume process, 4-1,

$= C_v . (T_1 - T_4)$

Heat rejected, to the cold body

= Heat rejected during, Constant volume process, 2-3

$= C_v . (T_2 - T_3)$

Work done = Heat supplied – Heat rejected

$= C_v . (T_1 - T_4) - C_v . (T_2 - T_3)$

and

$$\text{Efficiency, } \eta = \frac{\text{Work done}}{\text{Heat supplied}}$$

$$= \frac{C_v . (T_1 - T_4) - C_v . (T_2 - T_3)}{C_v . (T_1 - T_4)}$$

$$= 1 - \frac{T_2 - T_3}{T_1 - T_4} \qquad ...(i)$$

Now, considering, the adiabatic compression, 3-4

$$\frac{T_4}{T_3} = r^{\gamma - 1}$$

or

$$T_3 = \frac{T_4}{r^{\gamma - 1}} \qquad ...(ii)$$

Similarly, considering the adiabatic expansion, 1-2

$$\frac{T_1}{T_2} = r^{\gamma - 1}$$

or

$$T_2 = \frac{T_1}{r^{\gamma - 1}} \qquad ...(iii)$$

Now, substituting the values of T_3 and T_2 from (*ii*) and (*iii*) respectively in (*i*) we have;

$$\text{Efficiency, } \eta = 1 - \frac{\frac{T_1}{r^{\gamma-1}} - \frac{T_4}{r^{\gamma-1}}}{T_1 - T_4} = 1 - \frac{1}{r^{\gamma-1}}\left(\frac{T_1 - T_4}{T_1 - T_4}\right)$$

$$\text{Hence} \quad \text{Efficiency, } \eta = 1 - \frac{1}{r^{\gamma-1}}$$

Also, from eqn. (*i*)

$$\text{Efficiency, } \quad \eta = 1 - \frac{T_3\left(\frac{T_2}{T_3} - 1\right)}{T_4\left(\frac{T_1}{T_4} - 1\right)}$$

But from (*ii*) and (*iii*)

$$\frac{T_1}{T_2} = \frac{T_4}{T_3} \qquad (\text{each} = r^{\gamma-1})$$

or

$$\frac{T_2}{T_3} = \frac{T_1}{T_4}$$

$$\text{Efficiency, } \quad \eta = 1 - \frac{T_3\left(\frac{T_1}{T_4} - 1\right)}{T_4\left(\frac{T_1}{T_4} - 1\right)} = 1 - \frac{T_3}{T_4}$$

$$\text{or Efficiency, } \quad \eta = \frac{T_4 - T_3}{T_4}$$

Now, since T_4 is not the highest temperature in the cycle, the efficiency is less than carnot efficiency which would have been for the same range of temperature.

$$\text{Carnot } \eta = \frac{T_1 - T_3}{T_1} = 1 - \frac{T_3}{T_1}$$

Thus, we can achieve higher efficiency by increasing the compression ratio and lesser is the difference between T_1 and T_4 the more the efficiency reaches close to the carnot efficiency but in that case the power developed is reduced.

The various **assumptions** made in deriving the formula for the air standard efficiency of an Otto cycle are as follows :

(*i*) The working fluid is not subjected to chemical reactions.

(*ii*) There are no heat losses.

(*iii*) The working fluid is a perfect gas following the gas laws and has constant specific heats.

(*iv*) The heating and cooling is made to take place strictly at constant volume.

(*v*) The suction as well as exhaust take place at the atmospheric pressure.

9.3.2. Diesel Cycle or Constant Pressure Cycle

The cycle is shown on P–V diagram in Fig. 9.3. It consists of the following thermodynamic operations :

(*i*) Operation 1–2, an Adiabatic expansion.

(*ii*) Operation 2–3, heat rejection at constant volume.

(*iii*) Operation 3–4, an Adiabatic compression.

(*iv*) Operation 4 –1, Heat addition at constant pressure.

In case of Otto cycle, we have seen already that the heat is supplied to the working substance at constant volume, while in diesel cycle the heat is supplied at constant pressure. This is the main difference between these two cycles.

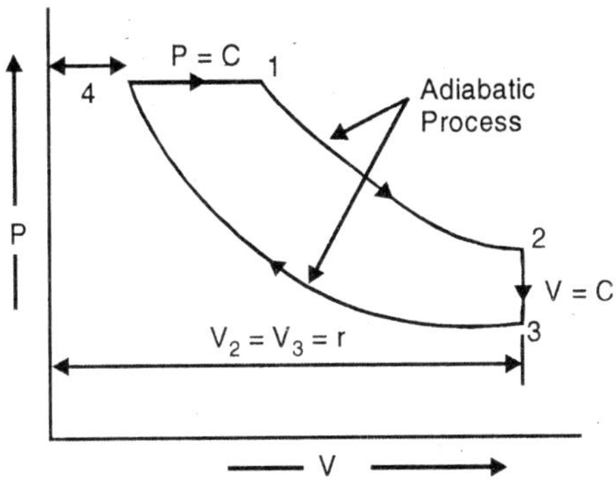

Fig. 9.3.

Let p_3, V_3 and T_3 be the pressure, volume and absolute temperature of the air in the cylinder at point 3. The piston compresses the air adiabatically from point 3 upto point 4. Let the condition of air at point 4 be p_4 V_4 and T_4. At point 4 the air occupies the clearance volume of the cylinder.

Now, a hot body is brought in contact with the cylinder end, to supply it heat at the point which the supply of heat is discontinued, is called cut off point. At this point T_1 be the abs., temperature and V_1 the volume which is represented by ρ. The air expands adiabatically, from point 1 to 2 and thus achieving condition p_2V_2 and T_2.

Now, a cold body is brought in contact with the cylinder end so that the pressure of air drops at constant volume till at point 3 the condition of air becomes p_3 V_3 and T_3, thus completing the cycle.

Let, the clearance volume, $V_4 = 1$

Ratio of expansion during the adiabatic process, 1–2

$$= \frac{V_2}{V_1} = \frac{r}{\rho}$$

Ratio of compression during the adiabatic process, 3–4

$$= \frac{V_3}{V_4} = \frac{r}{\rho} = r \qquad (\because\ V_4 = 1)$$

Now, consider one kg of air in the cylinder

Heat supplied by the hot body = heat supplied at constant pressure operation 4-1

$$= C_p(T_1 - T_4)$$

Heat rejected to the cold body = Heat rejected during constant volume operation, 2–3

and Work done = Heat supplied – Heat rejected

$$= C_p(T_1 - T_4) - C_v(T_2 - T_3)$$

Efficiency,

$$\eta = \frac{\text{Work done}}{\text{Heat supplied}}$$

or

$$\eta = \frac{C_p(T_1 - T_2) - C_v(T_2 - T_3)}{C_p(T_1 - T_4)}$$

or $$= 1 - \frac{C_p(T_2 - T_3)}{C_v(T_1 - T_4)}$$

$$= 1 - \frac{1}{\gamma}\left(\frac{T_2 - T_3}{T_1 - T_4}\right) \quad \left(\because \frac{C_p}{C_v} = \gamma\right)$$

Hence, efficiency, $$\eta = 1 - \frac{1}{\gamma}\left(\frac{T_2 - T_3}{T_1 - T_4}\right) \quad ...(i)$$

But, in actual practice, it is not easy to measure the instantaneous temperature although the volumes and pressures are mush easier to measure so, the efficiency is usually expressed in terms of compression ratio (r) and volumes at the point of cut-off (ρ)

Now, considering the adiabatic compression, 3–4

$$\frac{T_4}{T_3} = \left(\frac{V_3}{V_4}\right)^{\gamma - 1} = r^{\gamma - 1}$$

or $$T_3 = \frac{T_4}{r^{\gamma - 1}} \quad ...(ii)$$

Similarly, considering the constant pressure operation 4–1

$$\frac{p_4 . V_4}{T_4} = \frac{P_1 . V_1}{T_1}$$

or $$\frac{V_4}{T_4} = \frac{V_1}{T_1} \quad (\because p_4 = p_1)$$

or $$T_4 = T_1 . \frac{V_4}{V_1}$$

or $$T_4 = T_1 . \frac{1}{\rho} \quad ...(iii)$$

Now, substituting the value of T_4 from (*iii*) in (*ii*), we get :

$$T_3 = \frac{T_1 . \frac{1}{\rho}}{r^{\gamma - 1}}$$

or $$T_3 = \frac{T_1}{p . \gamma^{\rho - 1}} \quad ...(iv)$$

Considering the adiabatic expansion 1–2

$$\frac{T_1}{T_2} = \left(\frac{V_2}{V_1}\right)^{\gamma - 1} = \left(\frac{\gamma}{\rho}\right)^{\gamma - 1}$$

or $$T_2 = \frac{T_1}{\left(\frac{\gamma}{\rho}\right)^{\gamma - 1}} \quad ...(v)$$

So, substituting the value of T_4, T_3 and T_2 from (*iii*), (*iv*) and (*v*) respectively in (*i*), we get;

$$\text{Efficiency, } \eta = 1 - \frac{1}{p}\left[\frac{\dfrac{T_1}{\left(\frac{\gamma}{\rho}\right)^{\gamma - 1}} - \dfrac{T_1}{\rho . r^{\gamma - 1}}}{T_1 - \dfrac{T_1}{\rho}}\right]$$

or $$\eta = 1-\frac{1}{\gamma}\left[\frac{\dfrac{1}{\left(\dfrac{\gamma}{\rho}\right)^{\gamma-1}}-\dfrac{1}{\rho.\gamma^{\gamma-1}}}{1-\dfrac{1}{\rho}}\right]$$

or $$\eta = 1-\frac{1}{\gamma}\left(\frac{\rho}{\rho-1}\right)\frac{1}{r^{\gamma-1}}\left(p^{\gamma-1}-\frac{1}{\rho}\right)$$

or $$\eta = 1-\frac{1}{\gamma}\left(\frac{\rho}{\rho-1}\right)\frac{1}{r^{\gamma-1}}\left(p^{\gamma}-\frac{1}{\rho}\right)$$

or $$\eta = 1-\frac{r^{1-\gamma}}{r}\left(\frac{\rho^{\gamma}-1}{\rho-1}\right)$$

or $$= 1-\frac{1}{r^{\gamma-1}}\left(\frac{\rho^{\gamma}-1}{\rho-1}\times\frac{1}{r}\right)$$

Hence, Air standard efficiency for a diesel cycle.

$$\eta = 1-\frac{r^{1-\gamma}}{r}\left(\frac{\rho^{\gamma}-1}{\rho-1}\right)$$

or $$= 1-\frac{1}{r^{\gamma-1}}\left(\frac{\rho^{\gamma}-1}{\rho-1}\times\frac{1}{r}\right)$$

9.3.3. Dual Combustion Cycle or Limited Pressure Cycle

To achieve complete combustion of fuel in a diesel engine, without adversely affecting its efficiency it is often arranged in these engines of moderate power for the injection of the fuel to commence before the end of the compression stroke, so that the combustion proceeds, partly at constant volume and partly at constant pressure. Such a cycle of operation is known as the Dual combustion cycle. And an engine working on this cycle is known as *Semi-Diesel engine.*

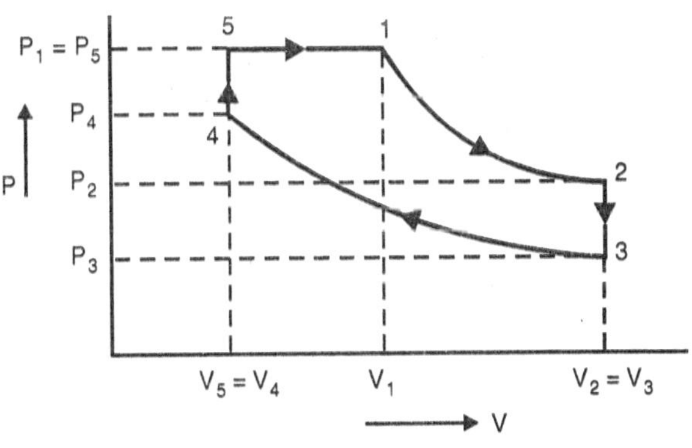

Fig. 9.4.

In fact, all the hot spot ignition engines and diesel engines, actually operate on this cycle, because of the necessity of injecting fuel before the end of compression stroke to take care of ignition lag characteristics of fuel.

This cycle as shown in Fig. 9.4 consists of the following thermodynamic operations;

(*i*) Operation 1–2, an Adiabatic expansion

(*ii*) Operation 2–3, Heat rejection at constant volume.

(*iii*) Operation 3–4, an Adiabatic compression.

(*iv*) Operation 4–5, Heat addition at constant volume.

(*v*) Operation 5–1, Heat addition at constant pressure.

Now, Total Heat supplied during the cycle

= Heat supplied during const. vol process 4–5

+ Heat supplied during const. pressure process 5–1

$= C_v (T_5 - T_4) + C_P (T_1 - T_5)$

and heat rejected during the cycle

= Heat rejected during const. vol. operation 2–3.

$= C_v (T_2 - T_3)$

Now, Work done = Heat supplied – Heat rejected

$$= [C_v(T_5 - T_4) + C_P (T_1 - T_5)] - C_v(T_2 - T_3)$$

$$\therefore \quad \text{Efficiency, } \eta = \frac{\text{Work done}}{\text{Heat supplied}}$$

$$= \frac{[C_v(T_5 - T_4) + C_p (T_1 - T_3)] - C_v (T_2 - T_3)}{C_v (T_5 - T_4) + C_p (T_1 - T_5)}$$

$$= 1 - \frac{C_v (T_2 - T_3)}{C_v(T_5 - T_4) - C_p (T_1 - T_5)}$$

$$= 1 - \frac{(T_2 - T_3)}{(T_5 - T_4) + \gamma (T_1 - T_5)} \quad \left(\because \gamma = \frac{C_p}{C_v}\right) \qquad \ldots(i)$$

Let, clearance volume, $V_4 = V_5 = 1$

and the volume, at cut off $= V_1 = \rho$

$$\therefore \quad \frac{V_1}{V_5} = \frac{\rho}{1} = \rho$$

Let, $V_2 = V_3 = r \qquad \therefore \quad \frac{V_2}{V_1} = \frac{r}{\rho}$

So, expansion ratio during adiabatic process, 1–2

$$\frac{T_4}{T_2} = \left(\frac{r}{\rho}\right)^{\gamma - 1} \qquad \ldots(ii)$$

For the adiabatic compression process, 3–4

$$\frac{T_4}{T_3} = r^{\gamma - 1} \qquad \ldots(iii)$$

Further, let pressure or explosion ratio

$$\frac{T_5}{T_4} = \alpha$$

For the constant volume operation, 4–5

$$\frac{p_4 \, . V_4}{T_4} = \frac{p_5 \, .V_5}{T_5}$$

or

$$\frac{p_4}{T_4} = \frac{p_5}{T_5} \qquad (\because V_4 = V_5)$$

or $$\frac{T_5}{T_4} = \frac{p_5}{p_4} = \alpha$$

or $$T_4 = \frac{T_5}{\alpha} \qquad ...(iv)$$

Substituting the value of T_4 from (iv) and (iii)

$$\frac{T_5}{\alpha / T_3} = r^{\gamma - 1}$$

or $$T_3 = \frac{T_5}{\alpha\, r^{\gamma - 1}} \qquad ...(v)$$

For constant pressure operation 5–1

$$\frac{p_1 V_1}{T_1} = \frac{p_5 V_1}{T_1}$$

or $$\frac{V_1}{T_1} = \frac{V_5}{T_5} \qquad (\because\ p_1 = p_s)$$

or $$\frac{T_1}{T_5} = \frac{V_1}{V_5} = \rho$$

or $$T_1 = \rho \,.\, T_5 \qquad ...(vi)$$

Substituting the value of T_1 from (vi) in (ii)

$$\frac{\rho \,.\, T_5}{T_2} = \left(\frac{r}{\rho}\right)^{\gamma - 1}$$

or $$T_2 = \rho \,.\, T_5 \left(\frac{\rho}{r}\right)^{\gamma - 1} \qquad ...(vii)$$

Now, substituting the values of T_4, T_3, T_1 and T_2 from (v), (vi) and (vii) respectively in (i) we get;

$$\text{Efficiency}, \eta = 1 - \frac{\rho \,.\, T_5 \left(\frac{\rho}{r}\right)^{\gamma - 1} - \frac{T_5}{\alpha . r^{\gamma - 1}}}{\left(T_5 - \frac{T_5}{\alpha}\right) + \gamma \left(\rho . T_5 - T_5\right)}$$

Dividing num. and deno. by T_5

or $$= 1 - \frac{\rho \,.\left(\frac{\rho}{r}\right)^{\gamma - 1} - \frac{1}{\alpha \,.\, r^{\gamma - 1}}}{\left(1 - \frac{1}{\alpha}\right) + \gamma\left(\rho - 1\right)}$$

or $$= 1 - \frac{\left(\frac{1}{r^{\gamma - 1}}\right) . \left(\rho^{r}\ \frac{1}{\alpha}\right)}{\left(\frac{\alpha - 1}{\alpha}\right) + \gamma\left(\rho - 1\right)}$$

or $$= 1 - \frac{1}{r^{\gamma - 1}} \cdot \frac{\left(\alpha \rho^{\gamma} - 1\right)}{\left(\alpha - 1\right) + \alpha\gamma\left(\rho - 1\right)}$$

Hence, Air standard efficiency of dual Cycle, η

or
$$= 1 - \frac{1}{r^{\gamma-1}} \cdot \frac{(\alpha\rho^{\gamma-1})}{(\alpha-1)+\alpha\gamma(\rho-1)}$$

9.3.4. Comparison of Otto, Diesel and Dual Combustion Cycles

Following are the *important variable factors which are used as a basis for comparison of the cycles* :

(*i*) Compression ratio
(*ii*) Maximum pressure
(*iii*) Heat supplied
(*iv*) Network

Some of the above mentioned variables are fixed when the performance of Otto, Diesel and dual combustion cycles is to be compared.

9.3.4.1. Efficiency Versus Compression Ratio

Figure 9.5 shows the comparison for the air standard efficiencies of the Otto, Diesel and Dual combustion cycles at various compression ratios and with given cut-off ratio for the Diesel and Dual combustion cycles. It is evident from the Fig. 9.5 that the air standard efficiencies *increase with the increase in the compression ratio. For a given compression ratio Otto cycle is the most efficient while the Diesel cycle is the least efficient.* ($\eta_{otto} > \eta_{dual} > \eta_{diesel}$).

Note. The maximum compression ratio for the petrol engine is limited by detonation. In their respective ratio rages, the Diesel cycles is more efficient than the Otto cycle.

9.3.4.2. For the Same Compression Ratio and the Same Heat Input

A comparison of the cycles (Otto, Diesel and Dual) on the *p-v* and *T-s* diagrams for the *same compression ratio and heat supplied* is shown in the Fig. 9.6.

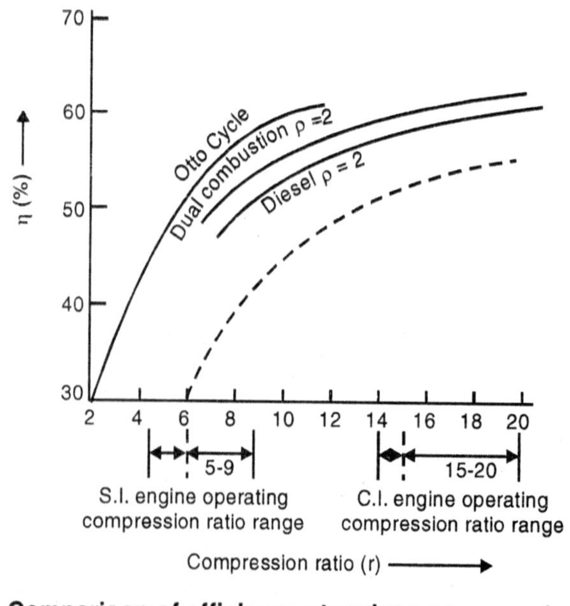

Fig. 9.5. Comparison of efficiency at various compression ratios.

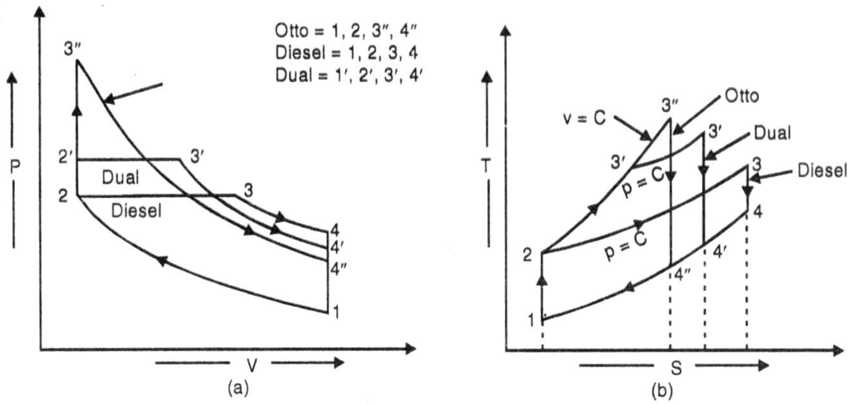

Fig. 9.6. (*a*) *P-V* diagram, (*b*) *T-S* diagram.

We know that, $$\eta = 1 - \frac{\text{Heat rejected}}{\text{Heat supplied}}$$

Since all the cycles reject their heat at the same specific volume, process line from state 4 to 1, the quantity of *heat rejected from each cycle is represented by the appropriate area under the line 4 to 1 on the T-S diagram*. As is evident from the above eqn. (1) the cycle which has the least heat rejected will have the highest efficiency. Thus, Otto cycle is the most efficient and Diesel cycle is the least efficient of the three cycles.

i.e., $$\eta_{otto} > \eta_{dual} > \eta_{diesel}$$

9.3.4.3. For Constant Maximum Pressure and Heat Supplied

Figure 9.7 shows the Otto and Diesel cycles on *P-V* and *T-S* diagrams for constant maximum pressure and heat input respectively.

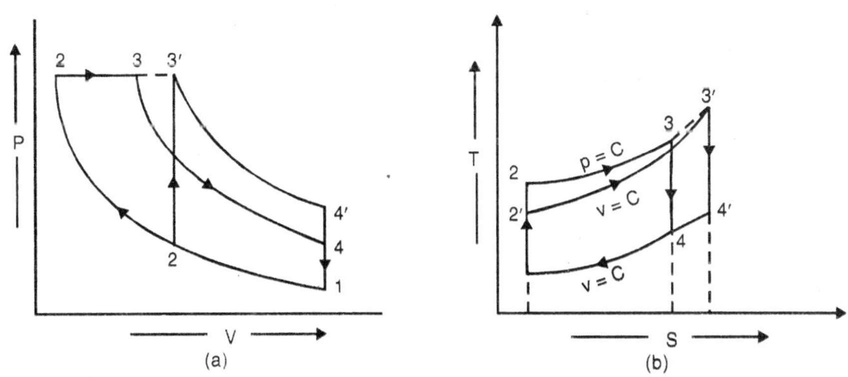

Fig. 9.7. (*a*) *P-V* diagram, (*b*) *T-S* diagram.

9.3.5. Reasons for Highest Efficiency of Carnot Cycle Than Other Cycles Working Between Same Temperature Limits

As we know that :

Carnot Efficiency, $$\eta_{Carnot} = \frac{T_1 - T_2}{T_1} \text{ or } 1 - \frac{T_2}{T_1} \quad \text{(From eqn. (9.1))}$$

and Otto Cycle Eff. $\eta_{Otto} = 1 - \dfrac{T_2 - T_3}{T_1 - T_4}$ (Eqn. (*i*) of Art. 9.3.1)

Whereas Diesel cycle Eff. $\eta_{Diesel} = 1 - \dfrac{1}{r}\left(\dfrac{T_2 - T_3}{T_1 - T_4}\right)$ (Eqn. (*i*) of Art. 9.3.2)

Now, if we analyse the above equations for various efficiencies we observe that :

(*i*) The 'η' of diesel cycle is different from 'η' of the Otto cycle by the factor $\left(\dfrac{1}{r}\right)$ which is to be multiplied with $\left(\dfrac{T_2 - T_3}{T_1 - T_4}\right)$. As the value of '$r$' is always more than one (unity), so $\dfrac{1}{r}$ will give value always less than one which after multiplication with $\left(\dfrac{T_2 - T_3}{T_1 - T_4}\right)$ will reduce the value of the term to be subtracted from one (unity) and thus will increase the diesel efficiency accordingly.

(*ii*) In both the cases of Otto and Diesel cycle the denominator $(T_1 - T_4)$ is not the highest temp. as in the case of carnot cycle where the denominator 'T_1' is the highest temp. of the cycle. So, as the value of denominator is max. in case of carnot cycle so the value of the term to be subtracted from unity to get 'η' of the cycle will reduce. Hence, the η_{Carnot} will be the max. possible efficiency.

NUMERICAL EXAMPLES

Example 9.1. Find the efficiency of an engine working on the Carnot cycle, if the heat is supplied at 500°C and rejected at 30°C.

Solution.

$$T_1 = 500 + 273 = 773 \text{ k}$$
$$T_2 = 30 + 273 = 303 \text{ k}$$

Because, efficiency of carnot cycle $= \dfrac{T_1 - T_2}{T_1}$

$$\therefore \quad \eta_{Carnot} = \frac{773 - 303}{773} = \frac{470}{773} = 0.608$$

or $\eta_{Carnot} = 60.8\%$ **Ans.**

Example 9.2. An engine working on Carnot cycle receives heat at 700° and rejects heat at 50°C. What is the efficiency of the engine. Now, if it absorbs 1000J of heat per minute, from the hot body, calculate the work done in J. min and the power of the engine in Watts.

Solution.

$$T_1 = 700 + 273 = 973 \text{ k}$$
$$T_2 = 50 + 273 = 323 \text{ k}$$

Because, the engine is working on the Carnot cycle,

$$\eta_{Carnot} = \frac{T_1 - T_2}{T_1} = \frac{973 - 323}{973} = \frac{650}{973} = 0.667$$

or $\quad \%\text{ age } \eta_{carnot} = 66.7\%$

Also, ∵ efficiency, $\eta = \dfrac{\text{Workdone / min}}{\text{Heat supplied / min}}$

$\therefore \quad \text{Work done/min} = \text{Heat supplied/min} \times \eta$

$= 1000 \times 0.667 = 667 \text{ J/min}$ **Ans.**

Also, power, in watt is the W.D. in J/sec

$\therefore \quad \text{Power} = \dfrac{667}{60} = 11.12 \text{ Watt}$ **Ans.**

Example 9.3. If the compression ratio of an engine working on otto cycle is increased from 5 to 6. Calculate the increase in efficiency of the engine. Assume air as the working fluid. Take γ for air to be 1.4.

Solution. (*i*) $r_1 = 5$

$$\therefore \quad \eta_1 = 1 - \frac{1}{r_1^{\gamma-1}} = 1 - \frac{1}{5^{1.4-1}} = 1 - \frac{1}{5^{0.4}}$$

$$= 1 - \frac{1}{1.903} = 1 - 0.525 = 0.475$$

Similarly, $r_2 = 6$

$$\eta_2 = 1 - \frac{1}{r_2^{\gamma-1}} = 1 - \frac{1}{6^{1.4-1}}$$

$$= 1 - \frac{1}{6^{0.4}} = 1 - \frac{1}{2.048} = 1 - 0.488$$

or $\quad \eta_2 = 0.512$

∴ Increase in efficiency $= \eta_2 - \eta_1 = 0.512 - 0.475 = 0.037$

or $\quad$ Increase in efficiency = 3.7% **Ans.**

Example 9.4. The efficiency of an otto cycle is 50% and $\gamma = 1.5$. What is the compression ratio, r ?

Solution. $\eta_{Otto} = 50\%$

$\gamma = 1.5$

We know; $\quad \eta_{Otto} = 1 - \dfrac{1}{r^{\gamma-1}}$

$$\therefore \quad 0.5 = 1 - \frac{1}{(r)^{1.5-1}}$$

or $\quad \dfrac{1}{r^{0.5}} = 1 - 0.5 = 0.5$

or $\quad r^{0.5} = \dfrac{1}{0.5} = 2$

or $\quad r = (2)^{1/0.5} = 4.$ **Ans.**

Example 9.5. In a Diesel engine the compression ratio is 13 : 8 : 1 and the expansion ratio is 6. Calculate the efficiency assuming heat is supplied at constant pressure and that compression and expansion and adiabatic.

Solution. $r = 13.8$

$$\frac{r}{\rho} = 6, \text{ or } \frac{13.8}{\rho} = 6$$

$$\therefore \quad \rho = 2.3$$

Assuming, $\gamma = 1.4$

We know Diesel cycle efficiency,

$$\eta_{\text{Diesel}} = \left[1 - \frac{r^{1-\gamma}}{\gamma}\left(\frac{\rho^{\gamma}-1}{\rho-1}\right)\right] \times 100$$

$$= \left[1 - \frac{(13.8)^{1-1.4}}{1.4}\left[\frac{(2.3)^{1.4}-1}{2.3-1}\right]\right] \times 100$$

$$= \left[1 - \frac{1}{1.4.(13.8)^{0.4}} \cdot \left(\frac{3.21-1}{2.3-1}\right)\right] \times 100$$

$$= \left[1 - \frac{2.21}{1.4 \times 2.86 \times 1.3}\right] \times 100 = 57.5\% \textbf{ Ans.}$$

Example 9.6. **Find the air standard efficiency of a Diesel engine having a compression ratio of 14 and in which cut-off takes place at 5.5% of the stroke.**

Solution. $r = 14$

Let clearance volume = 1

Now, $$\frac{\rho-1}{r-1} = \frac{5.5}{100}$$

$$\therefore \quad \frac{\rho-1}{14-1} = \frac{5.5}{100}$$

or $$\rho = 1 + \frac{13 \times 5.5}{100} = 1.715$$

$\therefore$ Air standard efficiency,

$$= 1 - \frac{r^{1-\gamma}}{\gamma}\left(\frac{\rho^{\gamma}-1}{\rho-1}\right)$$

$$= \left[1 - \frac{(14)^{1-1.4}}{1.4}\left(\frac{1.715^{1.4}-1}{1.715-1}\right)\right] \times 100$$

$$= \left[1 - \frac{1}{1.4.\times(14)^{0.4}}\left(\frac{2.13-1}{1.715-1}\right)\right] \times 100$$

$$= \left[1 - \frac{1.13}{1.4 \times 2.875 \times 0.715}\right] \times 100$$

$$= 60.74\% \textbf{ Ans.}$$

Example 9.7. **One kg of air is taken through a Diesel cycle. Initially the air is at 15°C and 1 atm. The compression ratio is 15 and the heat added is 1850 KJ. Calculate the ideal cycle efficiency.**

Solution. Initial volume of air,

$$V_1 = \frac{RT_1}{P_1}$$

$$= \frac{0.287 \times 288}{1.013 \times 100}$$

$$= 0.816 \text{ m}^3$$

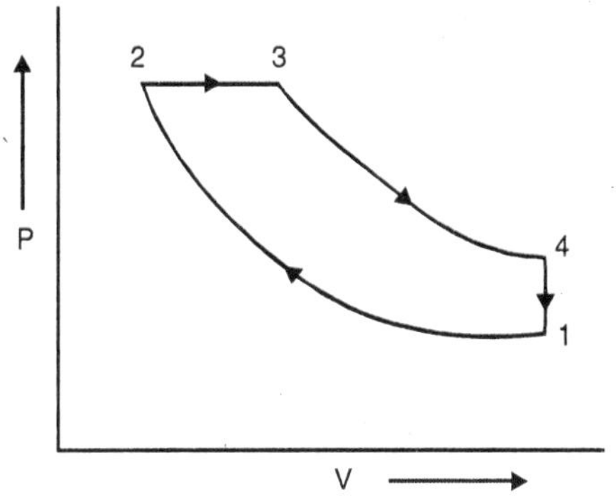

Fig. 9.8.

After compression

$$T_2 = T_1\left(\frac{V_1}{V_2}\right)^{\gamma-1}$$

$$= (273 + 15)\,(16)^{1.4-1} = 851°\text{k}$$

$$\therefore \quad V_2 = \frac{0.816}{15} = 0.054 \text{ m}^3$$

The temperature rise during heat addition is given by

$$Q = C_p(T_3 - T_2) = 1.005\,(T_3 - 851) = 1850 \text{ KJ/kg}$$

$$\therefore \quad T_3 = 2692 \text{ °k}$$

Also $\dfrac{V_4}{V_3} = \dfrac{V_4}{V_2} \cdot \dfrac{V_2}{V_3} = 15 \times \dfrac{851}{2692} = 4.742$

$$\therefore \quad T_4 = T_3\left(\frac{V_3}{V_4}\right)^{\gamma-1} = 2692\left(\frac{1}{4.742}\right)^{1.4-1} = 1444 \text{ k}$$

$$\text{Heat rejected} = C_p(T_1 - T_4) = 0.718\,(288 - 1444)$$

$$= -830 \text{ KJ/kg.}$$

$$\text{Net work done} = 1850 - 830 = 1020 \text{ KJ/kg}$$

$$\text{Cycle efficiency} = \frac{\text{Work done}}{\text{Heat supplied}} = \frac{1020}{1850} = 0.551 \; \textit{i.e.}, 55.1\% \quad \textbf{Ans.}$$

Example 9.8. (*a*) A Diesel cycle has compression ratio of 14 and heat supply upto 50% of the stroke. Calculate the air standard efficiency of the cycle.

(*b*) If the cut-off is increased from 5% to 7% with the compression ratio in unchanged, find the change in air standard efficiency.

(*c*) If the compression ratio is increased from 14 to 18 and the cut-off adjusted to give same air standard efficiency as in (*a*) find the required change in cut-off.

Take initial pressure of air as 1 bar and all processes is ideal.

Solution. Cut off is given at 5% and $r_c = 14$

$$\therefore \quad \rho = 1 + k(r_c - 1)$$

$$= 1 + \frac{5}{100}(14 - 1) = 1.65$$

$\therefore$ Air standard efficiency,

$$= 1 - \frac{1}{(r)^{\gamma-1}}\left[\frac{\rho^\gamma - 1}{\rho - 1} \times \frac{1}{\gamma}\right]$$

$$= 1 - \frac{1}{(14)^{1.4-1}}\left[\frac{1.65^{1.4}-1}{1.65-1} \times \frac{1}{1.4}\right]$$

$$= 0.611 \text{ i.e., } 61.1\%$$

(*b*) If the cut-off occurs at 7%

Then, $\rho = 1 + k\,(r - 1)$

$$= 1 + \frac{7}{100}(14-1) = 1.91$$

Air standard efficiency,

$$= 1 - \frac{1}{(4)^{1.4-1}}\left[\frac{(1.91)^{1.4}-1}{1.91-1} \times \frac{1}{1.4}\right]$$

$$= 0.5973 \text{ i.e., } 59.73\%$$

∴ Percentage decrease in air standard efficiency,

$$= 61.1 - 59.73 = 1.37\%$$

(*c*) As the air standard efficiency remains at 61.1% but,

Compression ratio, $r_c = 18$

$$\therefore \quad 0.611 = 1 - \frac{1}{(18)^{0.4}}\left[\frac{\rho^{1.4}-1}{\rho-1} \times \frac{1}{1.4}\right]$$

From which, $\rho = 1.95$

But, $\rho = 1 + k\,(r_c - 1)$

$\therefore \quad 1.95 = 1 + k\,(18 - 1)$

$\therefore \quad k = 0.0559$ *i.e.,* 5.59%

Hence, cut off occurs at 5.59% of the stroke.

Example 9.9. An ideal dual combustion cycle has a stroke volume of 0.01 m^3 of air at the commencement of adiabatic compression is one bar and 27°C respectively. The heat is added at constant pressure of 70 bar and for 5% of the stroke. Calculate,

(*i*) Pressure ratio

(*ii*) Cut-off ratio

(*iii*) Mass of air contained in the cylinder

(*iv*) Heat added per cycle

(*v*) Heat rejected per cycle

(*vi*) Work done per cycle

(*vii*) Internal efficiency of cycle

Take C_p = 1 KJ/kg°C, C_v = 0.714 KJ/kg°C

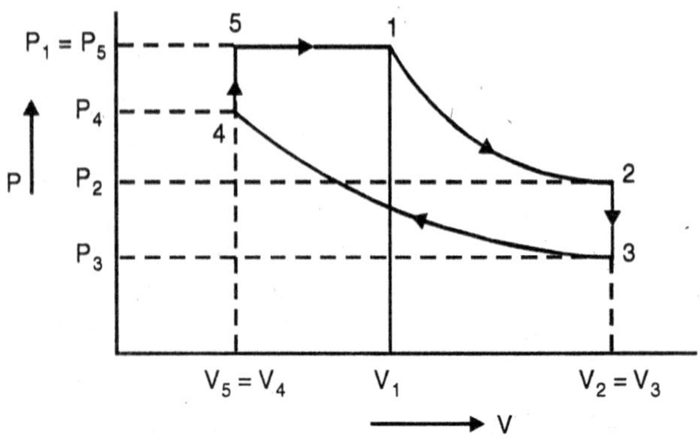

Fig. 9.9.

Solution.

The cycle is shown in the above figure.

$$\text{Stroke volume} = V_3 - V_4 = 0.01 \text{ m}^3$$

$$\text{Compression ratio} = \frac{V_3}{V_4} = 16$$

$$\therefore \quad V_3 = 16V_4$$

So, $$16V_4 - V_4 = 0.01$$

or $$V_4 = \frac{0.01}{15} = 0.00067 \text{ m}^3$$

$$\therefore \quad V_3 = 0.01 + 0.00067 = 0.01067 \text{ m}^3$$

Pressure before the adiabatic compression,

$$p_3 = 1 \text{ bar}$$

Temperature before adiabatic compression;

$$T_3 = 27 + 273 = 300 \text{ k}$$

Pressure at point 5 = p_5 = 70 bar

$$\gamma = \frac{C_p}{C_v} = \frac{1}{0.714} = 1.4$$

(*i*) Pressure ratio $$\frac{p_4}{p_3} = \left(\frac{V_5}{V_4}\right)^{\gamma} = (16)^{1.4} = 48.5$$

$$\therefore \quad p_4 = 48.5, \quad p_3 = 48.5 \times 1 = 48.5 \text{ bar}$$

and $$\frac{p_5}{p_4} = \frac{70}{48.5} = 1.44 \quad \textbf{Ans.}$$

(*ii*) Cut-off ratio $\rho = \frac{V_1}{V_5}$

But, $$V_1 = V_5 + 5\% \text{ of stroke volume}$$
$$= 0.00067 + 0.05 \times 0.01 = 0.00117 \text{ m}^3$$

$$\therefore \quad \rho = \frac{0.00117}{0.00067} = 1.75 \quad \textbf{Ans.}$$

(*iii*) Mass of air contained in the cylinder can be calculated from initial conditions,

$$p_3 . V_3 = mRT_3$$

$$\therefore \quad m = \frac{1 \times 10^5 \times 0.01067}{0.287 \times 1000 \times 300}$$
$$= 0.012 \text{ kg.} \quad \textbf{Ans.}$$

(*iv*) For adiabatic compression 3-4

$$T_4 = T_3\left(\frac{V_3}{V_4}\right)^{\gamma-1} = 300\ (16)^{1.4-1} = 9.0° \text{ k}$$

For constant volume process 4-5,

$$\frac{p_4}{T_4} = \frac{p_5}{T_5}$$

$$\therefore \quad T_5 = T_4\left(\frac{p_5}{p_4}\right) = \frac{909}{48.5} \times 70 = 1312° \text{ K}$$

During constant pressure process;

$$\frac{V_5}{V_5} = \frac{V_1}{T_1}$$

$$\text{or} \quad T_1 = T_5 . \frac{V_1}{V_5} = 1312 \times \frac{1.17}{0.67} = 2291° \text{K}$$

Heat added = Heat added during constant Vol. Process

+ Heat added during const. Pressure process

$= m[C_v(T_5 - T_4) + C_p(T_1 - T_5)]$

$= 0.012\,[0.714(1312 - 909) + (2291 - 1312)]$

$= 15.2$ KJ. **Ans.**

(*v*) From adiabatic expansion 1–2

$$\frac{T_1}{T_2} = \left(\frac{V_2}{V_1}\right)^{\gamma-1}$$

$$\therefore \quad \frac{2291}{T_2} = \left(\frac{0.01067}{0.00117}\right)^{0.4}$$

$$\text{or} \quad T_2 = 947 \text{ K}$$

Heat rejected during cycle

$= m \,.\, C_v \,.\, (T_2 - T_3)$

$= 0.012 \times 0.714(947 - 300)$

$= 5.54$ KJ **Ans.**

(*vi*) Work done per cycle = Heat added – Heat rejected

$= 15.20 - 5.54 = 9.66$ KJ. **Ans.**

(*vii*) Thermal Efficiency

$$= \frac{\text{Work done/cycle}}{\text{Heat added/cycle}}$$

$$= \frac{9.66}{15.20} = 0.7318 \text{ } i.e., \text{ } 73.18\% \text{ } \textbf{Ans.}$$

Example 9.10. With the help of *p-v* and T-s diagram compare the cold air standard Otto, diesel and dual combustion cycles for same maximum pressure and maximum temperature. (AMIE Summer, 1998)

Solution. Refer Fig. 9.8(*a*, *b*).

The air-standard Otto, Dual and Diesel cycles are drawn on common *P-V* and *T-S* diagrams for the same maximum pressure and maximum temperature, for the purpose of comparison.

Otto 1-2-3-4-1, Dual 1-2′-3′-3-4-1, Diesel 1-2″-3-4-1 (Fig. 9.10(*a*)).

Slope of constant volume lines on T-S diagram is higher than that of constant pressure lines (Fig. 9.10(*b*)).

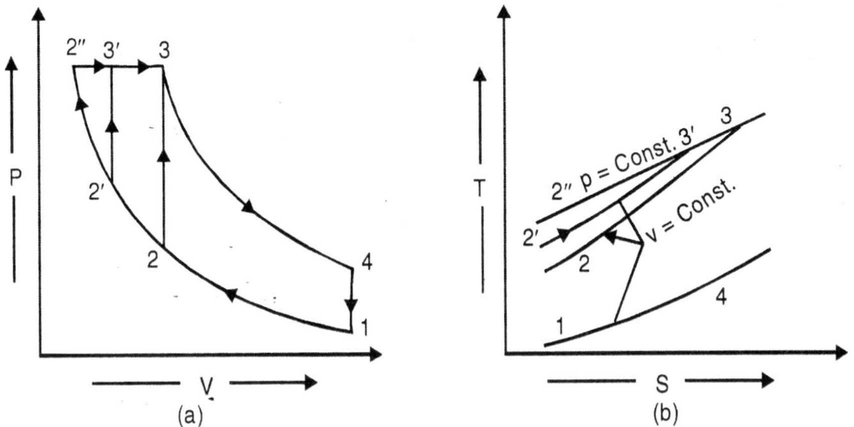

Fig. 9.10.

Here the otto cycle must be limited to a low compression ratio (r) to fulfill the condition that point 3 (same maximum pressure and temperature) is to be a common state for all the three cycles.

The construction of cycles as T-S diagram proves that for the given conditions the heat rejected is same for all the three cycles (area under process line 4-1). Since, by definition,

$$\eta = 1 - \frac{\text{Heat rejected}, Q_r}{\text{Heat supplied} Q_s} = 1 - \frac{\text{Const.}}{Q_s}$$

The cycle, with greater heat addition will be more efficient. From the T-s diagram,

$$Q_{s\,(diesel)} = \text{Area under } 2''\text{-}3$$
$$Q_{s\,(dual)} = \text{Area under } 2'\text{-}3'\text{-}3$$
$$Q_{s\,(otto)} = \text{Area under } 2\text{-}3$$

It can be seen that,

$$Q_{s\,(diesel)} > Q_{s\,(dual)} > Q_{s\,(otto)}$$

and thus $$\eta_{(diesel)} > \eta_{(dual)} > \eta_{(otto)}$$

- For the maximum pressure the points 3 and 3′ must lie on a constant pressure line.
- On T-S diagram the heat rejected from the Diesel cycle is represented by the area under the line 4 to 1 and this area is less than the Otto cycle area under the curve 4′ to 1. *Hence the Diesel cycle is more than the Otto cycle for the condition of maximum pressure and heat supplied.*

ASSIGNMENT

Objective Type Questions :

(A) Fill in the Blanks :

1. Reversible process is that process in which ______.
2. The efficiency of the Carnot cycle is ______ possible.
3. The efficiency of the Otto cycle is ______ than that of Diesel cycle under similar conditions.

4. Practical engines ______ work on carnot cycle.
5. The efficiency of the cycle increases with the ______ in compression ratio.
6. The best suited range of compression ratio in the diesel engines is ______.
7. The carnot cycle ______ operate in the reversed direction.
8. Severe detonation ______ the efficiency of the S.I. engine.
9. Relative efficiency may be defined as the ratio of ______.
10. Rate of combustion can be ______ with the increase of compression ratio is the S.I. engines.
11. In vapour power cycles, the working fluid usually employed is ______.
12. In gas power cycles, the working fluid usually employed is ______.
13. Carnot cycle consists of two adiabatic processes and two ______ processes.
14. Otto cycle consists of two adiabatic processes and two ______ process.
15. Diesel cycle essentially consists of two adiabatic processes one constant pressure process and two ______ process.
16. In Dual cycle the heat addition occurs partly at constant volume and partly at ______.
17. Practically the C.I. engines work on the ______ cycle.

(B) State True or False :

1. All the thermodynamic processes are reversible processes.
2. The efficiency of the reversible cycle is the maximum.
3. The Otto cycle is also known as constant volume cycle.
4. The Carnot cycle is also known as the constant temperature cycle.
5. The point of cut-off is a stage during the compression stroke.
6. In the diesel the heat is supplied to the working substance at constant pressure.
7. The Carnot efficiency decreases with the increase of hot body temperature.
8. Any thermodynamic process, which can be operated in the reverse direction is known as 'a reversible process'.
9. The Carnot cycle consists of two adiabatic processes and two isothermal processes.
10. The efficiency of the Carnot cycle increases with the decrease in the exhaust temp.
11. The efficiency of the Carnot cycle is the maximum possible that can be achieved.
12. Rankine cycle is not a vapour power cycle.
13. Rankine cycle is a modification of carnot cycle.
14. The work done during Rankine cycle is given by $\dfrac{n(P_1V_1 - P_2V_2)}{n-1}$, whereas letters have their usual meanings.
15. Otto cycle consists of two adiabatic processes and two constant volume processes.
16. Diesel cycle is also called constant pressure cycle.
17. Practically the diesel engine work on the Dual combustion cycle.

(C) Multiple Choice Questions :

1. Which does not constitute the theoretical cycle for the working of a practical *i.e.*, engine?

   (*a*) Otto cycle (*b*) Diesel cycle

   (*c*) Dual cycle (*d*) Ericsson cycle

2. A four stroke petrol engine theoretically operates on :
   (*a*) Otto cycle (*b*) Brayton cycle
   (*c*) Joule cycle (*d*) Bell-Coleman cycle
3. A slow speed diesel engine theoretically operates on the principle of :
   (*a*) heat addition at const. vol and heat rejection at const. vol.
   (*b*) heat addition at const. pressure and heat rejection at const pressure
   (*c*) heat addition at const. vol. and heat rejection at const. pressure
   (*d*) heat addition at const. vol. and heat rejection at const. pressure.
4. Which of the following processes is not associated with diesel cycle?
   (*a*) constant volume (*b*) constant pressure
   (*c*) isothermal (*d*) adiabatic
5. From the same compression ratio and heat input the cycles in decreasing order of thermal efficiency are:
   (*a*) otto, dual, diesel (*b*) diesel, otto, dual
   (*c*) dual, diesel, otto (*d*) otto, diesel, dual
6. What of the following has isentropic compression and expansion, but heat addition and heat rejection at constant volume.
   (*a*) Otto cycle (*b*) Diesel cycle
   (*c*) Brayton cycle (*d*) Atkinson cycle
7. Which one of the following cycles has adiabatic compression and expansion but addition at constant pressure and heat rejection at constant volume.
   (*a*) Otto cycle (*b*) Diesel cycle
   (*c*) Brayton cycle (*d*) Atkinson cycle
8. For a practical petrol engine working on Otto cycle, the compression ratio usually lies in the range.
   (*a*) 3-5 (*b*) 6-8
   (*c*) 10-15 (*d*) 16-22
9. The compression ratio for a practical diesel engine usually lies in the range.
   (*a*) 3-5 (*b*) 6-8
   (*c*) 10-15 (*d*) 16-22

ANSWERS

(A) Fill in the Blanks :

1. Thermal equilibrium exists between the source of heat and working fluid.
2. maximum **3.** less
4. do not **5.** increase
6. 15 to 18 **7.** can
8. decrease
9. actual cycle thermal efficiency to the air standard efficiency
10. increased **11.** Water or steam
12. a gas **13.** isothermal
14. constant volume **15.** constant volume
16. constant pressure **17.** Dual combustion or limited pressure.

(B) State True or False :

1. False **2.** True **3.** True **4.** False **5.** True
6. True **7.** False **8.** True **9.** True **10.** True

11. True 12. False 13. True 14. True 15. True
16. True 17. True

(C) Multiple Choice Questions :

1. (*d*) 2. (*a*) 3. (*b*) 4. (*c*) 5. (*a*)
6. (*a*) 7. (*b*) 8. (*b*) 9. (*d*)

(D) Short Answer Type Questions :

Q. 1. What are 'Power cycles'?

Ans. The thermodynamic cycles on which any device or systems, used to produce a net power output, called engines operates are known as 'power cycles'.

Q. 2. What are 'vapour power cycles'?

Ans. In 'Vapour Power Cycles' the working fluid used is in the vapour phase such as steam, but the working fluid does not remain in the vapour phase throughout the cycle but instead it remains in the vapour phase during one part of the cycle and in the liquid phase during the other part of the cycle.

Q. 3. What are the 'Gas power cycles'?

Ans. In 'Gas Power Cycles' the working fluid remains in the gaseous state throughout the entire cycle.

Q. 4. What is 'Reversible Process'?

Ans. Any thermodynamic process, which can be operated in the reverse direction is known as a 'Reversible Process'.

Q. 5. What are the two important conditions of a thermodynamically reversible process?

Ans. To obtain thermodynamically reversible process the following two conditions must be satisfied;

(*i*) There should be no temperature difference between the working fluid and the heat source.

(*ii*) There should be no friction between the working fluid and the walls of the container.

Q. 6. What do you know by the term 'Available Energy'?

Ans. Available energy is that energy which can be converted into work by processes, which are ideal in nature.

Q. 7. What do you know by the term 'Unavailable Energy'?

Ans. The unavailable energy is that portion of the energy which cannot be converted into work even by using processes which are ideal.

Q. 8. What is 'Carnot Cycle' ?

Ans. 'Carnot Cycle' is an ideal cycle and cannot be executed in actual practice because it has the maximum efficiency. It is the most efficient cycle and acts as the model, to which actual heat engines are compared or being approached.

Q. 9. What is the 'Importance of Carnot Cycle'?

Ans. The fact that we cannot use 'Carnot Cycle' practically does not mean that the cycle has no significance. But, it is very important cycle as well, which acts as the standard for all other actual cycles because it is 100% efficient cycle. Also, the 'Rankine Cycle' has been developed on the basis of this cycle.

Q. 10. What is the expression for the efficiency of a Carnot Cycle? How it can be maximized?

Ans. Efficiency of the Carnot cycle is given by the following expression.

$$\eta_{\text{Carnot}} = \frac{T_1 - T_2}{T_1}$$

where 'T_1' and 'T_2' are the two extremities of temperatures between which the cycle operates.

From this expression, it is clear that the efficiency increases with the decrease in the temperature 'T_2'. It becomes 100% when T_2 is zero, which is not practicable.

In order to obtain the maximum efficiency of the cycle, the temperature 'T_1' should be as high as possible and on the other hand temperature 'T_2' should be as low as possible. But, the temperature 'T_2' is generally the atmospheric temperature. Thereforem, the temperature 'T_1' is the only aim of the design engineer.

Q. 11. What is 'Rankine Cycle'? Is it the modification of 'Carnot Cycle'?

Ans. 'Rankine Cycle' is the earliest but the ideal vapour power cycle. Here the condensation process is continued until all the steam is converted/condensed into water.

Yes, the 'Rankine Cycle" is the modification of the 'Carnot Cycle'.

Q. 12. What is 'Brayton Cycle'?

Ans. The 'Brayton or Joule Cycle' is the air standard Cycle for the gas turbine power plant. Here air is first compressed reversibly and adiabatically, heat is added to it reversibly at constant pressure, air expands in the turbine reversibly at constant pressure to bring it to the initial state.

Q. 13. What is an 'Otto Cycle'?

Ans. The 'Otto Cycle' is the cycle presently being used in petrol of gasoline engines. In this cycle the air is simply heated and cooled. Also, the working fluid is assumed, to be a perfect gas having constant specific heat and the indicator diagram is strictly followed, *i.e.*, the heat losses are supressed and the suction and exhaust pressures are equal.

Q. 14. What are various assumptions which are to be made while deriving formula for the 'air standard efficiency' of an 'Otto Cycle'?

Ans. The various assumptions which are to be made while deriving formula for the air standard efficiency of an 'Otto Cycle' are follows :

(*i*) The working fluid is not subjected to chemical reactions.

(*ii*) There are no heat losses.

(*iii*) The working fluid is a perfect gas following the gas laws and has constant specific heats.

(*iv*) The heating and cooling is made to take place strictly at constant volume.

(*v*) The suction as well as exhaust take place at the atmospheric pressure.

Q. 15. What is a 'Diesel Cycle'? Where it is used?

Ans. The 'Diesel cycle' is a 'Power Cycle' generally being used in Diesel Engines (slow speed generally). In this cycle the heat is supplied constant pressure contrary to 'Otto Cycle' in which it is supplied at constant volume. Because of this reason it is also called 'Constant Pressure Cycle'.

Q. 16. What is the 'Dual Combustion Cycle'? Where it is used?

Ans. To achieve complete combustion of fuel in a diesel engine, without adversely affecting its efficiency it is often arranged in these engine of moderate power for the injection of the fuel to commence before the end of the compression stroke, so that the combustion proceeds partly at constant volume and partly at constant pressure. Such, cycle of operation is known as 'Dual combustion Cycle'. And, the engine working on this cycle is known as 'Semi-Diesel engine' of high speed diesel engine.

Q. 17. What is the effect of increase in compression ratio on the cycle efficiency?

Ans. The efficiency of the cycle increases with the increase in the compression ratio.

Q. 18. What is the effect increase of hot body temperature on the 'Carnot efficiency'?

Ans. The 'Carnot efficiency' increases with the increase of hot body temperature.

Q. 19. What is the cycle which we obtain by reversing the 'Brayten Cycle' or 'Joule Cycle' and where it is used?

Ans. 'Brayton Cycle' or 'Joule Cycle' if reversed is called the 'Bell-Coleman Cycle' which is used in air refrigeration.

Q. 20. Are the practical high speed diesel engines work on 'Diesel Cycle'. Why?

Ans. No, the practical high speed diesel engines do not work on the 'Diesel Cycle' although they work on the 'Dual Combustion Cycle' because to achieve complete combustion of fuel in a diesel engine without adversely affecting the efficiency it is often arranged in these engines of moderate power for the injection of the fuel to commence before the end of the compression stroke so that the combustion proceeds partly at constant volume and partly at constant pressure.

REVIEW QUESTIONS

1. What is meant by 'thermodynamic reversibility'? Sketch the T-ϕ diagram of carnot cycle and the determine the maximum efficiency of a reversible heat engine operating between two given heat reservoirs. **(SBTE Pb., 1974)**
2. (*i*) Enlist the conditions of reversibility.
   (*ii*) Define vapour power cycle and a gas power cycle.
   (*iii*) What are the implications of carnot cycle?
3. Find an expression for the air standard efficiency of a gas engine working on the Otto cycle. **(SBTE Pb., 1975)**
4. Derive an expression for the air standard efficiency of a Diesel cycle. State clearly the assumptions involved.
5. Derive an expression for the air standard efficiency of an engine working on the dual combustion cycle. **(SBTE Pb., 1976)**
6. What are the assumptions made while determining the air standard efficiency? **(SBTE Pb., 1976)**
7. Difference between Otto and Diesel cycle. **(Pb. Dec., 99)**
8. Note on concept of reversibility. **(Pb. Dec., 99)**

PROBLEMS FOR PRACTICE

1. Find the air standard efficiency of the Carnot cycle when the upper and lower limits of temperatures are 200°C and 30°C. [**Ans.** 35.94%]
2. A Carnot cycle steam engine used steam between pressure of 30 bar and 0.04 car. Calculate the heat and work transfers and cycle efficiency. [**Ans.** Q = 724 KJ.kg, W = 724 KJ/kg, η = 40.4%]
3. In an engine working on Otto cycle, what will be the percentage increase in efficiency if the compression ratio is increased from 7 to 9? [**Ans.** 8%]
4. An engine working on an Otto cycle, using air as its working fluid has compression ratio 7:1. The maximum temp. in the cycle is six times the min. temp. Assuming the pressure at the beginning of the adiabatic compression to be 1 bar abs. Calculate the work done during the cycle and the A.S.E. Assume initial temperature to be 35°C. [**Ans.** 91.8 kg cm, 54%]

5. Find the air standard efficiency of a Diesel engine having a compression ratio of 14 and in which cut-off takes place at 5.5% of The stroke.

[**Ans.** η = 60.74%]

6. An indicator diagram taken on a Diesel engine shows that the compression curve follows the law, $PV^{1.4}$ = constant. At two points lying on the compression curve at 1/8th and 7/8th of the strokes the pressure are respectively 1.6 bar and 16 bar. Find the compression ratio of The engine.

If the cut off occurs at 6% the stroke, calculate The air standard efficiency of the engine. [**Ans.** η = 64%]

7. An oil engine working on the dual combustion cycle has a cylinder of 20 cm bore and a stroke of 40 cm. The compression ratio is 14 and The explosion ratio obtained from an indicator card is 1.4. From The indicator card it was also found that the cut off occurred at 5% of stroke. Find the ideal efficiency.

[**Ans.** 62.2%]

8. An ideal dual combustion cycle has a stroke volume of 0.01 m^3 of air with a compression ratio of 16. The pressure and temperature of air at the commencement of adiabatic compression is 1 bar and 27°C respectively. The heat is added at constant pressure of 70 bar and for 5% of the stroke. Calculate

(*i*) Pressure ratio
(*ii*) Cut off ratio
(*iii*) Mass of air contained in the cylinder
(*iv*) Heat added per cycle
(*v*) Heat rejected per cycle
(*vi*) Work done per cycle
(*vii*) Thermal efficiency of the cycle
(*viii*) Mean effective pressure.

Take C_p = 1 KJ/kg k, C_v = 0.714 KJ/kg k

[**Ans.** (*i*) 1.44 (*ii*) 1.75
(*iii*) 0.012 kg (*iv*) Q_A = 15.2 KJ
(*v*) Q_B = 5.54 KJ (*vi*) W = 9.66 KJ
(*vii*) η_{th} = 73.18% (*viii*) p_m = 0.966 KN/m^2]

9. An engine working on Carnot cycle receives heat at 800°C and rejects heat at 30°C. Find the efficiency of the cycle. If the engine receives 5000 kJ/min for the source, calculate the power developed by the engine. [**Ans.** 59.8 KW]

10. A Carnot engine having 15 KW power requires 121000 KJ/h from a source of heat at 300°C. Calculate the temp. of the sink. [**Ans.** 44.33°C]

11. An engine working on Otto cycle has a cylinder dia of 200 mm and stroke of 300 mm. Clearance vol. is 1500000 mm^3. Determine the air standard efficiency of the engine. [**Ans.** 54.8%]

12. In a diesel engine, compression ratio is 20. If fuel cut-off ratio is 2.5. Find the air standard efficiency of the cycle. [**Ans.** 77.53%]

13. Find the air standard efficiency of an I.C. Engine working on Diesel Cycle, if the compression ratio is '17' and the cut-off takes place at 8% of the stroke. Take, the value of '*r*' for air as 1.4. [**Ans.** 62.4%]

14. An I.D. Engine working on Otto Cycle has diameter of the cylinder equal to 200 mm, stroke length of 400 mm, with a clearance volume of 0.005 m^3. Find the thermal efficiency of the engine. take, r = 1.4 for air. [**Ans.** 12.5%]

❑❑❑

10

CHAPTER

Air Compressors

INSIDE THIS CHAPTER

10.1. INTRODUCTION

An '*Air Compressor*' is a machine, which receives air from atmosphere, reduces its volume by compression and increases its pressure with the aid of some mechanical energy and then delivers it to some vessel for storage. The high pressure air thus obtained may be supplied, through a pipe line to the site, wherever, it is required. Compressed air is today, finding many chemical and industrial applications.

The compressed air stores the mechanical energy, which when put into air in the form of work in a reversible process, is returnable as work. However, a small amount of it is lost in friction or as heat.

10.1.1. Industrial Uses of Compressed Air

The compressed air may be used in the different industrial fields which are given below :

(*i*) It is used to produce an air blast for a workshop.

(*ii*) It is used for spraying fuel into the cylinder of an I.C. engine.

(*iii*) It is used in case of pneumatic drills, pneumatic brakes for locomotives, rolling stocks and automobiles.

(*iv*) It is used for the pumping of water.

(*v*) It is used for compressed air engines in coal mines.

(*vi*) In coalmines the machinery is driven by compressed air because no electrical or fuel motor can be used due to the danger of fire.

(*vii*) It is also used for cooling large buildings.

(*viii*) It is used to operate reciprocating tools such as riveting hammer, chipping hammer.

(*ix*) It is used to operate rotating tools, such as grinders, screw drivers, drills, reamers, air motors and air operated pumps.

(*x*) It is used to operate air piston for presses, opening doors, lifting, clamping etc.

(*xi*) It is used to operate air jet for cleaning, moving packages through tubes, conveying materials like sand, concrete etc.

(*xii*) It is used to create blast in blast furnaces.

(*xiii*) It is used to inflate tyres, to charge shock absorbers.

(*xiv*) It is used to start diesel engines and hoisting I.C. engines and running gas turbines.

(*xv*) It is used to clean the heavy castings by mixing with it small steel balls.

(*xvi*) It is also used to spray paints in case of paintings.

(*xvii*) It is used in earth moving machinery.

10.2. CLASSIFICATION - DESCRIPTION OF RECIPROCATING AND ROTARY AIR COMPRESSOR

10.2.1. Classification of Air Compressors

The air-compressors may be classified as follows :

1. According to design and principle of operation :

(*i*) Reciprocating compressors :

(*a*) Single acting

(*b*) Double acting

(*ii*) Rotary Compressors :

(*a*) Steady flow compressor

(*b*) Displacement compressor

These can also be classified as :

(*a*) Fans

(*b*) Blowers

(*c*) Axial flow compressor

(*d*) Mixed flow compressor

(*e*) Centrifugal compressor

2. According to the number of stages :

(*i*) Single stage compressor

(*ii*) Multi-stage compressor

3. According to the Pressure limit :

(*i*) Low pressure compressor (Delivery pressure less than 1 MN/m^2)

(*ii*) Medium pressure compressor (Delivery pressure from 1 to 8 MN/m^2)

(*iii*) High pressure compressor (Delivery pressure from 8 to 10 MN/m^2)

(*iv*) Super high pressure compressor (Delivery pressure above 10 MN/m^2)

4. According to the capacity :

(*i*) Low capacity compressor (Delivers comp. air upto 0.15 m^3/sec)

(*ii*) Medium capacity compressor (Delivers comp. air from 0.15 to 5 m^3/sec)

(*iii*) High capacity compressor (Delivers comp. air more than 5 m^3/sec)

5. According to the method of cooling :

(*i*) Air cooled compressor

(*ii*) Water cooled compressor

6. According to the installation :

(*i*) Portable compressor

(*ii*) Fixed compressor

7. According to the application :

(*i*) Pneumatic hand tool compressor

(*ii*) Rock drill compressor

(*iii*) Sand blasting compressor

(*iv*) Spray painting compressor

(*v*) Mine prospecting compressor

(*vi*) Refrigerator compressor etc.

8. Rotary compressors can be further classified according to I.S.I. standard as :

(*i*) Positive displacement compressor

(*ii*) Single shaft vane type compressor

(*iii*) Multi-shaft compressor

(*iv*) Roots blower or compressor

(*v*) Screw compressor

(*vi*) Turbo blower

(*vii*) Turbo compressor

(*viii*) Axial blowers :

(*a*) Single stage

(*b*) Multi stage

10.2.2. Description of Reciprocating Air Compressor

10.2.2.1. Single Stage Reciprocating Air Compressor

(*i*) Single Stage, Single Acting Reciprocating Air Compressor. A single stage single acting reciprocating compressor has been shown in Fig. 10.1.

It is a positive displacement compressor from which the delivery is intermittent. It essentially consists of a piston and cylinder. The piston reciprocates inside the cylinder. It is driven by crankshaft through a connecting rod. There are two valves mounted in the head of the cylinder. These valves are the suction valve and the delivery valve. Both the valves, usually operate as a result of pressure difference across them.

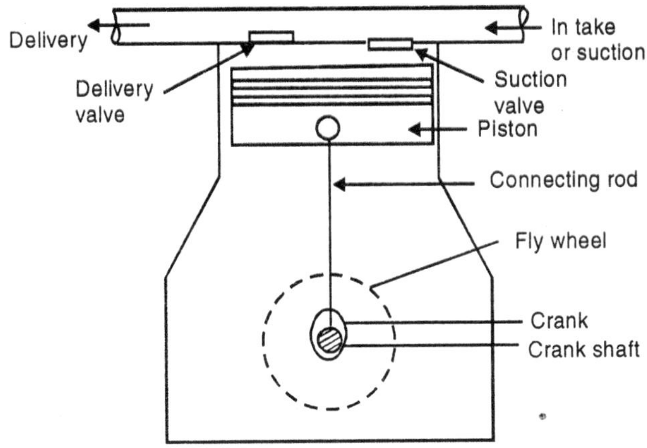

Fig. 10.1. Single stage single acting reciprocating air compressor.

(*ii*) **Single Stage Double Acting Reciprocating Air Compressor.** When the atmospheric air is compressed to the desired pressure in a single operation and the compression is carried out by the two sides of the piston, then the compressor is known as 'single stage, double acting reciprocating compressor'. A single stage double acting reciprocating compressor has been shown in Fig. 10.2.

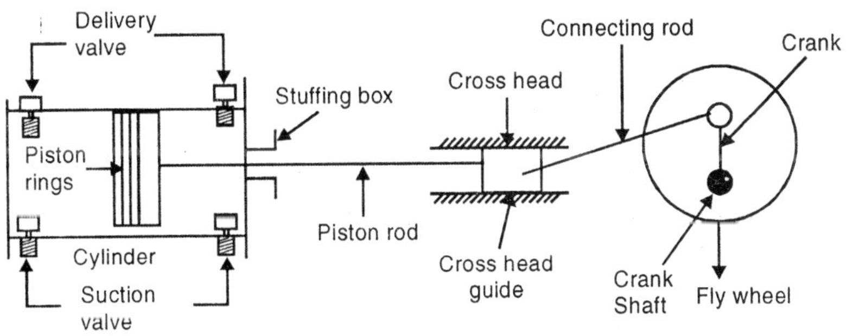

Fig. 10.2. Single stage double acting reciprocating air compressor.

In this compressor, air is compressed by both the faces of the piston as it reciprocates inside the cylinder. The crank end side of the cylinder is provided with a stuffing box through which the piston rod reciprocates. The function of stuffing box is to make the crank end side of the cylinder air tight. The piston rod is connected to the connecting rod through a cross-head.

10.2.2.2. Multistage Reciprocating Air Compressor

When the required delivery pressure of compressed air is achieved by the use of two or more than two stages/cylinders, then the compressor is said to be the 'multistage compressor'.

Multistage compressor is employed whenever it is not possible to achieve the required very high pressure of air in a single stage compressor.

A two stage reciprocating compressor with intercooler has been shown in Fig. 10.3.

In this the first cylinder is known as the low pressure cylinder and it is the first stage of compression. The second cylinder receives the compressed air through the intercooler and is known as high pressure cylinder. It is the second stage of compression. The dimensions of the H.P. cylinder is smaller than that of L.P. cylinder, because the volume of air is decreased as it is compressed. The intercooler saves amount of work required during the second stage of compression. The L.P. and H.P. cylinders are coupled and are driven by the same prime-mover.

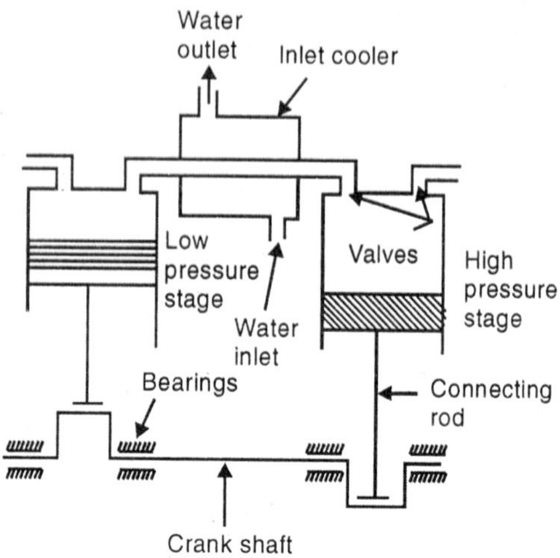

Fig. 10.3. Two stage reciprocating air compressor.

10.2.2.3. Advantages of Multistage Compression

(*i*) **Reduced Driving Power.** For a given compression (pressure ratio), the requirement of driving power is reduced.

(*ii*) **Improved Volumetric Efficiency.** For a given compression, the volumetric efficiency is improved by multistage compression.

(*iii*) **High Delivery Pressure.** Multistage compression makes possible even a very high delivery pressure.

(*iv*) **Improved Cooling.** Multi-cylinder arrangement with intercooling of a multistage compression gives better mechanical balance and improved cooling during compression.

On the other hand, in single stage compression, it is difficult to transfer (reject) heat from the air in the small time available during compression. Hence, the process cannot be approached to isothermal and then the work required is large.

(*v*) **Reduced Flammability Danger.** Due to intercooling in a multi-cylinder compression, there is no danger of formation of flammable mixture of compressed air with lubricating oil. But, this danger remains in case of single stage high compression.

(*vi*) **Reduced Cylinder Size.** The single cylinder size, in case of single stage compression, will be too large if used for high compression ratio. This can be avoided through multistage compression. Then, size and strength of the cylinders can be adjusted to suit volume and pressure of air.

10.2.3. Description of Rotary Air Compressors

The rotary air compressors run at very high speeds. These are used for supplying large quantity of air at relatively low pressures. Due to their high speed these can be directly coupled with steam turbine or gas turbine shafts. Pressures upto nearly 11 bar can be obtained with a multi stage turbo compressor. Rotary compressors may be subdivided into.

(*i*) Fans

(*ii*) Rotary displacement blowers

(*iii*) Turbo blowers

(*iv*) Turbo compressors

(*i*) **Fans.** The fan consists of a single rotating ring of curved vanes rotating at high speed and producing a flow of air at a comparatively low pressure. The fan produces a gauge pressure of less than 0.07 bar. It is generally used for air circulation, production of air cooling currents low pressure blast air and ventilation.

(*ii*) **Rotary displacement blowers.** In this type of compressor the air is compressed in the space formed by two sets of mutually engaging cam surfaces. As the two surfaces engage, the space between them is reduced in volume, so that the air between them is compressed. One type of rotary displacement blower (the root compressor) is shown in Fig. 10.4.

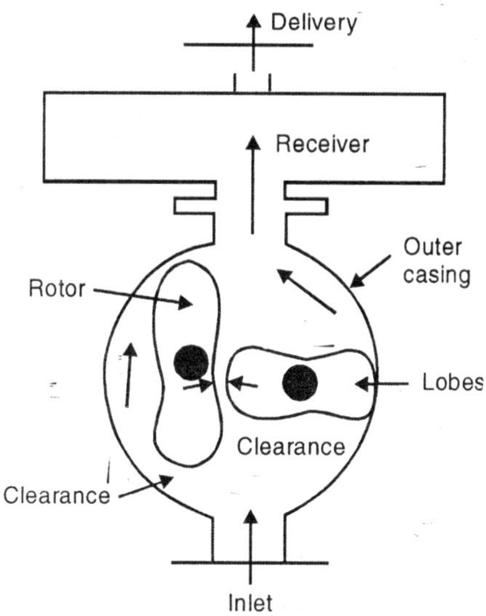

Fig. 10.4. Rotary displacement blower (Root compressor).

The Pressure Ratio upto 3.6 : 1 can be obtained in this type of blower. Higher pressure may be obtained by placing two or more blowers in series, having intercooler between each stage. The blower can be coupled directly to a turbine shaft and it occupies a very small space compared with other compressors for the same output. It is used for super charging I.C. engines, for producing scavenging air for two stroke diesel engine and as a compressor in gas turbine plants.

(*iii*) **Turbo blowers.** This is a low pressure type of rotary compressor and it consists of one or more rotating rings of curved blades. The complete blade ring is known as an impeller as shown in Fig. 10.5.

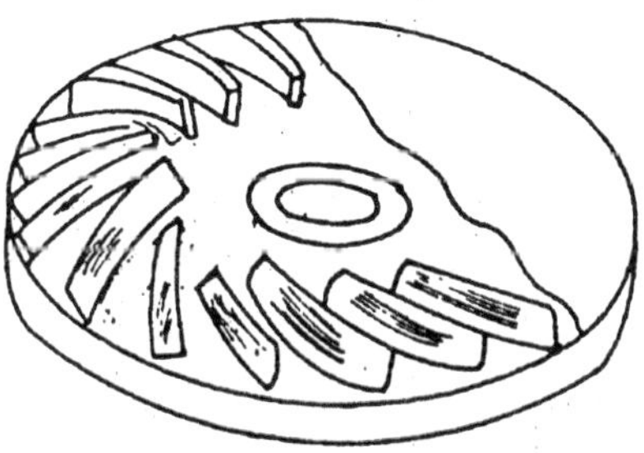

Fig. 10.5. Impeller of turbo blower.

Several impellers may be keyed on to a common shaft. Each impeller is surrounded by a diffuser which converts, the high velocity of the air leaving the impeller into pressure energy. A vane in the outer casing then guides the air stream into the next impeller. Each impeller and its diffuser combined is known as a stage and each stage of blower may increase the pressure of the air by a ratio of 1.6 : 1. The turbo blowers produce air pressure upto 2.5 bar. These are used for ventilators for producing blast air in work shops and for engine super chargers.

(*iv*) **Turbo compressors.** The working principle of turbo compressor is the same as that of turbo blower. But the turbo compressor usually has more number of stages. Delivery air pressure upto 12 bar is obtained and as many as 20 impellers in series may be used. Turbo compressors are designed to run at one speed only. They further can be classified as follows :

(*a*) **Vane Blower.** It consists of rotor located eccentrically in a cylindrical outer casing, a set of vanes mounted on the rotor and rotating eccentrically in the outer casing, inlet and delivery ports as shown in the Fig. 10.6.

Each vane will start collecting air after it leaves the portion marked by position 1 in the figure and no more air will be sucked after vane moves to position 2.

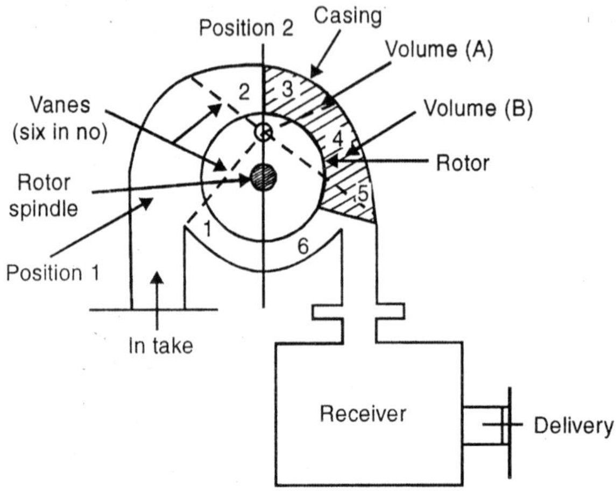

Fig. 10.6. Vane blower.

Then compression of trapped air could commences due to decreasing space volume, between the rotor and the casing. When the vane travels from position 3 to 4, the air occupying volume A is compressed, into a smaller volume B. This type of blower is used for air capacities upto 150 m^3/min and pressure ratio upto 8 : 5 : 1.

(*b*) **Centrifugal Compressor.** A centrifugal compressor is shown in the Fig. 10.7.

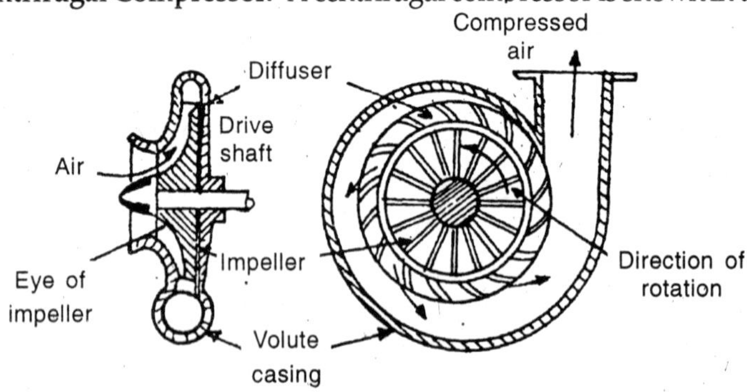

Fig. 10.7. Centrifugal compressor.

This type of compressor mainly consists of the following elements :

(*i*) **Series of Inlet Buckets.** These are attached to the shaft and rotate with it.

(*ii*) **Rotating Impeller.** The impeller is a disc fitted with radial vanes. The air supplied to the these vanes by the inlet buckets.

(*iii*) **Volute Casing.** The casing surrounds the rotating impeller.

(*iv*) **Diffuser.** The diffuser is housed in radial portion of the casing.

The air enters the eye of the impeller with a low velocity and at atmospheric pressure. Depending upon the centrifugal action of the impeller, the air moves radially outwards and during its movement, it is guided by the impeller vanes. The impeller traverses energy of the drive to the air causing a rise both in static pressure and temperature and an increase in velocity. The air now enters the diverging passage called diffuser, where it is efficienctly slowed down. The kinetic energy of the air is converted into pressure energy with the result that there is further rise in static pressure. In case of multi stage centrifugal compressor a number of single stage centrifugal compressors are arranged on one shaft. Where the delivery of one unit leads to the intake of the next for further compression. Delivery from the last unit is made use of for doing the required work.

(*c*) **Axial Flow Compressors.** If the air flowing through the compressor is forced to flow in a direction parallel to its axis, then this compressor is called as an 'axial flow compressor'.

It consists of several moving blade rings attached to a common rotating shaft which are separated by rings of fixed blades attached to the casing as shown in the Fig. 10.8.

The circular area is usually reduced from inlet to outlet of the compressor. This is done to keep the flow of velocity constant throughout the compressor length. In the diverging passage, there is a rise in pressure due to diffusion. The fixed blades serve two purposes;

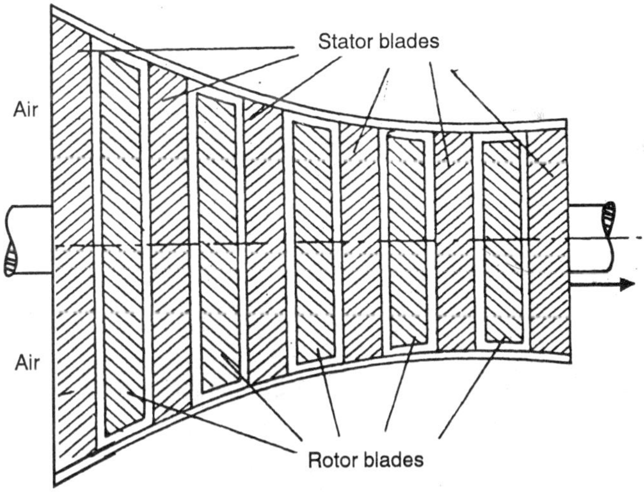

Fig. 10.8. Axial flow compressor.

(*i*) They convert a part of the kinetic energy of the air into pressure energy and this conversion is achieved by diffusion process carried out in the diverging blade passage.

(*ii*) Guide and redirect the air flow so that entry to the next stage is without shock. Axial flow compressors are used to produce pressure upto 5.5. bar and delivering about 1,00,000 kg of air per hour.

10.2.4. Comparison of Reciprocating and Rotary Air Compressors

Various points of comparison between reciprocating and rotary air compressors are given below :

Reciprocating Air Compressor	Rotary Air Compressor
1. Generally suitable for low discharge of air at high pressure.	Suitable for handling large volumes of air at low pressures.
2. Generally operate at low speeds.	The speed of operation is usually high.
3. Compressed air discharge is pulsating. To overcome this problem receiver tanks are installed.	The air supply is continuous.
4. The balancing is a problem resulting in cyclic vibration.	Due to absence of cyclic vibrations there are less vibrations.
5. Lubricating system is generally complicated as reciprocating as well as rotating parts are to be lubricated.	Generally simple lubrication systems are required.
6. Air delivered is generally contaminated with oil as it comes in contact with lubricating oil during compression.	Air delivered is relatively more clean.
7. The size of the air compressor is large for the given discharge.	The size of the compressor is small for same discharge.
8. Generally, available in sizes which can handle upto 250-300 m^3 per minute of free air.	Can handle 2000–3000 m^3 per minute of free air.
9. Can be used for air delivery pressures upto 800–1000 bar.	Normally restricated to pressures below 10 bar.
10. Isothermal compression is the usual standard of comparison for reciprocating compressors.	Isentropic compression is the normal standard of comparison for the rotary compressors.

10.3. FANS, BLOWERS AND SUPERCHARGER

The fans and blower have already been described in the previous section.

10.3.1. Supercharger

The process of increasing the pressure of the fuel air mixture (in S.I. engines) or that of air only (in I.C. engines) entering an engine during suction, to a pressure greater than that of atmospheric air, is known as 'supercharging' and sometimes 'boosting'.

A small compressor or blower which is employed for super charging is known as 'super charger'. But an air or gas compressor which is employed to raise the pressure of already compressed air or gas, is known as 'booster'.

The loss of power produced by an engine, when operating at high altitudes or when operating at a very high speed can be prevented by the use of supercharger. It is generally driven by crank shaft of the engine. Thus, it absorbs some of the power produced by the engine. It is found that the power absorption by the supercharger increases with the increase of altitude. It is generally expressed in terms of percentage of brake power of the engine.

There are two main types of superchargers :

(*i*) **Reciprocating Supercharger.** It has a piston reciprocating inside the cylinder. It is generally used on large and slow speed, stationary engines. The main disadvantages of this type of supercharger are large space occupied, more weight and more lubrication required.

(*ii*) **Rotary Supercharger.** It has a rotary impeller or rotor, rotating inside the casing and is quite successful for the supercharging of I.C. engines, in particular two stroke I.C. engines. The rotary superchargers further have the following two types :

(*a*) Roots blower

(*b*) Vane blower

These two have already been explained in the previous sections.

10.4. WORKING PRINCIPLE OF RECIPROCATING SINGLE AND TWO STAGE COMPRESSORS

10.4.1. Working Principle of a Single Stage Reciprocating Compressor

Figure 10.9 shows a theoretical *p-V* diagram of operations for an ideal single stage compressor.

Assumptions made for the derivation of expressions for work to be done during a cycle and power required to drive the compressor :

(*i*) The clearance volume is zero.

(*ii*) There is no pressure drop in the valves, *i.e.*, suction and delivery valves.

(*iii*) There is no pressure drop in the suction and delivery pipings.

(*iv*) The entire compression takes place in the same cylinder.

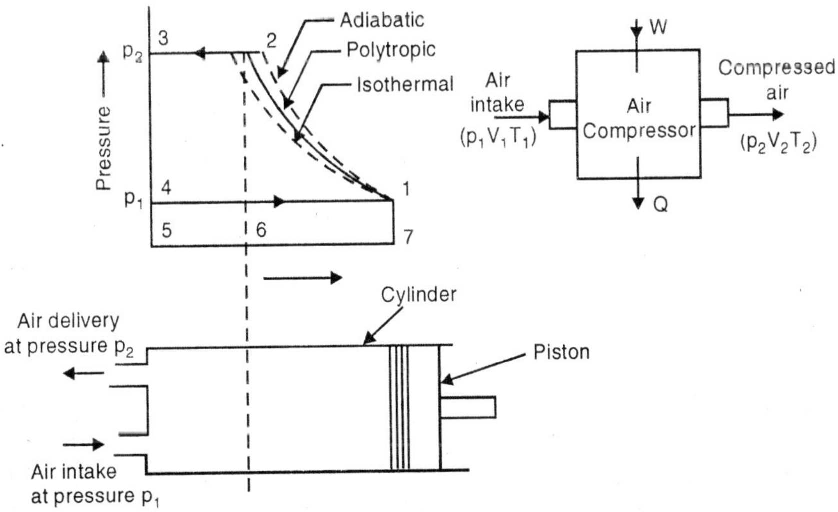

Fig. 10.9. Ideal *p-V* diagram for single stage reciprocating compressor.

At the beginning of the outward stroke of the piston the intake valve opens at a pressure P_1 and air is admitted in the cylinder. Because, there is no clearance volume, V_1 represents the volume of air admitted in the cylinder. This operation is represented by the line 4-1.

The intake valve closes instantly, as the motion of the piston is, towards the cylinder head. The compression of the air is carried out in the compressor cylinder from pressure P_1 to pressure P_2 according to the law PV^n = Constant. In this process the mass of the air in the cylinder remains constant. This operation is represented by line 1-2.

The discharge valve opens, when the pressure of compressed air becomes equal to the receiver pressure. The air is discharged at a constant pressure into the receiver. Because, there is no clearance so whole of the air is discharged into the receiver. This process is represented by the line 2-3.

Now, there is no air in the cylinder and pressure becomes again equal to intake pressure P_1. Again the compressor is ready for another cycle when the piston moves outwards.

Now, work to be done or work required per cycle during polytropic, isothermal and adiabatic compression can be calculated as follows :

(*a*) When the compression is polytropic *i.e.*, following the law PV^n = Constant :

Work required/cycle, W = Area of the p-V diagram

or W = Area 1 – 2 – 3 – 4

or W = Area (3 – 2 – 6 – 5) + Area (2 – 1 – 7 – 6) – Area (4 – 1 – 7 – 5)

$$\text{or } W = P_2 V_2 + \frac{P_2 V_2 - P_1 V_1}{n-1} - P_1 V_1$$

$$\text{or } W = \frac{nP_2 V_2 - P_2 V_2 + P_2 V_2 - P_1 V_1 - nP_1 V_1 + P_1 V_1}{n-1}$$

$$\text{or } W = \frac{n(P_2 V_2 - P_1 V_1)}{n-1} \qquad ...(10.1)$$

It can be noted, from this expression *i.e.*, 10.1, that the work required/cycle is equal to the work required during the polytropic compression multiplied by the polytropic index, 'n'

From eqn. (10.1)

$$\text{Work required/Cycle } W = \frac{n}{n-1} P_1 V_1 \left(\frac{P_2 V_2}{P_1 V_1} - 1\right) \qquad ...(i)$$

Also, during polytropic process, $P_1 V_1^n = P_2 V_2^n$

$$\text{or} \qquad \frac{V_2}{V_1} = \left(\frac{P_1}{P_2}\right)^n$$

Now, substituting the value of $\frac{V_2}{V_1}$ in eqn. (*i*), we have work required/cycle,

$$W = \frac{n}{n-1} . P_1 V_1 \left[\frac{P_2}{P_1} \times \left(\frac{P_1}{P_2}\right)^n - 1\right]$$

or $$W = \frac{n}{n-1} . P_1 V_1 \left[\left(\frac{P_2}{P_1} \right)^{1-\frac{1}{n}} - 1 \right]$$

or $$W = P_1 V_1 . \frac{n}{n-1} \left[\left(\frac{P_2}{P_1} \right)^{\frac{n-1}{n}} - 1 \right] \quad ...(10.2)$$

(*b*) When the compression is isothermal :

Work required/cycle W= Area (1 – 2 – 3 – 4)

$$= P_2 V_2 + P_2 V_2 . \log_e \frac{V_1}{V_2} - P_1 V_1 \quad ...(ii)$$

But, during isothermal process, $P_1 V_1 = P_2 V_2$

∴ From eqn. (*ii*)

Work required/cycle, $W = P_2 V_2 . \log_e \frac{V_1}{V_2}$

(Cancelling ($P_2 V_2$) and ($P_1 V_1$) with each other)

Now V_1/V_2 = Compression ratio, r

Hence work required/cycle, $W = P_2 V_2 \log_e r$ ...(10.3)

(*c*) When the compression is adiabatic *i.e.*, following the law PV^γ = constant

Work required/cycle, W during adiabatic compression can be calculated by replacing the 'n' by 'γ' in eqn. (10.2). *i.e.*, for polytropic compression.

Hence, work required/cycle :

$$W = P_1 V_1 . \frac{\gamma}{\gamma - 1} \left[\left(\frac{P_2}{P_1} \right)^{\frac{\gamma - 1}{\gamma}} - 1 \right] \quad ...(10.4)$$

The power required to drive the compressor :

If P_1 and P_2 are in N/m² or P_a and V_1 in m³

Then, work reqd./cycle, W will be in N-m or Joule

If n = Number of cycles/minute

and N = Number of revolutions per minute of the crank shaft,

Then $n = N$ (for single acting compressor)

and n = Number of strokes/min *i.e.*, 2N (for double acting compressor)

Now, work done/minute = $W \times n$ N-m/min or J/min

Hence, power required to drive the compressor,

$$P = \frac{W \times n}{60,000} \text{ kW} \quad ...(10.5)$$

Mean Effective Pressure :

As, work required/cycle = $P_1 V_1 . \frac{n}{n-1} \left[\left(\frac{P_2}{P_1} \right)^{\frac{n-1}{n}} - 1 \right]$ ...(*iii*)

In case of single acting reciprocating compressor,

Work required/cycle = Work required/stroke

Also, Work required/stroke = $P_m \times V_s$ ...(*iv*)

(where P_m is the mean effective pressure and V_s the swept volume)

∴ From eqn. (*iii*) and (*iv*)

$$P_m \times V_s = P_1 V_1 \frac{n}{n-1}\left[\left(\frac{P_2}{P_1}\right)^{\frac{n-1}{n}} - 1\right] \quad ...(v)$$

If we neglect the clearance volume, then :

$$V_s = V_1$$

and the eqn. (*v*) becomes,

$$P_m = P_1 \frac{n}{n-1}\left[\left(\frac{P_2}{P_1}\right)^{\frac{n-1}{n}} - 1\right] \quad ...(10.6)$$

which gives the value of mean effective pressure, P_m directly.

10.4.2. Working Principle of a Reciprocating Two Stage Air Compressor

The *p-V* diagram of a two stage air compressor is shown in Fig. 10.10.

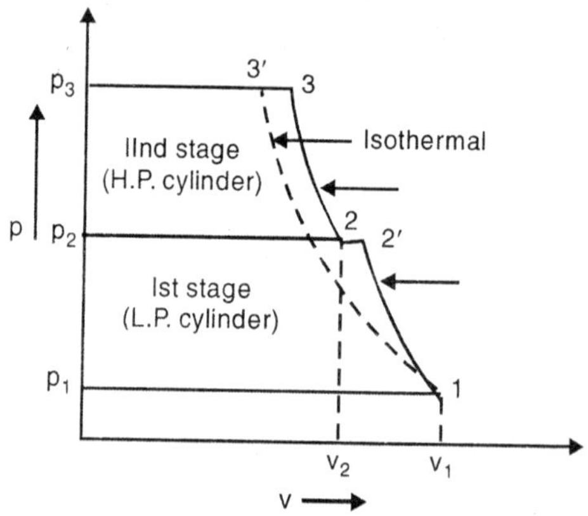

Fig. 10.10. Ideal *p-V* diagram for two stage reciprocating air compressor.

Assumptions made for the derivation of expressions for work and the power required to drive the compressor :

(*i*) The clearance volume is zero.

(*ii*) The compression is assumed to be polytropic.

(*iii*) The suction and delivery takes place at constant pressure.

(*iv*) The process of cooling is at constant pressure.

(*v*) The temperature of the air entering into the second stage is the same as the temperature of the air leaving the intercooler.

(*vi*) The inter cooling is not complete, *i.e.*, the temperature of air after cooling does not become equal to its temperature before compression.

In this case, it is assumed that there is no pressure drop in the intercooler and the intercooling is incomplete, so that the temperature of air entering high pressure cylinder is greater than that of air entering low pressure cylinder. Let P_1 be the pressure and V_1 the volume of air entering the low pressure cylinder, $P_2 V_2$ represents the pressure and volume of air entering high pressure cylinder and P_3 be the final delivery pressure of air. The total work required to drive the compressor per cycle will be the sum of the work required in each cylinder.

Work required per cycle in low pressure cylinder,

$$= P_1 V_1 \frac{n}{n-1}\left[\left(\frac{P_2}{P_1}\right)^{\frac{n-1}{n}} - 1\right] \text{ Joule} \quad ...(i)$$

Similarly, work required per cycle in high pressure cylinder,

$$= P_2 V_2 \frac{n}{n-1} \cdot \left[\left(\frac{P_3}{P_2} \right)^{\frac{n-1}{n}} - 1 \right] \text{Joule} \quad ...(ii)$$

Total work required per cycle,

$$W = \frac{n}{n-1} \left[P_1 V_1 \left\{ \left(\frac{P_2}{P_1} \right)^{\frac{n-1}{n}} - 1 \right\} + P_2 V_2 \left\{ \left(\frac{P_3}{P_2} \right)^{\frac{n-1}{n}} - 1 \right\} \right] \text{J} \quad ...(10.7)$$

If the intercooling is perfect, we have, $P_1 V_1 = P_2 V_2$

So, work required will be,

$$W = \left(\frac{n}{n-1} \right) P_1 V_1 \left[\left(\frac{P_2}{P_1} \right)^{\frac{n-1}{n}} + \left(\frac{P_3}{P_2} \right)^{\frac{n-1}{n}} - 2 \right] \text{Joule} \quad ...(10.8)$$

Now, if 'n' is the number of cycles per minute then,

$$\text{Power required} = \frac{W.n}{60{,}000} \text{ kW} \quad ...(10.9)$$

Let, T_1 = Abs. temperature of air at point '1'

T_2 = Abs. temperature of air at point '2'

T'_2 = Abs. temperature of air at point 2'

m = Mass of air in the cylinder

and C_p = Sp. heat of air at constant pressure.

$\therefore$ Heat rejected to intercooler = $m \times C_p \times (T'_2 - T_2)$ ...(10.10)

The temperatures T_2 and T'_2 and m can be calculated as follows :

(*i*) From Gas equation, we know, that;

$$P_1 V_1 = m.RT_1 \quad ...(10.11)$$

As P_1, V_1, T_1 and R are known, so 'm' can be found out.

(*ii*) During the polytropic process 1 – 2'

$$P_1 V_2^n = P_2 V_2'^n \quad ...(10.12)$$

As P_1, V_1, P_2 and the polytropic index 'n' are known so, V_2' can be found out.

(*iii*) Also, $P_2 V_2' = m RT_2'$ ...(10.13)

Here, P_2, V'_2, m and R known, so T_2' can be found out.

(*iv*) During constant pressure process 2' – 2

$$\frac{V_2'}{T_2'} = \frac{V_2}{T_2} \quad ...(10.14)$$

Here, V'_2, V_2 and T'_2, are known, So T_2 can be found out.

10.4.2.1. Conditions for Maximum Efficiency

Let us consider a reciprocating, two-stage compressor. The work required to drive the compressor will be maximum when the point of start of second stage of compression lies on the isothermal line shown in Fig. 10.10. The expression for the work required per cycle is given by the equation;

$$W = P_1 V_1 \times \frac{n}{n-1} \left[\left(\frac{P_2}{P_1}\right)^{\frac{n-1}{n}} + \left(\frac{P_3}{P_2}\right)^{\frac{n-1}{n}} - 2 \right] \qquad ...(10.15)$$

If we take initial pressure, P_1 and the final pressure P_3, the best value of intermediate pressure P_2 can be determined by differentiating the eqn. (10.15) with respect to P_2.

$$\therefore \frac{dW}{dP_2} = P_1 V_1 \frac{n}{n-1}\left[\left(\frac{1}{P_1}\right)^{\frac{n-1}{n}} \cdot \left(\frac{n-1}{n}\right) P_2^{-1/n} - (P_3)^{\frac{n-1}{n}} \cdot \left(\frac{n-1}{n}\right) P_2^{-1/n} - 2\right]$$

Multiplying throughout by $\frac{n-1}{n} P_2^{-1/n}$ and equating to zero.

$$\therefore \left(\frac{1}{P_1}\right)^{\frac{n-1}{n}} - \frac{(P_3)^{\frac{n-1}{n}}}{P_2^2\left(\frac{n-1}{n}\right)} = 0$$

Thus, $$P_2^2 = P_1 . P_3$$

or $$\boldsymbol{P_2 = \sqrt{P_1 . P_3}} \qquad ...(10.16)$$

Hence, for maximum efficiency, the intermediate pressure is the geometric mean of the initial and final pressures.

Now, substituting this value of P_2 in the eqn. (10.15), we have

$$W = P_1 V_1 \left(\frac{n}{n-1}\right)\left[\left(\frac{\sqrt{P_1 P_3}}{P_1}\right)^{\frac{n-1}{n}} + \left(\frac{P_3}{\sqrt{P_1 P_3}}\right)^{\frac{n-1}{n}} - 2\right]$$

or $$W = P_1 V_1 \left(\frac{2n}{n-1}\right)\left[\left(\frac{P_3}{P_1}\right)^{\frac{n-1}{2n}} - 1\right] \text{ Joules} \qquad ...(10.17)$$

10.4.2.2. Heat Rejected in the Intercooler

Heat rejected in the intercooler can be determined if the temperature at the end of compression in the low pressure cylinder and at suction of the high pressure cylinder are known.

Let, T_1 = Temp. of air at the time of suction in L.P. cylinder

T'_2 = Temp. of air at the end of compression.

T_2 = Temp. of air at the time of entering the H.P. cylinder

m = Mass of air taken per cycle.

Now, $$\frac{T'_2}{T_1} = \left(\frac{P_2}{P_1}\right)^{\frac{n-1}{n}} \qquad ...(10.18)$$

From this eqn. (10.18) the value of T_2' can be calculated

Also, $$\frac{P_2 V'_2}{T'_2} = \frac{P_2 . V_2}{T_2} \qquad ...(10.19)$$

From eqn. (10.19) the value of T_2 can be calculated.

So, the heat rejected to the intercooler per cycle

$$= m \cdot C_p \cdot (T'_2 - T_2) \quad (10.20)$$

10.4.2.3. Effect of Intercooling

In two stage compression, with perfect intercooling the intermediate pressure ratio P_2/P_1 and P_3/P_2 are equal and is equal to $\sqrt{P_3/P_1}$. So, in general if the compression has 'N' number of stages the intermediate pressure ratios will all be equal to 'N' the root of the overall pressure ratio.

The work required per kg of air in 'N' stage compressor with perfect intercooling to compress air from P_1 to P^{n+1} is given by;

$$W = \left(\frac{n}{n-1}\right).N.RT_1\left[\left(\frac{P_{N+1}}{P_1}\right)^{\frac{n-1}{n-N}} - 1\right] \text{J/kg} \quad ...(10.21)$$

10.4.2.4. Effect of Type of Compression

The operation of compression may be adiabatic, isothermal or polytropic as represented in Fig. 10.11.

The process line 1-2 represents the adiabatic compression. If the compression is polytropic, it is represented by process line 1-5, whereas the isothermal compression is represented by process line 1-6.

It is clear from the figure that the isothermal compression process line (1-6) has less slope, as compared to the other two processes. This means that if the compression is isothermal, the work required per cycle is reduced considerably, which is given by the area (1-6-3-4).

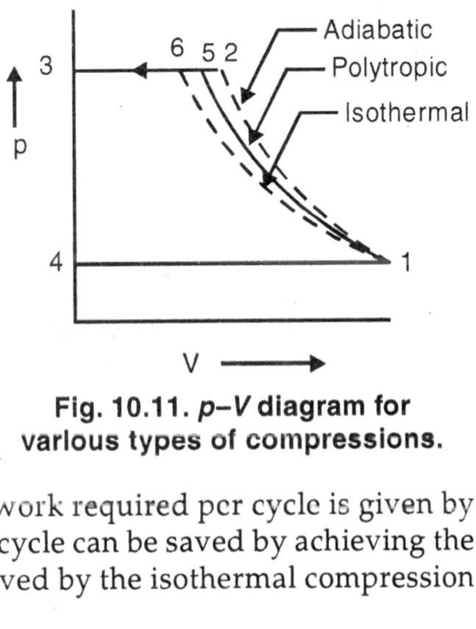

Fig. 10.11. *p–V* diagram for various types of compressions.

When the compression is adiabatic, the work required per cycle is given by the area (1-2-3-4). So, the work required per cycle can be saved by achieving the compression as isothermal one. The work saved by the isothermal compression is shown in Fig. 10.11.

But, in actual practice it is not possible to compress the air isothermally (at constant temperature), because an isothermal compression would mean that the compressor should run at a very slow speed. Whereas, the compressor is required to run at high speeds, so as to compress as much air as possible in a given period. Still every attempt is made to get the compression which is approaching isothermal, to save a considerable amount of work required to drive the compressor.

10.4.2.5. Methods of Saving the Work Required

There are generally two methods in use to get the isothermal compression approximately, when the compressor is running at a high speed. These are by using;

(*i*) Cold water spray, and

(*ii*) Multistage compression

(*i*) **Cold Water Spray.** When the air is being compressed, the increasing temperature of the air is reduced by spraying cold water. With this, although the actual compression may not be the isothermal yet, it is approaching the isothermal compression.

It is shown in Fig. 10.12 that a large amount of work is saved (shown shaded).

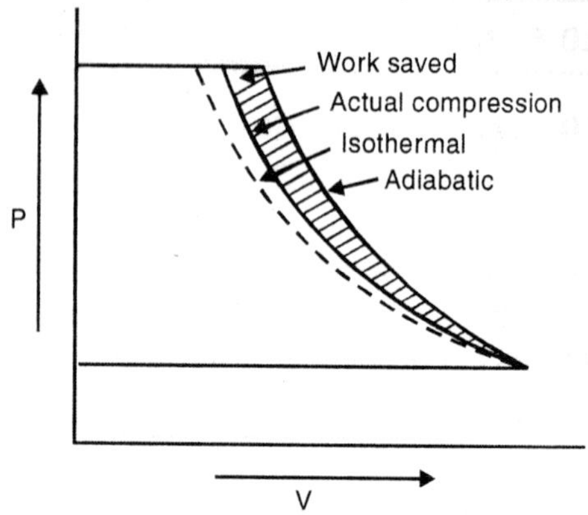

Fig. 10.12. ***p–V*** **diagram showing actual and isothermal compression.**

(*ii*) **Multi-Stage Compression.** In multistage compression the air is compressed in the cylinder of compressor, in two or more stages in separate cylinders. The air after being compressed in the L.P. cylinder is cooled and fed into the next H.P. cylinder and so on. The *p-V* diagram of a four stage air compressor is shown in Fig. 10.13.

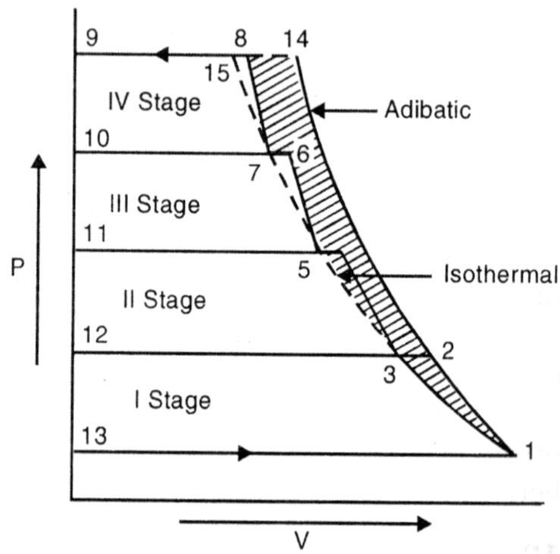

Fig. 10.13. ***p-V*** **diagram for four stage reciprocating air compressor.**

The process lines (1-14) and (1-15) represents the adiabatic and isothermal compression lines respectively. In order to save the work required, the actual compression in each cylinder should be as nearly as approaching the isothermal.

Let the air is compressed adiabatically in the L.P. cylinder. It is represented by the line 1-2. At point 2, the compressed air is cooled down at constant pressure to its initial temperature, if possible. Cooling is represented by line 2-3 and generally known as 'Intercooling'. If the intercooling is complete, *i.e.*, the temperature of the compressed air becomes equal to its initial temperature then the point 3 should lie on the isothermal line.

This cooled air corresponding to point 3 is drawn into the next cylinder, where it is again compressed. It is represented by line 3-4. The air is again cooled by means of the intercooler and is further drawn by the subsequent H.P. cylinder for the third stage of its compression and so on the process is continued.

It can be noted that the actual compression during each stage is very close to the isothermal. It is made possible by compressing the air in a number of stages and providing intercoolers in between the stages. If this total compression had taken place in a single stage, then the compression would have followed the adiabatic line (1-14). Hence, a lot of work required is saved, which is shown shaded in the Fig. 10.13.

10.5. INTERCOOLING, VOLUMETRIC EFFICIENCY

10.5.1. Intercooling

The process of exchanging heat from the hot compressed air in between the various stages of an air compressor to that of a coolent by means of a simple heat exchanger intercooler is known as 'Intercooling'.

Whereas the process of exchanging heat from the hot compressed air after the last stage of an air compressor to that of a coolant by means of a simple heat exchanger (here, aftercooler) is known as 'After cooling'.

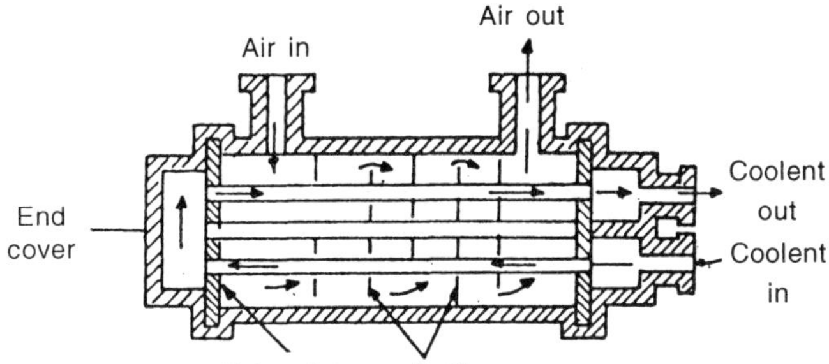

Fig. 10.14. Intercooler.

Figure 10.14 shows the sectioned view of an intercooler or aftercooler (as the case may be).

The inter cooler mainly has the two paths, one for the air and the other for the coolant. The coolant which may be water or any other fluid passes through the tubes secured, between two end plates and the air circulates over the tubes through a system of baffles. These are provided to ensure intimate contact between the tubes and the air by passing the air as shown in the Fig. 10.14.

10.5.2. Effect of Clearance Volume

The volume in between the cylinder head and the piston top, when it is at its top most position, is called the 'clearance volume'.

The clearance volume is necessary to accommodate the intake and delivery valve opening during operation and to allow for wear and mechanical freedom.

It is clear from Fig. 10.15 that during the first part of the suction stroke, the compressed air left in the clearance space V_3 at pressure P_2 at the end of compression and delivery stroke air will first expand to volume V_4 at pressure P_1. The expanssion of air may be assumed polytropic, *i.e.*, to follow the law PV^n = constant.

During expansion no fresh air from atmosphere is admitted into the cylinder even though the piston has travelled through a certain distance during its suction stroke so long as the pressure inside the cylinder is more than the atmospheric pressure.

The air from the atmosphere is admitted during the remaining part of the suction stroke, when the pressure inside the cylinder becomes equal or less than atmospheric pressure. It will decrease the volumetric efficiency of the compressor to a great extent. Therefore, in order to increase the volumetric efficiency of the compressor, the clearance volume should be as small as possible. In high class high pressure compressors, the clearance volume is limited to 3% of the swept volume.

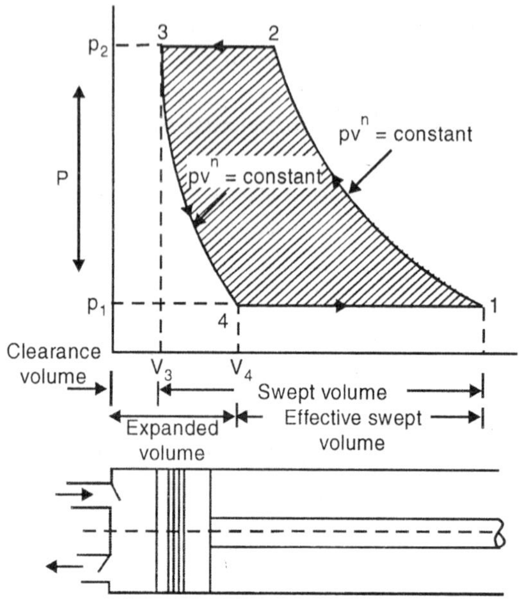

Fig. 10.15. ***p–V*** **diagram showing polytropic compression.**

As the compressed air in the clearance space expands during the initial part of the suction stroke, some work is produced, which reduces the work required to run the compressor.

The net work required to run the compressor can be calculated as follows :

Refer to Fig. 10.15 and neglecting the clearance volume.

Work required during compression/cycle,

$$= P_1V_1 \frac{n}{n-1}\left[\left(\frac{P_2}{P_1}\right)^{\frac{n-1}{n}} - 1\right] \quad ...(i)$$

and work obtained during expansion/cycle,

$$= P_1V_4 \frac{n}{n-1}\left[\left(\frac{P_2}{P_1}\right)^{\frac{n-1}{n}} - 1\right] \quad ...(ii)$$

Subtracting (*ii*) from (*i*)

∴ Net, work required/cycle,

$$= P_1V_1 \frac{n}{n-1}\left[\left(\frac{P_2}{P_1}\right)^{\frac{n-1}{n}} - 1\right] - P_1V_4 \frac{n}{n-1}\left[\left(\frac{P_2}{P_1}\right)^{\frac{n-1}{n}} - 1\right]$$

$$= P_1(V_1 - V_4).\left(\frac{n}{n-1}\right)\left[\left(\frac{P_2}{P_1}\right)^{\frac{n-1}{n}} - 1\right] \quad ...(10.22)$$

In eqn. (10.22) $(V_1 - V_4)$ is called the 'effective swept volume'.

10.5.3. Volumetric Efficiency

The ratio of actual volume of air taken in, during the suction stroke, at standard atmospheric conditions, to the swept volume of the piston, is known as 'Volumetric efficiency' of the compressor.

The actual volume of the air taken in during suction stroke at S.T.P. will be less than the swept volume of the piston, because of the reasons already discussed in the previous section.

$$\therefore \quad \text{Volumetric efficiency} = \frac{\text{Actual volume of air taken in at S.T.P.}}{\text{Swept volume of the piston}}$$

Generally, the volumetric efficiency varies from 70% to 90% depending upon the type of the compressor.

For the derivation of expression for volumetric efficiency again refer Fig. 10.15.

$$\text{Thus, Volumetric efficiency} = \frac{V_1 - V_4}{V_1 - V_3} \quad \text{...}(i)$$

Considering the value of polytropic index 'n' same for compression and expansion.

$$\therefore \quad \frac{V_4}{V_3} = \left(\frac{P_3}{P_4}\right)^{\frac{1}{n}} \quad \text{or} \quad V_4 = V_3 \times \left(\frac{P_3}{P_4}\right)^{\frac{1}{n}}$$

$$\text{or} \quad = V_3 \times \left(\frac{P_2}{P_1}\right)^{\frac{1}{n}} \quad (\because P_2 = P_3 \text{ and } P_1 = P_4)$$

$$\therefore \text{Volumetric efficiency} = \frac{V_1 - V_3 \times \left(\frac{P_2}{P_1}\right)^{\frac{1}{n}}}{V_1 - V_3}$$

$$= \frac{V_1 - V_3 \times \left(\frac{P_2}{P_1}\right)^{\frac{1}{n}}}{\frac{V_1 - V_3}{V_3} \times V_3} \quad \text{...}(ii)$$

$$\text{But} \quad \frac{V_3}{V_1 - V_3} = \text{Clearance ratio} = k \text{ (say)}$$

$$\therefore \quad \frac{V_1 - V_3}{V_3} = \frac{1}{k}$$

$$\text{or} \quad V_1 = \frac{V_3}{k} + V_3 = V_3\left(\frac{1}{k} + 1\right)$$

Now, substituting the value of $\frac{V_1 - V_3}{V_3}$ and V_1 in eqn. (ii), we have

$$\text{Volumetric Efficiency} = \frac{V_3 \times \left(\frac{1}{k} + 1\right) - V_3\left(\frac{P_2}{P_1}\right)^{\frac{1}{n}}}{\frac{1}{k} \times V_3} = \frac{\left(\frac{1}{k} + 1\right) - \left(\frac{P_2}{P_1}\right)^{\frac{1}{n}}}{\frac{1}{k}}$$

$$\text{or} \quad \text{Volumetric efficiency, } \eta_{vol} = 1 + k - k\left(\frac{P_2}{P_1}\right)^{\frac{1}{n}} \quad \text{...}(10.23)$$

From eqn. (10.23) it can be seen that if $P_1 = P_2$ the volumetric efficiency will be 100%. On the other hand the volumetric efficiency will be zero if $V_1 = V_4$ [from eqn. (*i*)].

For 100% volumetric efficiency, delivery pressure should be equal to the atmospheric pressure, *i.e.*, there is no compression.

Also in case of the multistage air compression, the swept volume of the piston refers only to the low pressure cylinder.

10.5.3.1. Factors Affecting the Volumetric Efficiency

The volumetric efficiency of an air compessor is affected by the following factors;

(*i*) **Increase in Delivery Pressure.** The volumetric efficiency is lowered, when the delivery pressure of the compressor under consideration is increased. It is shown clearly in the Fig. 10.16.

By increasing the delivery pressure, the actual volume of the air taken in during suction stroke is decreased, whereas the swept volume of the piston remains the same.

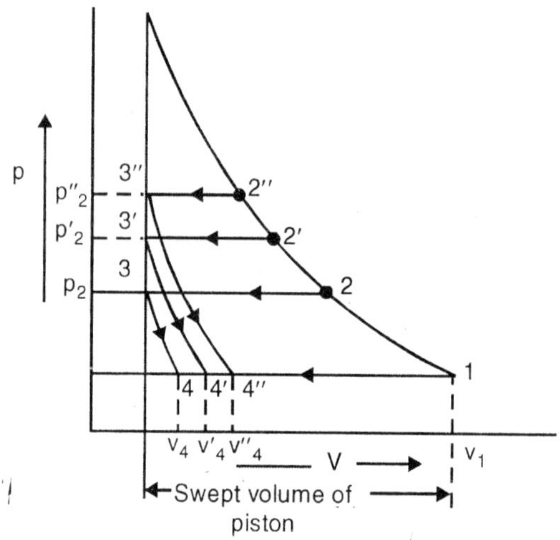

Fig. 10.16. Effect of delivery, pressure on volumetric efficiency.

(*ii*) **Very High Speed.** The volumetric efficiency is lowered when the speed of the compressor is increased to very high. It is because of the fact that the actual volume of the air taken in is considerably reduced, because of less time available for the suction of fresh air, during suction stroke.

(*iii*) **Leakage Past the Piston.** During the compression and delivery, some air leaks past the piston and cylinder walls thus decreasing the amount of air delivered. This tendency increases with the increase of delivery pressure.

(*iv*) **Very Large Clearance Volume.** The increase in clearance volume lowers the volumetric efficiency as discussed already in the preceding section.

(*v*) **Restricted Air Intake.** Any obstruction at the inlet valve decreases the actual volume of the air taken in during the suction stroke and thus lowering the volumetric efficiency of the compressor.

(*vi*) **Overheating of Inlet Air.** During the suction of air into the cylinder, it is heated and expanded, because of the heat taken by it from the cylinder walls and temperature of the expanded remaining air is more than that of the entering fresh air. So, the volume of air taken in during the suction stroke is decreased considerably, thus lowers the volumetric efficiency of the compressor.

(*vii*) **Turbulence in the Air Passages.** The pressure inside the cylinder during the suction stroke is decreased, due to the frictional resistance in the air itself and turbulence in air passages.

This decreases the actual quantity of air taken in during suction. Thus the volumetric efficiency is decreased.

Thus volumetric efficiency can be improved by taking into consideration the above factors, during the designing of the compressor.

10.5.4. Efficiencies of a Reciprocating Compressor

To have the clear understanding of the various compressor efficiencies, the term indicated power should be explained first :

The indicated power, $ip = \dfrac{P_m \cdot LAn}{60000}$ kW ...(10.24)

where P_m = Mean effective pressure (N/m^2 or P_a)
L = Length of stroke (m)
A = Area of cross-section of the piston (m^2)
n = Number of working strokes per minute
= N (for single acting compressor)
= $2N$ (for double acting recicompressor)

where 'N' is the speed of crank shaft in rev. per min.

Mathematically,

$$P_m = \frac{\text{Area of indicator diagram}}{\text{Length of indicator diagram}} \times \text{Spring constant}$$

or $P_m = \dfrac{a}{l} \times S$ ...(10.25)

where a = Area of the indicator diagram (cm^2)
l = Length of the indicator diagram (cm)
and S = Spring constant (N/m^2 per cm)

Some of the efficiencies concerned with the air compressor are defined and calculated as follows :

(*i*) **Compressor Efficiency.** It may be defined as the ratio of the isothermal power to the indicated power of the compressor.

i.e., $$\text{Compressor Efficiency} = \frac{\text{Isothermal power}}{\text{Indicated power}}$$

(*ii*) **Isothermal Efficiency.** It may be defined as the ratio of isothermal power to the brake power of the compressor. It is generally about 70%.

i.e., $$\text{Isothermal Efficiency} = \frac{\text{Isothermal power}}{\text{Brake power}}$$

(*iii*) **Adiabatic Efficiency.** It may be defined as the ratio of adiabatic power to the brake power of the compressor.

i.e., $$\text{Adiabatic Efficiency} = \frac{\text{Adiabatic power}}{\text{Brake power}}$$

(*iv*) **Mechanical Efficiency.** It may be defined as the ratio of indicated power to the brake power of the compressor.

i.e., $$\text{Mechanical Efficiency} = \frac{\text{Indicated power}}{\text{Brake power}}$$

(*v*) **Overall Efficiency.** It may be defined as the ratio of indicated power of the compressor to the indicated power of the prime mover, which drives it.

i.e. $$\text{Overall Efficiency} = \frac{\text{Indicated power of compressor}}{\text{Indicated power of prime mover}}$$

10.6. OPERATION AND MAINTENANCE OF RECIPROCATING COMPRESSORS

Following points must be kept in mind during the operation and for the maintenance of a reciprocating compressor :

(*i*) **Care.** It involves ensuring that the machine is not mishandled, overloaded and overrun. It also involves observing that the machine possesses no unwanted noise, vibrations, temperatures and leakages in its running. It also ensures that the machine is not allowed to be subjected to unfavourable weather conditions and it is properly protected during idle or out of use periods.

(*ii*) **Cleaning.** It involves, daily as well as after regular time intervals, the removal of dust, sticky materials etc. So that they do not accumulate on or between the moving parts. The foreign matter including abrasive particles, if be allowed to enter between moving parts due to poor cleaning, may lead to excessive wear or jamming of such parts.

(*iii*) **Lubrication.** Lubrication should always be maintained properly, both as regards grade of lubricant and quantity of lubricant. The lubricant should be replaced from crank case after specific interval of use. It helps in reducing wear of moving parts and loss of power in friction.

(*iv*) **Periodic Inspection.** Periodic inspection is the best way to ascertain necessary adjustments, replacements, repairs and overhauls needed by the machine.

(*v*) **Adjustments, Repairs and Replacements.** It is essential that minor adjustments, repairs and replacements of parts should be at once attended to.

It is quite helpful in avoiding the development of major faults or breakdown of the machinery. It also helps in maintaining good working efficiency of the machine and increasing the working life of the machine.

NUMERICAL EXAMPLES

Example 10.1. An air compressor takes in air at 1 bar and 20 °C compresses it according to the law $PV^{1.2}$ = constant, and delivers it to a receiver at a constant pressure of 10 bar. Find the temperature at the end of compression. Calculated per kg of air;

(*i*) The work done and the heat rejected during compression.

(*ii*) The work done during delivery.

Take $R = 287$ J/kg °K and $\gamma = 1.4$.

Solution. Intake temperature, $T_1 = 20 + 273 = 293$ °K

$$\because \quad \frac{T_2}{T_1} = \left(\frac{P_2}{P_1}\right)^{\frac{n-1}{n}} = (10)^{\frac{0.2}{1.2}}$$

$$\therefore \quad T_2 = T_1 \times 10^{0.2/1.2}$$

$$T_2 = 293 \times 10^{0.1/1.2} = 430.4 \text{ K.}$$

Thus, temperature at the end of compression,

$$T_2 = 430.4 - 273 = 157.4 \text{ °C. } \textbf{Ans.}$$

(*i*) Work done on air during compression 1-2

$$= \frac{P_2 V_2 - P_1 V_1}{n-1}$$

$$= \frac{R}{n-1}(T_2 - T_1) \text{ per kg of air}$$

$$= \frac{287}{1.2-1}(430.4 - 293)$$

$$= \frac{287 \times 137.4}{0.2} = 197169 \text{ Joule}$$

$$= 197.17 \text{ KJ}$$

Heat rejected during compression

$$Q = \frac{\gamma - n}{\gamma - 1} \times W = \frac{1.4 - 1.2}{1.4 - 1} \times 197.17$$

$$= \frac{0.2}{0.4} \times 197.17 = 98.58 \text{ KJ/kg. } \textbf{Ans.}$$

(*ii*) The work done during delivery

$$= RT_2 \text{ KJ/kg of air}$$

$$= 0.287 \times 430.4$$

$$= 123.52 \text{ KJ/kg of air. } \textbf{Ans.}$$

Example 10.2. A single stage single cylinder reciprocating air compressor has a piston displacement of 0.05 m^3. Air is sucked by the compressor at 1 bar and 17 °C. The air after compression to 6 bar is delivered to the receiver at constant pressure. Three alternatives of compression are;

(*i*) Isentropic compression, $\gamma = 1.5$

(*ii*) Polytropic compression, $n = 1.3$

(*iii*) Isothermal compression.

Determine for each alternative :

(*a*) Temperature at the end of compression.

(*b*) Work done by air during suction.

(*c*) Work done on the air and heat transferred during compression process alone.

(*d*) Work done on air during delivery.

(*e*) Net work done on the air per cycle.

Solution. Mass of air handled,

$$m = \frac{P_1 V_1}{RT_1} \qquad (\because\ P_1V_1 = mRT_1)$$

$$= \frac{1 \times 10^5 \times 0.05}{1000 \times 0.287 \times (273 + 17)}$$

$$= 0.0601 \text{ kg.}$$

(*i*) **Isentropic Compression**

The law of compression is

$$P_1 V_1^{1.4} = P_2 V_2^{1.4}$$

(*a*) Temperature T_2 at the end of compression is given by the equation,

$$T_2 = T_1\left(\frac{P_2}{P_1}\right)^{\frac{\gamma-1}{\gamma}} = (273 + 17)\left(\frac{6}{1}\right)^{\frac{1.4-1}{1.4}} = 484 \text{ K} = 211\ °\text{C. } \textbf{Ans.}$$

(*b*) Work done by air during suction

$= (P_1 \times 10^5) \times V_1$ Nm

$= 1 \times 10^5 \times 0.05 = 5000$ Nm or J. **Ans.**

(*c*) Work done on air during compression

$$= \frac{m.R}{\gamma - 1}(T_2 - T_1)$$

$$= \frac{0.0601 \times 287}{1.4 - 1}(484 - 290)$$

$= 9099$ Nm or J. **Ans.**

During isentropic process the heat neither flows into the system nor flows out of the system. Hence, heat transferred during the process is zero.

(*d*) Work done on air during delivery,

$= P_2 V_2 = m \,.\, RT_2$

$= 0.0601 \times 287 \times 484$

$= 8348$ Nm or J. **Ans.**

(*e*) Net work done on air during cycle,

= Work done on air during compression + Work done on air during delivery – Work done by air during suction.

$= 9099 + 8348 - 5000 = 12447$ Nm or J

$= 12.45$ KJ. **Ans.**

(*ii*) **Polytropic Compression :**

The law of compression is, $P_1 V_1^{1.3} = P_2 V_2^{1.3}$

(*a*) Temperature T_2 at the end of compression is given by,

$$T_2 = T_1 \left(\frac{P_2}{P_1}\right)^{\frac{n-1}{n}}$$

$$= 290\,(6)^{\frac{1.3-1}{1.3}} = 438.5 \text{ K} = 165.5\ °\text{C}. \textbf{ Ans.}$$

(*b*) Work done by air during suction $= P_1 V_1$

$= 1 \times 10^5 \times 0.05 = 5000$ N-m. **Ans.**

(*c*) Work done on air during compression

$$= \frac{mR}{n-1}(T_2 - T_1)$$

$$= \frac{0.0601 \times 287}{1.3 - 1}(438.5 - 290) = 8538 \text{ N-m}$$

Heat transfered to cylinder walls,

$$= \frac{r - n}{n - 1} \times \text{Work done during compression}$$

$$= \frac{1.4 - 1.3}{1.3 - 1} \times 8538 = 2846 \text{ N-m or J. } \textbf{Ans.}$$

(*d*) Work done on air during delivery

$= P_2 V_2 = mRT_2$

$= 0.0601 \times 287 \times 438.5$

$= 7564$ N-m or J. **Ans.**

(*e*) Net work done during the cycle,

$$= 7564 + 8538 - 5000$$

$$= 11100 \text{ Nm or J} = 11.1 \text{ KJ. } \textbf{Ans.}$$

(*iii*) **Isothermal Compression :**

(*a*) During isothermal compression temperature remains constant. Hence, temperature at the end of compression,

$$T_2 = T_1 = 290 \text{ K} = 17\ °\text{C. } \textbf{Ans.}$$

(*b*) Work done by the air during suction

$$= P_1V_1 = 5000 \text{ N-m. } \textbf{Ans.}$$

(*c*) Work done on the air during compression

$$= P_1V_1 \log_e \frac{P_2}{P_1} = 1 \times 10^5 \times 0.05 \log_e (6)$$

$$= 8959 \text{ N-m or J. } \textbf{Ans.}$$

During isothermal compression there is no change in internal energy. The heat transfered to the cylinder walls is equal to the work done.

∴ Heat transfered to the cylinder walls = 8959 N-m or J.

(*d*) Work done on air during delivery,

$$= P_2V_2 = P_1V_1 = 5000 \text{ N-m or J. } \textbf{Ans.}$$

(*e*) Net work done during the cycle,

$$= 8959 + 5000 - 5000 = 8959 \text{ N-m or J. } \textbf{Ans.}$$

Example 10.3. Free-air is to be compressed from 1 bar to 8 bar. There are two alternatives :

(*i*) Single stage compression

(*ii*) Two stage compression with perfect intercooling.

The law of compression in both cases is $PV^{1.3}$ = Constant. Find the percentage saving in second alternative over the first.

Solution. Work required to be done in case of single stage compression,

$$W_1 = P_1 V_1 \frac{n}{n-1}\left[\left(\frac{P_2}{P_1}\right)^{\frac{n-1}{n}} - 1\right]$$

$$= (1 \times 10^5 \times V_1)\frac{1.3}{1.3-1}\left[(8)^{\frac{1.3-1}{1.3}} - 1\right]$$

$$= 2.669 \times 10^5\ V_1 \text{ N-m or J} \qquad ...(i)$$

In case of two stage compression with perfect intercooling the work required to be done,

$$W_2 = 2 \times P_1 V_1 \frac{n}{n-1}\left[\left(\frac{P_2}{P_1}\right)^{\frac{n-1}{2n}} - 1\right]$$

where, P_2 = final discharge pressure, 8 bar

$$\therefore \quad W_2 = 2 \times (1 \times 10^5)\ V_1\left(\frac{1.3}{1.3-1}\right)\left[(8)^{\frac{1.3-1}{2 \times 1.3}} - 1\right]$$

$$= 2.35 \times 10^5\ V_1 \text{ N-m or J} \qquad ...(ii)$$

From (*i*) and (*ii*)

Percentage saving in second alternative over the first

$$= \frac{W_1 - W_2}{W_1} \times 100$$

$$= \left(\frac{2.669 \times 10^5 V_1 - 2.35 \times 10^5 V_1}{2.669 \times 10^5 V_1} \right) \times 100$$

$$= 11.95\% \textbf{ Ans.}$$

Example 10.4. Determine the size of cylinder for a double acting compressor of 37.5 kW (*ip*) in which the air is drawn in at 1 bar pressure and 15 °C, and compressed according to the law $PV^{1.2}$ = Constant to 6 bar, r.p.m = 100, average piston speed = 150 m/min. Neglect clearance volume.

Solution. As we know that :

$$\text{Work done per stroke} = P_1 V_1 \frac{n}{n-1}\left[\left(\frac{P_2}{P_1}\right)^{\frac{n-1}{n}} - 1\right] \quad ...(i)$$

Also, Work done per stroke = $P_m \times V_s$ ...(*ii*)

where P_m = Mean effective pressure

and V_s = Swept volume

Equating (*i*) and (*ii*)

$$\therefore \quad P_m \cdot V_s = P_1 V_1 \frac{n}{n-1}\left[\left(\frac{P_2}{P_1}\right)^{\frac{n-1}{n}} - 1\right] \quad ...(iii)$$

It is given that, clearance volume is zero

$$\therefore \quad V_s = V_1$$

So, eqn. (*iii*) becomes

$$P_m = P_1 \frac{n}{n-1}\left[\left(\frac{P_2}{P_1}\right)^{\frac{n-1}{n}} - 1\right]$$

Substituting the values

$$P_m = 1 \times \frac{1.2}{1.2-1}\left[\left(\frac{6}{1}\right)^{\frac{1.2-1}{1.2}} - 1\right]$$

$$= \frac{1.2}{0.2} \times 0.348 = 2.088 \text{ bar.}$$

If A is the cross-sectional area of the piston in m^2. Then, Indicated power,

$$ip = \frac{P_m L A n}{60000}$$

or

$$37.5 = \frac{2.088 \times 150 \times A \times 10^5}{60000} \quad (\text{Here } n = 2N)$$

$$(2LN = 150 \text{ m/min})$$

$$\therefore \quad A = \frac{37.5 \times 1000 \times 60000}{2.088 \times 150 \times 10^5} = 0.072 \text{ m}^2 = 720 \text{ cm}^2$$

Since, $$A = \frac{\pi}{4} \times d^2 = 720$$

$\therefore$ $$d = \sqrt{\frac{720 \times 4}{\pi}} = 30.25 \text{ cm. } \textbf{Ans.}$$

Now, $$2LN = 150 \text{ m/min.}$$

$\therefore$ $$L = \frac{150}{2 \times 100} \text{ m} = 0.75 \text{ m} = 75 \text{ cm. } \textbf{Ans.}$$

Example 10.5. A two stage air compressor with complete intercooling delivers air to the mains at a pressure of 30 bar, the suction conditions being 1 bar and 27 °C. If both the cylinders have the same stroke, find the ratio of the cylinder diameters, for the efficiency of compression to be a maximum. Assume, n = 1.3.

Solution. Volume of HP cylinder, $V_a = V_2$

Volume of LP cylinder, $V_b = V_1$

If d_1 and d_2 are the diameters of the LP and HP cylinders respectively, then,

$$\frac{V_1}{V_2} = \frac{\frac{\pi}{4} \times d_1^2}{\frac{\pi}{4} \times d_2^2}$$

and $$\frac{d_1}{d_2} = \sqrt{\frac{V_1}{V_2}}$$

From $$PV^{1.3} = \text{Constant}$$

$$P_1 V_1^{1.3} = P_2 V_c^{1.3}$$

or $$\frac{V_1}{V_c} = \left(\frac{P_2}{P_1}\right)^{1/1.3}$$

But for maximum efficiency,

$$P_2 = \sqrt{P_1 . P_3} = \sqrt{1 \times 30} = 5.477 \text{ bar}$$

$\therefore$ $$\frac{V_1}{V_c} = \left(\frac{5.477}{1}\right)^{\frac{1}{1.3}} = 3.699 \qquad ...(i)$$

Now, temperature, T_c at the end of compression in the LP cylinder

$$T_c = T_b \times \left(\frac{P_2}{P_1}\right)^{\frac{n-1}{n}} = 300 \times \left(\frac{5.477}{1}\right)^{\frac{1.3-1}{1.3}} = 444 \text{ K} \quad (\because\ T_b = 273 + 27 = 300 \text{ K})$$

From constant pressure process,

$$\frac{V_d}{T_d} = \frac{V_c}{T_c}$$

or $$\frac{V_2}{T_d} = \frac{V_c}{T_c}$$

$\therefore$ $$\frac{V_c}{V_2} = \frac{T_c}{T_d} \qquad (\because \text{ cooling being perfect So, } T_b = T_d)$$

$$= \frac{444}{300}$$

or $$\frac{V_c}{V_2} = 1.48 \quad ...(ii)$$

Now, From (*i*) and (*ii*)

$$\frac{V_1}{V_2} = \frac{V_1}{V_c} \times \frac{V_c}{V_2}$$

$$= 3.699 \times 1.48 = 5.464$$

Thus diameter ratio, $\frac{d_1}{d_2} = \sqrt{\frac{V_1}{V_2}} = \sqrt{5.464} = 2.337$ **Ans.**

Example 10.6. In a test of a single cylinder single acting air compressor of 11 cm bore and 14 cm stroke, air is taken from the atmosphere at 1 bar and delivered through a valve which maintains a delivery pressure of 8 bar to a reservoir of 1.5 m^3 capacity. A motor giving 87 bp, drives the compressor at 400 rpm and *i.m.e.p.* is 2 bar. The reservoir is initially at atmospheric pressure (1 bar) and 17 °C and after 19 minutes running, reaches 7.5 bar and 75 °C. Calculate the delivery in m^3 of free air per minute at 17 °C. Calculate the delivery in m^3 of free air per minute at 17 °C, the volumetric efficiency at atmospheric conditions, the isothermal and mechanical efficiency.

Solution. Swept Volume $= \frac{\pi}{4} \times d^2 \times l$

$$= \frac{\pi}{4} \times (11)^2 \times 14 \times 10^{-6} \text{ m}^3$$

$$= 0.00133 \text{ m}^3$$

Initial mass of air in the reservoir,

$$m_1 = \frac{P_1 V_1}{RT_1}$$

Final mass of air after 19 minutes

$$m_2 = \frac{P_2 V_2}{RT_2}$$

∴ Mass of air delivered in 19 minutes

$$= \frac{P_2 V_2}{RT_2} - \frac{P_1 V_1}{RT_1}$$

$$= \frac{V}{R}\left[\frac{P_2}{T_2} - \frac{P_1}{T_1}\right] \quad [\because V_1 = V_2 = V]$$

and volume delivered at initial conditions,

$$= \frac{RT_1}{P_1} \times \frac{V}{R}\left[\frac{P_2}{T_2} - \frac{P_2}{T_1}\right]$$

$$= \frac{T_1}{P_1} \times V\left[\frac{P_2}{T_2} - \frac{P_1}{T_1}\right]$$

$$= \frac{290}{1} \times 1.5 \times \left[\frac{7.5}{348} - \frac{1}{290}\right]$$

$$= 8.0735 \text{ m}^3.$$

$$\text{Volume delivered/min} = \frac{8.0735}{19} = 0.4243 \text{ m}^3 \textbf{ Ans.}$$

$$\text{Volume delivered/stroke} = \frac{0.4243}{400} = 0.001062 \text{ m}^3$$

$$\therefore \text{ Volumetric Efficiency} = \frac{0.001062}{0.001331} \times 100 = 79.6\%. \textbf{ Ans.}$$

$$\text{Also, Indicated Power, } ip = \frac{P_m LAn}{60000}$$

$$ip = \frac{2 \times 0.001331 \times 400 \times 10^5}{60000} = 1.77 \text{ kW}$$

($\because$ LA = Swept volume = 0.001331 m^3 and $n = N = 400$)

$$\therefore \quad \text{Mechanical Efficiency, } \eta_m = \frac{ip}{bp} \times 100$$

$$= \frac{1.77}{1.87} \times 100 = 94.6\% \textbf{ Ans.}$$

$$\text{Isothermal Power} = \frac{P_1 V_1 \log_e \frac{P_2}{P_1}}{60000}$$

$$= \frac{10^5 \times 0.4243}{60000} \log_e \frac{8}{1} = 1.47 \text{ kW}$$

$$\therefore \text{ Isothermal Efficiency} = \frac{1.47}{1.77} \times 100 = 82.8\% \textbf{ Ans.}$$

Example 10.7. A single stage single acting air compressor delivers 15 m^3/min of free air from 1 bar abs. to 8 bar abs. at 300 rpm. The clearance volume is 6% of the stroke volume and compression and expansion follow the law $PV^{1.3}$ = constant. Calculate, the diameter and stroke of the compressor. Take L/D = 1.5. The temperature and pressure of air at suction are the same as atmospheric air.

Solution. As we know that;

$$\text{Clearance Ratio,} \quad K = \frac{6}{100} = 0.06$$

$\therefore$ Volumetric Efficiency,

$$\eta_{vol} = 1 + K - K\left(\frac{P_2}{P_1}\right)^{\frac{1}{n}}$$

$$= 1 + 0.06 - 0.06\left(\frac{8}{1}\right)^{\frac{1}{1.3}}$$

$$= 1 + 0.06 - 0.2969$$

$$= 0.763 = 76.3\%$$

The free air delivered per minute is given by

$$V_s \times \eta_{vol} \times 300 = 15$$

or

$$V_s \times 0.763 \times 300 = 15$$

$$\therefore \quad V_s = \frac{15}{0.763 \times 300} = 0.06553 \text{ m}^3$$

$$\therefore \quad V_s = \frac{\pi}{4} D^2 \times L = 0.06553$$

or
$$\frac{\pi}{4} \times D^2 \times 1.5D = 0.06553 \qquad (\because L = 1.5\, D)$$

$$\therefore \quad D^3 = 0.06553 \times \frac{4}{\pi} \times \frac{1}{1.5} = 0.556$$

$$\therefore \quad D = 3\sqrt{0.0556} = 0.3817 \text{ m} = 38.17 \text{ cm} \textbf{ Ans.}$$

and $L = 0.3817 \times 1.5 = 0.58255$ m = 58.26 cm. **Ans.**

ASSIGNMENT

Objective Type Questions

(A) Fill in the Blanks :

1. There are mainly _______ types of the compressors.
2. Pneumatic tools are driven by mean of _______ air.
3. The pressure ratio in the case of a reciprocating compressor is _______ than that of a rotary compressor.
4. The size of the L.P. cylinder is _______ than that of H.P. cylinder.
5. The compression should be _______ for the less power required.
6. Intercooler is provided in the _______ stage compression.
7. For the maximum efficiency, the intermediate pressure should be _______ .
8. The volumetric efficiency increases with the _______ of clearance volume.
9. Continuous and uniform flow of compressed air can be achieved with the help of _______ compressor.
10. The diabatic efficiency of the compressor is _______ .
11. Turbo blowers are used to compress air or gas upto a pressure of _______ bar.
12. The creterion of thermodynamic efficiency of reciprocating compressors is _______ while that of rotary compressors is _______.
13. A 5 m^3/min reciprocating compressor means that it compresses 5 m^3/min of _______ .
14. Free air is the atmospheric air at any _______ .
15. The performance of reciprocating air compressors at high altitudes, as compared to that at sea level will be _______ .
16. An intercooler is basically a _______ which may be either air cooled or water cooled.
17. A _______ in compressor installation is used to damp pressure waves in compressor discharge.
18. The flow of air is perpendicular to the machine axis in a _______ compressor.
19. _______ work on a cycle which is the reverse of the reciprocating air compressor cycle.
20. In a single stage reciprocating air compressor, the compression process is carried out isothermally. The calculations for work input are made by using the relation _______ .

(B) State True or False :

1. In mines, the compressed air is used to drive the pneumatic tools.
2. The reciprocating compressor is used when large volume of air at lesser pressure is required.
3. The compressors are always double acting.
4. The intercooler is employed to decrease the pressure and temperature of the air.
5. Centrifugal compressor is a positive displacement compressor.
6. Reciprocating compressors are used to compress air upto a pressure of more than 100 bar.
7. The rotary compressor which compresses air upto a pressure of 2.5 bar, is known at, turbo-blower.
8. A pair of a moving and fixed blades is known as a stage of the compressor.
9. Multistage compression helps in reducing the work required.
10. The intercooler is a type of heat-exchanger.
11. Rotary compressors are used to handle large quantities of air at low pressure.
12. Rotary compressors cannot be coupled directly to turbines or high speed I.C. engines.
13. Rotary compressors give cleaner supply of air than reciprocating compressors.
14. Turbo compressors are invariably water cooled.
15. Compressed air is used for spray painting, conveyance of cement, sand etc.
16. Intercooling is essential in multistage air compressor.
17. In a two stage compressor, for maximum efficiency the pressure at the end of the first stage should be the geometrical mean of the pressure at the initial and final stage.
18. Compressed air is used in hydraulic machinery.
19. If the pressure at the suction is more than atmospheric the blower is known as booster.
20. In a two stage compressor the size of the L.P. cylinder is kept smaller than the size of the H.P. cylinder.

(C) Multiple Choice Questions :

1. Identify the non-positive displacement compressor :
   (*a*) axial flow compressor (*b*) roots blower
   (*c*) vane blower (*d*) reciprocating compressor.
2. Reciprocating compressors are usually preferred for ;
   (*a*) high pressure and high discharge
   (*b*) high pressure the low discharge
   (*c*) low pressure and high discharge
   (*d*) lowe pressure and low discharge.
3. Reciprocating compressors are used to compress air upto pressure of :
   (*a*) 10 bar (*b*) 50 bar
   (*c*) 100 bar (*d*) more than 100 bar
4. The capacity of a reciprocating air compressor expressed in m^3/min refers to volume of ;
   (*a*) receiver tank (*b*) compressor displacement
   (*c*) free air delivered (*d*) compressed air discharged.

5. Free air is the air at;
   (*a*) 1 bar pressure and 0 °C temperature
   (*b*) 1 bar pressure and 15 °C temperature
   (*c*) 1 bar pressure, 20 °C temperature and 36% relative humidity
   (*d*) atmospheric conditions of pressure and temperature at any specific location.
6. Standard air is the air at ;
   (*a*) 1 bar pressure and 0 °C temperature
   (*b*) 1 bar pressure and 15 °C temperature
   (*c*) 1 bar pressure, 20 °C temperature and 36% relative humidity
   (*d*) atmospheric conditions of pressure and temperature at any specific location.
7. Reciprocating compressors have a maximum capacity of ;
   (*a*) 50 m^3/min (*b*) 100 m^3/min
   (*c*) 200 m^3/min (*d*) 300 m^3/min.
8. The work input to air compressor will be least if the exponent 'n' in the compression process PV^n = constant equals ;
   (*a*) $n = 1$ (*b*) $n = 1.2$
   (*c*) $n = \gamma = 1.4$ (adiabatic compression)
   (*d*) n takes the infinite value.
9. Isothermal compression though most efficient cannot be realised in practice because that necessitates;
   (*a*) Cylinder of large dimension
   (*b*) placing the cylinder in a stream of water
   (*c*) engine operation at slow speeds
   (*d*) compression of air to low pressure only.
10. Isothermal efficiency of a compressor is defined as the ratio of;
    (*a*) volume of free air delivered per stroke to the swept volume of the piston
    (*b*) indicated power to shaft power of the motor or engine required to drive the compressor
    (*c*) adiabatic power to the power required to drive the compressor
    (*d*) isothermal work to the actual work required to compress the air for the same pressure ratio.
11. The clearance volume in a reciprocating compressor is provided to;
    (*a*) create turbulence in the air to be delivered
    (*b*) accommodate valves in the cylinder head of the cylinder
    (*c*) increase the volumetric efficiency of the compressor
    (*d*) reduce the work input per kg of air delivered.
12. With increase in clearance volume, the specific work input (work input per kg of air);
    (*a*) increase (*b*) decrease
    (*c*) remains constant
    (*d*) unpredictable, may increase or decrease depending on compressor capacity.
13. The volumetric efficiency of an air compressor is the ratio of;
    (*a*) displacement volume to clearance volume
    (*b*) displacement volume to volume of air sucked inside the cylinder

(c) actual free air delivered to the displacement volume
(d) volume of air before compression to the volume of air after compression.

14. The volumetric efficiency of a reciprocating compressor usually lies in the ranges;
(a) 15 to 25% (b) 25 to 45%
(c) 50 to 65% (d) 70 to 85%.

15. An intercooler is used with multistage reciprocating air compressors to cool;
(a) The air before, it is sucked into the compressor cylinder
(b) The air before it is discharged to the receiver
(c) The air after one stage of compressor before it is sent to the next stage
(d) The compressor cylinder.

16. For a two stage reciprocating air compressor with perfect intercooling, the work input will be minimum if the supply pressure 'P_1' the intercooler pressure P_2 and the delivery pressure P_3 conform to the relation;
(a) $P_2 = \sqrt{P_1 . P_3}$ (b) $P_2 = \sqrt{P_1 + P_3}$
(c) $P_2 = \frac{P_1 + P_2}{2}$ (d) $P_2 = \frac{\sqrt{P_3}}{P_1}$.

17. Maximum delivery pressure for rotary bladed compressors is usually restricted to;
(a) 5 bar (b) 10 bar
(c) 20 bar (d) 50 bar

18. The main loss in a centrifugal compressor is due to
(a) inlet losses (b) diffuser losses
(c) impeller channel losses (d) all of the above.

19. The function of diffusion passage in a compressor is to;
(a) make the flow stream line
(b) increase the degree of reaction
(c) convert kinetic energy into pressure energy
(d) convert pressure energy into kinetic energy.

20. Compared to centrifugal compressor, an axial flow compressor;
(a) requires a low starting torque
(b) has high pressure ratio per stage
(c) has large air handling ability per unit of frontal area
(d) wide range of operation between surging and choking limits.

ANSWERS

(A) Fill in the Blanks :

1. two (reciprocating and rotary) **2.** compressed
3. higher **4.** bigger
5. isothermal **6.** multi
7. the geometric mean of the initial and final delivery pressure
8. decrease **9.** rotary
10. the ratio of adiabatic power to the brake power
11. (2.5) **12.** isothermal compression, isentropic compression
13. free air **14.** specific location **15.** inferior

16. heat-exchanger	**17.** receiver	**18.** centrifugal	
19. air motors	**20.** $p_1 \cdot v_1 \log_e \frac{v_1}{v_2}$.		

(B) State True or False :

1. True	**2.** False	**3.** False	**4.** False	**5.** False
6. True	**7.** True	**8.** False	**9.** True	**10.** True
11. True	**12.** True	**13.** True	**14.** False	**15.** True
16. False	**17.** False	**18.** False	**19.** True	**20.** False

(C) Multiple Choice Questions :

1. (*a*)	**2.** (*b*)	**3.** (*d*)	**4.** (*c*)	**5.** (*d*)
6. (*c*)	**7.** (*a*)	**8.** (*a*)	**9.** (*c*)	**10.** (*d*)
11. (*b*)	**12.** (*c*)	**13.** (*c*)	**14.** (*d*)	**15.** (*c*)
16. (*a*)	**17.** (*b*)	**18.** (*d*)	**19.** (*c*)	**20.** (*c*)

(D) Short Answer Type Questions :

Q 1. What is an 'Air-compressor' ?

Ans. An 'Air-compressor' is a machine, which receives air from atmosphere, reduces its volume by compression and increases its pressure with the aid of some mechanical energy and then delivers it to some vessel for storage.

Q 2. How air-compressors are classified, according to the design and principle of operation ?

Ans. (*a*) Reciprocating compressors :

(*i*) Single acting

(*ii*) Double acting

(*b*) Rotary compressor :

(*i*) Steady flow compressors

(*ii*) Displacement compressors

These can also be classified as;

(*i*) Fans.

(*ii*) Blowers

(*iii*) Axial flow compressor

(*iv*) Mixed flow compressor

(*v*) Centrifugal compressor.

Q 3. How air-compressor are classified according to the number of stages ?

Ans. (*i*) Single-stage compressor

(*ii*) Multistage compressor.

Q 4. How air compressor are classified according to the pressure limit ?

Ans. (*i*) Low pressure compressors (Del. pressure less 1 MN/m^2)

(*ii*) Medium pressure compressors (Del. pressure 1 to 8 MN/m^2)

(*iii*) High pressure compressors (Del. pressure 8 to 10 MN/m^2)

(*iv*) Super high pressure compressor (Del. pressure above 10 MN/m^2)

Q 5. How compressors are classified according to the capacity ?

Ans. (*i*) Low capacity compressor (Del. comp. air upto to 0.15 m^3/sec)

(*ii*) Medium capacity compressor (Delivers air 0.15 to 5 m^3/sec)

(*iii*) High capacity compressor (Delivers comp. air more than 5 m^3/sec)

Q 6. How compressors are classified acc. to the method of cooling ?

Ans. (*i*) Air-cooled compressor.

(*ii*) Water-cooled compressor.

Q 7. How compressors are classified acc. to the installation ?

Ans. (*i*) Portable compressor.

(*ii*) Fixed compressor.

Q 8. How compressor are classified acc. to the application ?

Ans. (*i*) Pneumatic hand tool compressor.

(*ii*) Rock drill compressor

(*iii*) Sand blasting compressor

(*iv*) Spray painting compressor

(*v*) Mine prospecting compressor

(*vi*) Refrigerator compressor.

Q 9. How air-compressor are classified acc to I.S.I. standaqrd ?

Ans. (*i*) Positive displacement compressor.

(*ii*) Single shaft vane type compressor.

(*iii*) Multi-shaft compressor.

(*iv*) Roots Blower or compressor.

(*v*) Screen compressor.

(*vi*) Turbo-blower.

(*vii*) Turbo-compressore.

(*viii*) Axial blower.

(*a*) Single stage (*b*) Multi stage.

Q 10. Enlist the main advantages of multi stage compression.

Ans. (*i*) Reduced driving power.

(*ii*) Improved volumetric efficiency.

(*iii*) High delivery pressure.

(*iv*) Improved cooling.

(*v*) Reduced flammability danger.

(*vi*) Reduced cylinder size.

Q 11. What is supercharging ?

Ans. The process of increasing the pressure of the fuel air mixture (in S.I. engine) or that of air (in C.I. engine) extering an engine cylinder during suction, to a pressure greater than that of atmospheric air is known as 'supercharging'.

Q 12. What is the condition for 'maximum-efficiency' of a multi stage compressor ?

Ans.

$$P_2 = \sqrt{P_1 . P_3}$$

In other words; for maximum efficiency, the intermediate pressure is the geometric mean of the initial and final pressures.

Q 13. Write the formula for heat rejected in the intercooler for an air-compressor.

Ans. The heat rejected to the inter-cooler per cycle

$$= m \cdot C_p \cdot (T'_2 - T_2)$$

where, 'm' is the mass of air

'C_p' the sp. heat at constant pressure.

'T'_2' the temperature at the end of compression.

'T_2' temperature of air at the entrance to H.P. cylinder.

Q 14. Enlist the methods of saving the work required for compression in an air-compressor.

Ans. (*i*) Cold water spray.

(*ii*) Multi-stage compression.

Q 15. Write the formula for volumetric efficiency of an air compressor.

Ans. Volumetric efficiency $= \dfrac{\text{Actual volume of air taken in at S.T.P.}}{\text{Swept volume of the piston}}$.

Q 16. What is the value of volume efficiency of an air compressor ?

Ans. Generally, the volume efficiency varies from 70% to 90% depending upon the type of air compressor.

Q 17. What are various factors affecting the volume efficiency of an air-compressor ? Name them.

Ans. (*i*) Increase in delivery pressure.

(*ii*) Very high speed.

(*iii*) Leakage part the piston.

(*iv*) Very large clearance volume.

(*v*) Restricted air-leakage.

(*vi*) Overheating of inlet-air.

(*vii*) Turbulence in the air-passages.

Q 18. What is the 'compressor-efficiency' ?

Ans. Compressor efficiency $= \dfrac{\text{Isothermal power}}{\text{Indicated power}}$.

Q 19. What is 'Mech. efficiency' ?

Ans. Mech. efficiency $= \dfrac{\text{Brake power}}{\text{Indicated power}}$.

Q 20. What is the 'Overall efficiency' of air-compressor ?

Ans. Overall efficiency $= \dfrac{\text{Indicated Power of Compressor}}{\text{Indicated Power of Prime - Mover}}$.

REVIEW QUESTIONS

1. Give the various industrial uses of compressed air.
2. Discuss the classified of air compressors.
3. Discuss the difference between the reciprocating and rotary compressors.
4. Describe the construction and working of reciprocating compressor with sketch.
5. Discuss the working of centrifugal compressor and axial flow compressor with sketch.
6. What is the function of an intercooler ?
7. Write a note on turbo blower, turbo compressor, and vane compressor.
8. Why intercooling is essential in multistage compression ?
9. What are the advantages of 'Multistage compression' ?
10. What do you mean by the volumetric efficiency of a compressor ? **(PSBTE, May 2003)**
11. Explain the functions of fans, blowers and booster. Where are they used ? **(PSBTE, May 2003)**

12. Derive an expression for the work required to drive a single stage compressor.
13. Derive an expression for the volumetric efficiency in a single stage compressor neglecting the clearance volume and explain the meaning of it.
14. What is the effect of clearance volume on the volumetric efficiency of the compressor ?
15. Name and define the various efficiencies you know in case of reciprocating compressor.
16. Describe the construction and working of a Rotary air compressor with sketch. **(PSBTE, May 2003)**

PROBLEMS FOR PRACTICE

1. A single stage air compressor is required to compress 72 cubic metre of air per minute from 15 °C and one atmosphere to 8 atmospheric pressure. Find the temperature at the end of compression, work done, power and heat rejected during each of the following processes (*i*) isothermal (*ii*) adiabatic (*iii*) polytropic compression following $PV^{1.25}$ = constant.

   [**Ans.** (*i*) 15°C, 150×10^5 J/min, 250 kW, 15000 KJ/min ; (*ii*) 248°C, 204.75×10^5 J/min, 341.26 kW ; (*iii*) 162°C, 185.76×10^5 J/min, 309.6, 6960 KJ/min]

2. A two stage air compressor with perfect intercooling takes in air at 1 bar pressure and 27°C. The law of compression in both the stages is $PV^{1.3}$ = constant. The compressed air is delivered at 9 bar from the HP cylinder to an air receiver. Calculate per kg of air.
   (*i*) The minimum work done.
   (*ii*) The heat rejected to intercooler.
   (*iii*) The minimum work done in a three stage compressor working under the same conditions. Take C_p = 1 KJ/kg K.

   [**Ans.** 213.29 KJ. 85.8 KJ/kg of air, 205.56 KJ]

3. An air compressor driven directly by a diesel engine handle 350 m^3 of free air per minute, the induction pressure being 1 bar, the temperature 15 °C and the compression pressure 12 bar. If the overall isothermal efficiency of the compressor is 70% and the brake thermal efficiency of the engine is 25%, find the litres of oil required per 24 hours if each litre liberates heat equivalent to 29400 KJ. [**Ans.** 23916.38 litres]

4. Estimate the power of compound air compressor necessory to compress 420 m^3 per hour of dry air at 1 bar abs. pressure and 10°C to a final pressure of 34 bar abs. The intermediate receiver cools the air to 30° and 7.5 bar. Assume mechanical efficiency of 90% and $\gamma = 1.4$. [**Ans.** 1911 KJ, 616 kW]

5. A single acting air compressor draws in 10 m^3/min of air at 1 bar and 17°C and compresses it to 8 bar. The law of compression is $PV^{1.3}$ = constant. Calculate the indicated power required in the following alternatives of compresion.
   (*i*) Single stage compression.
   (*ii*) Two stage compression with intercooling to 22 °C.

   [**Ans.** 44.5 kW, 39.50 kW]

❑❑❑

11

CHAPTER

Introduction to Heat Transfer

INSIDE THIS CHAPTER

11.1. INTRODUCTION

From the first law of thermodynamics and its corollary heat energy is obtained from other forms of energy. The limit and direction of the heat flow is due to the second law of thermodynamics. It states that the heat energy flows only from a body at higher temperature to a body at lower temperature. Uptil now no mention has been made of actual mechanism of heat transfer namely conduction, convection and radiation which are explained below.

In the previous chapters, time has never been considered a limiting factor for the transfer of heat, but in this chapter we are concerned with the rate of heat transfer *i.e.* the heat transfer per unit time. The rate of heat transfer may be constant or variable, depending on whether conditions are such that the temperature remain the same or change continuously with time. The former is known as the *'steady state of heat transfer'* while the latter, *'unsteady state of heat transfer'*. Here, we are dealing with *'steady state heat transfer'* only.

11.2. CONDUCTION

'Heat Conduction' is due to the elastic impacts of the molecules, in gases, due to longitudinal oscillations, in solid non-conductors of electricity and due to the motions of the electrons in metals. This means exchange of heat or heat transfer, is between contiguous bodies or parts of a body which are at different temperatures. For example the heat-transfer in a metal rod is by conduction. The basic law of heat conduction is the Fourier's law which is in complete analogy to the law of electrical conduction.

11.2.1. Fourier's Law of Heat Conduction

The basic law of heat conduction is due to Biot, but is called Fourier's law because he used it as a fundamental equation in his analytical theory of heat.

Fourier's law is based on the empirical observation of one dimensional steady state heat flow through a solid. One dimensional flow implies that the temperature is uniform over surfaces perpendicular to the direction of heat conduction and steady state heat flow implies that the temperature at any point does not change with time. It is easy to appriciate that in one dimentional steady-state heat transfer the rate of heat flow through successive surfaces is constant.

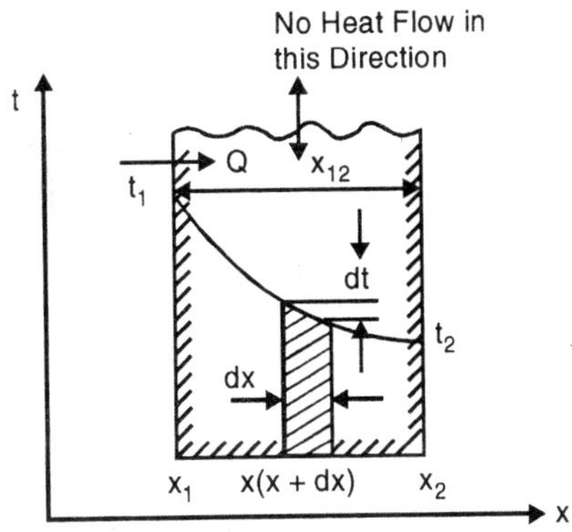

Fig. 11.1.

Consider a plane layer of thickness 'dx' (as shown in the Fig. 11.1) its one surface at a temperature 't' and the other surface at temperature $(t + dt)$.

According to Fourier' law, the heat flow :

$$Q = -kA\left(\frac{dt}{dx}\right) \quad \text{...(11.1)}$$

where
Q = Rate of heat flow
A = Area measured transversely to the direction of heat flow
$x_2 - x_1$ = x_{12} = Thickness of the plate
k = Constant of proportionality called the co-efficient of thermal conductivity.
dt = Difference of temperature between two faces of the body.
dx = Thickness of the body along the direction of heat.
$\frac{dt}{dx}$ = Temperature of gradient.

The –ve sign in the above equation is used, because the temperature decreases with the increasing value of x, making $\left(\frac{dt}{dx}\right)$, essentially negative and the –ve sign makes 'Q' positive for convenience.

The thermal conductivity, 'k' is defined as *the heat conducted in unit time, across unit area, through unit thickness when a temperature graidient of unity is maintained between the opposite faces.* The thermal conductivity is a property of a material and for most substances it varies with temperature. However, for simplicity of calculations the effect temperature on conductivity is generally ignored.

For steady flow, the constant value of thermal conductivity the temperature gradient will also be a constant *i.e.* the temperature profile will be linear.

Now eqn. (11.1) can be integrated between the limits x_2 and x_1 to enable the heat-flow to be expressed in terms of the surface temperature t_2 and t_1 by.

$$Q = -kA\frac{t_2 - t_1}{x_2 - x_1}$$

or $$Q = -kA\frac{t_2 - t_1}{x_2} \qquad \text{...(11.2)}$$

or $$Q = -kA\frac{\Delta t}{\Delta x}$$

The units in this equation are

$$\frac{\text{KJ}}{\text{hr}} = \frac{\text{KJ}}{\text{m.k.hr}} \times \text{m}^2 \times \frac{\text{k}}{\text{m}}$$

11.2.2. Conduction Through a Composite Wall

Walls of several materials constructed in layers are called 'composite walls'. The example of this is the wall of a building in which stones or bricks are covered with a layer of plaster on both sides or the walls of an insulated room having layers of insulation, brick, plaster etc.

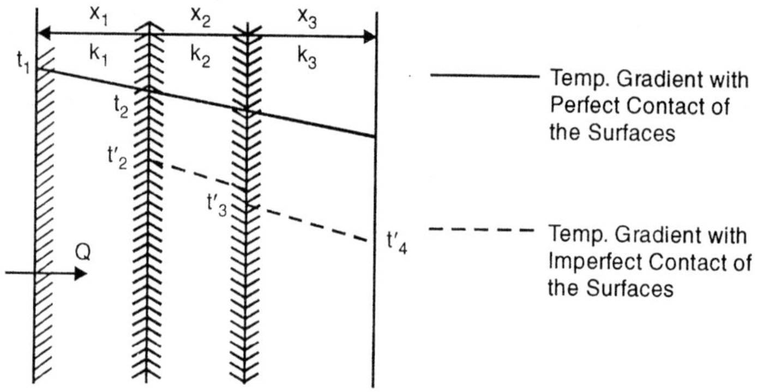

Fig. 11.2.

A wall of three different materials is shown in the Fig. 11.2. The outer surface temperature of the wall are 't_1' and 't_4'. Let the interfaces be at temperatures 't_2' and 't_3'.

Now assuming perfect conduction between the surfaces and steady state conditions,

$$Q = -k_1 . A\frac{(t_1 - t_2)}{x_1} = -k_2 . A\frac{(t_2 - t_3)}{x_2} = -k_3 . A\frac{(t_3 - t_1)}{x_3}$$

$$\therefore \quad (t_1 - t_2) = \frac{Q.x_1}{k_1 . A}$$

$$(t_2 - t_3) = \frac{Q.x_2}{k_2 . A}$$

and $$(t_3 - t_4) = \frac{Q.x_3}{k_3 . A}$$

Adding all these we get

$$(t_1 - t_4) = \frac{Q}{A}\left(\frac{x_1}{k_1} + \frac{x_2}{k_2} + \frac{x_3}{k_3}\right)$$

$$Q = \frac{(t_1 - t_4)}{\frac{x_1}{k_1 . A} + \frac{x_2}{k_2 . A} + \frac{x_3}{k_3 . A}} \qquad \text{...(11.3)}$$

or $$= \frac{(t_1 - t_4)}{\Sigma \frac{x}{kA}}$$

or $$Q = \frac{\Delta t}{R_1 + R_2 + R_3} = \frac{\Delta t}{\Sigma R}$$

where 'R' is the thermal resistance of wall.

∴ For a composite wall as a whole we can write

$$Q = U.A.(t_1 - t_2) \quad ...(11.4)$$

where 'U' is the 'overall heat transfer co-efficient' for the composite wall.

The units of U are $\frac{KJ}{m^2.h.k}$ and $\frac{1}{U} = \frac{x_1}{k_1} + \frac{x_2}{k_2} + \frac{x_3}{k_3}$.

In general, for any number of layers

$$\frac{1}{U} = \sum \frac{\Delta x}{k} = \sum R \quad ...(11.5)$$

$\frac{1}{U}$ is the overall resistance to the heat flow and is equal to the sum of individual resistance in series.

11.2.3. Radial Conduction Through Wall of a Tube

In the long tubes when the end effects are neglected heat flows only in radial direction, thus it is an one dimensional conduction problem. Unlikes plates the area through which heat flows is not constant in tubes. It is therefore convenient to consider a given length of the tube rather than a given area.

Consider a long tube of inside radius 'r_1', outside radius 'r_2' and length 'l' as shown in Fig. 11.3. Inside and outside temperatures are assumed constant as 't_1' and 't_2' respectively. The heat flow for given length 'l' is constant through successive layers of the tube, but area increases with radius, therefore the temperature gradient decreases with radius as shown in Fig. 11.3.

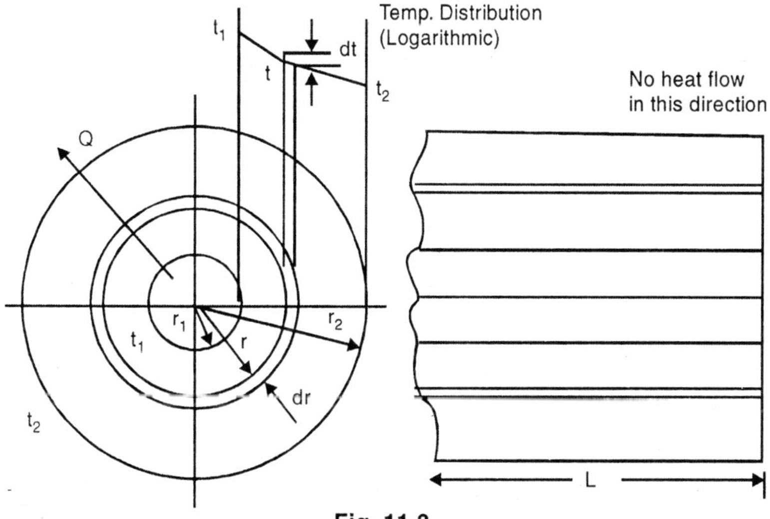

Fig. 11.3.

Consider an elemental shell of thickness 'dr'. The heat flow, by Fourier's law, is

$$Q = -kAr\frac{dt}{dr}$$

or
$$Q = -k2\pi rl\frac{dt}{dr}$$

Integrating, with the boundary conditions as temperature 't_1' and 't_2' at radius 'r_1' and 'r_2' respectively, the heat flow per unit length is,

$$Q\int_{r_1}^{r_2}\frac{dr}{r} = -k2\pi\int_{t_1}^{t_2}.dt$$

or
$$Q = -\frac{2\pi k(t_2 - t_1)}{\log_e \frac{r_2}{r_1}} = -\frac{(t_1 - t_2)}{\frac{\log_e \frac{r_2}{r_1}}{2\pi k}} \quad ...(11.6)$$

Defining a mean radius by 'r_m'

$$Q = -2\pi r_m k.\frac{t_2 - t_1}{r_2 - r_1} \quad ...(11.7)$$

The value of 'r_m' is found comparing eqns. (11.6) and (11.7)

$\therefore$
$$r_m = \frac{r_2 - r_1}{\log_e \frac{r_2}{r_1}} \quad ...(11.8)$$

Similarly, the equivalent mean area per unit length of tube

$$A_m = \frac{r_2 - r_1}{\log_e \frac{r_2}{r_1}} \times 2\pi = \frac{A_2 - A_1}{\log_e \frac{r_2}{r_1}} \quad ...(11.9)$$

Conduction Through Thick Spherical Shell

The Fig. 11.4 shows a thick spherical shell of inside radius 'r_1' and outside radius 'r_2' with temperature 't_1' and 't_2' respectively. Considering an element shell of thickness 'dt', the heat flow by Fourier's law is,

$$Q = -kAr.\frac{dt}{dx}$$

or
$$Q = -k4\pi r^2.\frac{dt}{dx}$$

Integrating and substituting the boundary conditions,

$$Q\int_{r_1}^{r_2}\frac{dr}{r^2} = -4\pi k\int_{t_1}^{t_2} dt$$

$$\text{or}\left[\frac{r_2 - r_1}{r_1 r_2}\right] = -4\pi k(t_2 - t_1)$$

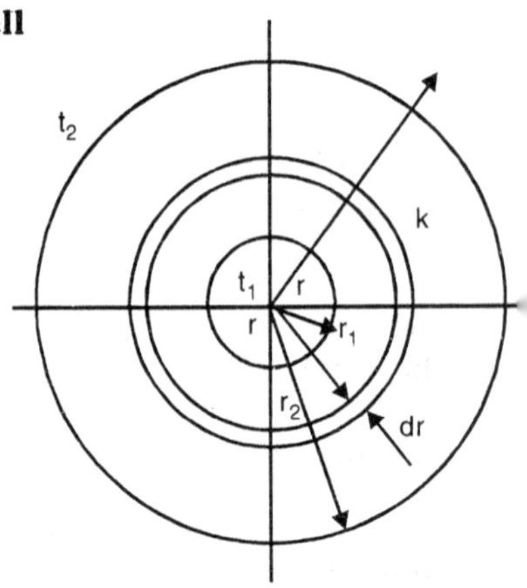

Fig. 11.4.

$$\therefore \quad Q = \frac{4\pi k r_1 r_2 (t_2 - t_1)}{r_2 - r_1} \quad ...(11.10)$$

Introducing, an overall heat transfer co-efficient, U

$$Q = U_a A_a (t_1 - t_2) = U_b . A_b .(t_1 - t_2) \quad ...(11.11)$$

where 'U_a' and 'U_b' are the overall heat transfer coefficients referred to the inside and outside areas of the shell respectively.

11.2.4. Critical Insulation

It is generally expected that the heat flow from an insulated pipe should decrease with the increasing thickness of insulation. This is however not always true as in certain cases the heat flow may increase, with the increasing thickness of insulation. This interesting phenomenon can be explained as follows.

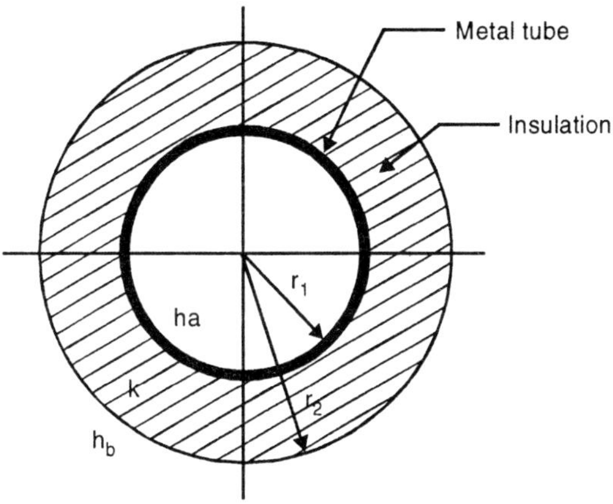

Fig. 11.5.

Consider a small diameter tube with insulation as shown in Fig. 11.5. Assuming negligible thermal resistance of the tube material, the heat flow equation can be written as follows :

$$Q = \frac{t_b - t_a}{\frac{1}{A_a . h_a} + \frac{\log_e \frac{r_2}{r_1}}{2\pi . lk} + \frac{1}{A_b . h_b}} \quad ...(11.12)$$

The heat flow will be maximum when the denominator becomes the minimum. Differentiating the denominator with respect to the variable outer radius 'r_2' and equating to zero, we get

$$\frac{1}{k_1 . r_2} - \frac{1}{h_b . r_2^2} = 0 \quad ...(11.13)$$

From which :

$$r_2 = r_{\text{critical}} = \frac{k_1}{h_b} \quad ...(11.14)$$

where 'critical radius' is defined as the outer radius, which would give maximum heat flow.

Referring to the original equation the term $\frac{1}{h_a . r_1}$ is constant with respect to outer radius, the term $\frac{1}{k}\log_e \frac{r_2}{r_1}$ (thermal resistance due to insulation) increases with 'r_2' while the term $\frac{1}{h_b . r_2}$ (thermal resistance due to film outside) decreases with increasing value of 'r_2'. At the critical radius, the increase of insulation resistance is equal to the r_1 of decrease in film resistance, thus giving minimum value of the resistance. If the value of r_0 is below or above the critical radius value the net resistance would be more. This phenomenon of critical thickness of insulation may happen with small tube diameter, relative large thermal conductivity of the insulation and small heat transfer coefficient. It has practical application in both cases where heat conduction is desired and when it is to be prevented. In electrical cable the objective is to provide adequate electrical insulation and also the maximum cooling of wire. This is achieved by providing insulation critical thickness. In steam and refrigeration pipes where heat insulation is the objective, the loss of heat can actually be decreased by decreasing the insulation thickness, if the outer radius of insulation near critical radius.

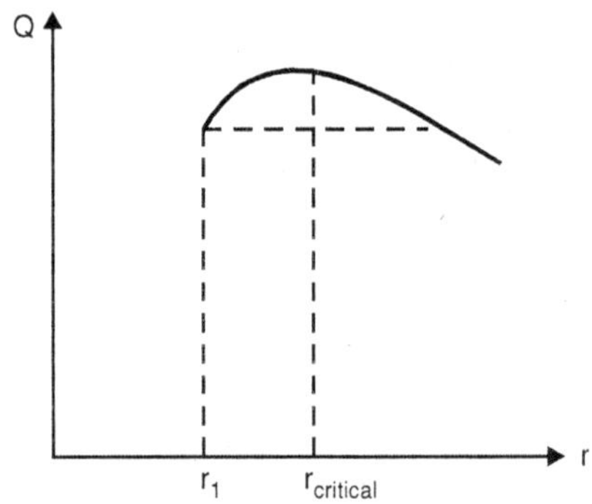

Fig. 11.6.

The phenomenon of variation of heat loss with thickness of the insulation is shown in Fig. 11.6.

For spheres critical thickness of insulation can be similarly calculated as,

$$r_2 = \frac{2k}{h} \quad ...(11.15)$$

11.3. HEAT-CONVECTION

It is the transportation and exchange of heat due to the mixing motion of different parts of a fluid.

It is of two types as follows :

(*i*) Natural or free convection,

(*ii*) Forced convection.

When the motion of the fluid is caused by the difference in density due to temperature, it is called *natural or free convection*, but when the motion of the fluid is caused by some mechanical means it is called *forced convection*.

Heating of rooms by room heaters is an example of free convection whereas heating of circulating water in condenser tubes is an example of forced convection. The heat transfer from a solid surface to a gas or liquid over it, is partly by convection. A thin boundary layer of the fluid always exists above the plate where the heat flow by conduction is dominant. Away from the plate where there is appreciable movement of fluid, the heat flow by convection is dominant. The heat transfer by convection is governed by the laws of fluid dynamics in combination with the laws of heat conduction.

The Newton's heat convection equation is;

$$Q = h \cdot A \cdot \Delta t \text{ Watt}$$

where h = Convective heat transfer co-efficient in W/m^3k

A = Area in m^2

Δt = Temperature gradient in k

11.4. HEAT RADIATION

The term *'radiation'* is applied to all kinds of processes which transmit energy by means of electromagnetic waves. The amount of radiation mainly depends upon the nature of the body, its temperature and the kind and extent of its surface. If there are two bodies at different temperatures, each body receives the energy radiated from the other. The body at the lower temperature will receive more energy and hence its internal energy will increase where as the body at higher temperature will receive less energy and hence its internal energy will decrease. Thus, the net transfer of energy between two radiating surfaces is from the body at higher temperature to the body at lower temperature which satisfies the second law of thermodynamics. Radient energy, being electromagnetic in characteristic, needs no medium for its propagation and will pass through vacuum.

In practice the heat transfer may be by combination of conduction, convection and radiation. In such a case the amount of heat transfer by each mode is calculated separately and the result is obtained by summing up. In engineering there are two main groups of problems, first, the desirable transfer of heat with the minimum possible surface area and temperature difference as in heat exchanges, boilers and condensers and second, the prevention of transfer of heat *i.e.* with thermal insulation, as in pipes carrying steam, air conditioning rooms etc.

11.4.1. Absorption Reflection and Transmission of Radiation

When heat is radiated, a second body intercepting some of the waves, may absorb, reflect or transmit the energy carried by the waves through the body. It is shown in the Fig. 11.7.

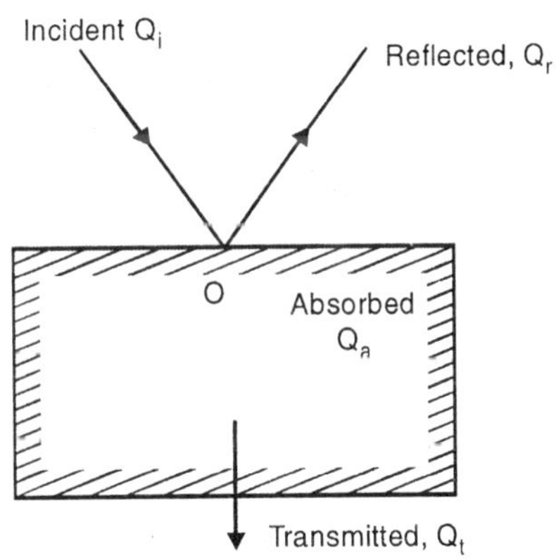

Fig. 11.7.

Mathematically,

$$Q_i = Q_a + Q_r + Q_t \quad ...(11.16)$$

or

$$1 = \frac{Q_a}{Q_i} + \frac{Q_r}{Q_i} + \frac{Q_t}{Q_i} \quad ...(11.17)$$

or

$$1 = \alpha + \rho + \tau \quad ...(11.18)$$

where Q_i = Total incident radiation

Q_a = Energy absorbed

Q_r = Energy radiated

Q_t = Energy transmitted,

α = The ratio of energy absorbed to the total incident energy and is called 'absorptivity'.

ρ = The ratio of the energy reflected to the total incident energy and is called 'reflectivity'.

and τ = The ratio of the energy transmitted to the total incident energy and is called 'transmissivity'.

For most of the solid materials used in engineering the transmissivity is zero. Thus for solids,

$$\alpha + \rho = 1$$

Depending on the radiating properties the bodies can be classified as follows :

(*i*) Black body (all radiation absorbed) $\rho = 0, \tau = 0$ and $\alpha = 1$

(*ii*) White body (all radiation reflected) $\alpha = 0, \tau = 0$ and $\rho = 1$

(*iii*) Transparent or Diathermanous body (all radiation transmitted) $\alpha = 0, \rho = 0$ and $\tau = 1$

(*iv*) Opaque body $\tau = 0$ and $\rho + \alpha = 1$ (partly reflected and partly absorbed)

11.4.2. Concept of Black Body

The *'black body'* is defined as the body, which absorbed all radiation incident upon it. In thermal sense it has nothing to do with colour. Actually no material known in nature is perfectly black body but some bodies as lampback and platinum black, reflect only a very small fraction of the incident radiation and consequently are nearly black bodies. *Snow is also nearly black to thermal radiation.*

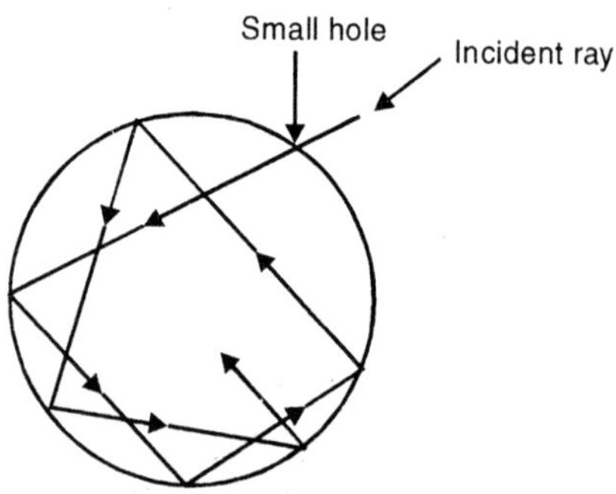

Fig. 11.8.

A black body may be approximated by a hollow space with uniform interior temperature and a small hole as shown in Fig. 11.8. An incident ray of energy enters it through the hole is partly absorbed and partly reflected at the interior surface. With each reflection there is a fraction of the energy absorbed and when the radiation beam finally escape from the hole again, the energy with it is negligible. Thus, the cavity with a small hole acts very nearly as a black body.

11.4.3. Total Emissive Power

The total amount of radiation emitted by a body per unit time is defined as the 'total emissive power'.

11.5. STEFAN BOLTZMANN'S LAW

In 1879, Joseph Stefan from experimental data concluded that *'The total emission from a black body per unit area and per unit time is proportional to the fourth power of the absolute temperature of the body.* It was established theoratically by Boltzmann with the help of the second law of thermodynamics in 1884.

Mathematically, Stefan Boltzmann's Law is given by equation.

$$E_b = \sigma \,.\, T^4 \qquad \text{...(11.19)}$$

where 'σ' is a constant called stefan Boltzmann's constant and is equivalent to 5.67×10^{-8} W/m^{2}K^4, which reduces it to the same equation as given by Plank.

NUMERICAL EXAMPLES

Example 11.1. The inner surface of a 325 mm brick wall of a furnace is kept at 800 °C and it is found that the outer surface temperature is 170 °C. Calculate the heat loss per square metre of wall area. Given that the conductivity is 0.865 W/mk.

Solution. We know that:

$$Q = \frac{k.A.\Delta t}{\Delta x}$$

Given:

$$k = 0.865 \text{ W/mk}$$
$$A = 1 \text{ m}^2$$
$$\Delta t = 800 - 170 = 630 \text{ k}$$
$$\Delta x = 235 \text{ mm} = 0.235 \text{ m}$$

Substituting the values in above equation

$$\therefore \quad Q = \frac{0.865 \times 630}{0.235}$$

or $$Q = 2319 \text{ W/m}^2 \textbf{ Ans.}$$

Example 11.2. The wall of a furnace is made up of 250 mm of fire brick, k = 1.05 W/m K, 120 mm of insulation brick, k = 0.15 W/m °k and 200 mm of red brick, k = 0.85 W/m °k. The inner and outer surface temperatures of the wall are 850 °C and 65 °C respectively. Neglecting the resistance of morter joints, compute the temperatures at the contact surfaces.

Solution. The rate of heat flow per square metre is given by

$$Q = \frac{1 \times (850 - 65)}{\frac{0.250}{1.05} + \frac{0.120}{0.15} + \frac{0.200}{0.85}} = 616.5 \text{ W/m}^2$$

As this quantity of heat flows through all the layers, hence

$$\frac{Q}{A} + \frac{t_1 - t_2}{\frac{\Delta x_{12}}{k_{12}}}$$

or $$616.5 = \frac{850 - t_2}{\frac{0.250}{1.05}}$$

$$\therefore \quad t_2 = 703 \text{ °C } \textbf{Ans.}$$

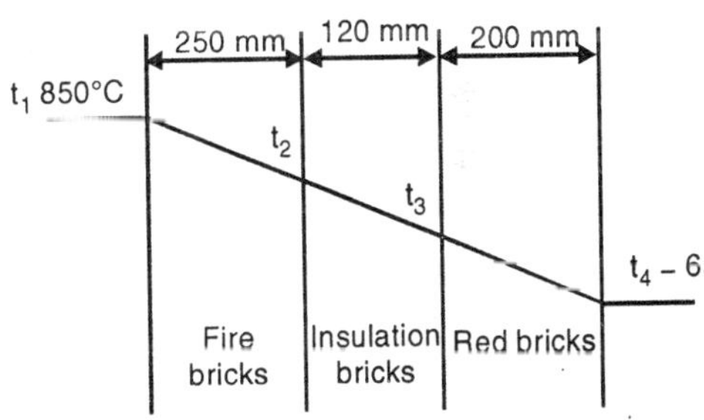

Fig. 11.9.

Similarly, t_3 can be determined from the relation,

$$\frac{Q}{A} = \frac{t_2 - t_3}{\Delta x_{23}/k_{23}}$$

or $$616.5 = \frac{703 - t_3}{\frac{0.120}{0.15}}$$

$$\therefore \quad t_3 = 210 \text{ °C. } \textbf{Ans.}$$

Example 11.3. A 25 mm steam main, 225 metres long is covered with 5 cm of high temperature insulation (k = 0.095 W/m k). The inner and outer surface temperatures as measured are 400 °C and 50 °C respectively. Compute;

(*i*) The total heat loss per hour.

(*ii*) The total heat loss per square metre of outer surface.

(*iii*) The heat loss per square metre of pipe surface.

(*iv*) The temp. conduction through pipe insulation.

Neglect heat conduction through pipe material.

Solution.

(*i*) Outside diameter of the pipe,

$$d_1 = 25 \text{ cm.}$$

Outside diameter of first layer,

$$d_2 = 25 + 2 \times 5 = 35 \text{ cm.}$$

Outside diameter of second layer.

$$d_3 = 35 + 2 \times 4 = 43 \text{ cm.}$$

Also,

$$k_1 = 0.095 \text{ W/m °k}$$

$$k_2 = 0.065 \text{ W/m °k}$$

Total heat loss,

$$Q = \frac{2\pi L(t_1 - t_3)}{\log_e \frac{(d_2/d_1)}{k_1} + \log_e \frac{(d_3/d_2)}{k_2}}$$

or

$$Q = \frac{2\pi \times 225\,(400 - 50)}{\log_e \frac{35/25}{0.0.95} + \log_e \frac{43/35}{0.065}} = 73754 \text{ W.}$$

$$= \frac{73754 \times 3600}{1000} \text{ KJ/hr.} = 265514 \text{ KJ/hr.}$$ **Ans.**

(*ii*) Total heat loss per square metre of outer surface

$$= \frac{265514}{\pi \times 0.43 \times 225} = 873.5 \text{ KJ/hr.}$$ **Ans.**

(*iii*) Heat loss per square metre of pipe surface

$$= \frac{265514}{\pi \times 0.25225} = 1502.5 \text{ KJ/hr.}$$ **Ans.**

(*iv*) For temperature between two layers.

$$\frac{Q}{L} = \frac{2\pi K(t_1 - t_2)}{\log_e\left(\frac{d_2}{d_1}\right)}$$

$$\therefore \quad \frac{73754}{225} = \frac{2\pi \times 0.095\,(400 - t_2)}{\log_e\left(\frac{35}{25}\right)}$$

$$\therefore \quad t_2 = 215 \text{ °C.}$$ **Ans.**

Example 11.4. **A spherical shaped vessel of 1.2 m diameter is 100 mm thick. Find the rate of heat leakage, if the temperature differences between the inner the outer surfaces is 200 °C. Thermal conductivity of the material of sphere is 0.3 KJ/m-hr °C.**

Solution. Outside diameter of sphere, = 1.2 m

So,

$$r_2 = 0.6 \text{ m}$$

Inside diameter of sphere, = 1.2 – 2 × 0.1 = 1.0 m

So,

$$r_1 = 0.5 \text{ m}$$

Given : $t_1 - t_2 = 200\ °C$

$k = 0.3$ KJ/m-hr °C

Using the relation for heat transfer,

$$Q = \frac{4\pi k r_1 r_2 (t_1 - t_2)}{(r_0 - r_1)}$$

$$= \frac{4\pi \times 0.3 \times 0.5 \times 0.6 \times 200}{0.1} = 2262 \text{ KJ/hr.} \quad \textbf{Ans.}$$

Example 11.5. The temperature of a black surface 0.25 m^2 in area is 650 °C. Calculate the total rate of energy transmission. Take, Steam Boltzman's Constant, $\sigma = 56.7 \times 10^{-12}$ KW/m^2 °k^4.

Solution. From Stefan Boltzmann's law :

$$Q = \sigma . \Delta T°$$

where, Q = Energy emission rate in KW

$\sigma = 56.7 \times 10^{-12}$ KW/m^2 k^4

A = Area in m^2

T = abs. temperature in k

$$\therefore \quad Q = 56.7 \times \left(\frac{273 + 650}{1000^4}\right)$$

or $$Q = 10.288 \text{ KW.} \quad \textbf{Ans.}$$

ASSIGNMENT

Objective Type Questions :

(A) Fill in the Blanks :

1. Three modes of heat transfer are _______ .
2. The unit of thermal conductivity in S.I. system of units is _______ .
3. The rate of heat transfer is proportional to the reciprocal of _______ .
4. In forced convection of movement of fluid is obtained by _______ .
5. Pipe lagging is done to _______ .
6. The transfer of heat from Sun to the earth is by _______ .
7. In the case of black body, the absorptivity is _______ .
8. Total emissive power is defined as _______ .
9. Stefan Boltzman's law states that _______ .
10. The basic law of heat conduction is the _______ law.
11. The heat-transfer in a metal rod is by _______ .
12. The two modes of heat convection are _______ and _______ .
13. Heating of rooms by room heaters is an example of _______ convection.
14. Heating of circulating water in condenser tubes is an example of _______ convection.
15. The amount of heat radiation mainly depends upon the nature of the body, the kind and extent of its surface and its _______ .
16. In practice the heat transfer may be by _______ and _______.
17. The ratio of energy absorbed to the total incident energy is called _______ .

18. The ratio of energy reflected to the total incident energy is called ______ .
19. The ratio of the energy transmitted to the total incident energy is called ______ .
20. The black body is defined as the body, which absorbed all ______ incident upon it.

(B) State True or False :

1. Solids are heated by conduction.
2. Liquids are heated by radiation.
3. Rate of heat flow depends upon the nature of material of the body.
4. Heat transfer is more in the case of natural convection than forced convection.
5. In opaque body, the transmissivity is unity.
6. Pipe lagging is a coating of insulation material to avoid heat transfer.
7. The value of Stefan Boltzmann's constant is 56.7×10^{-12} KW/m^2 $°k^4$.
8. Heat radiation can occur through vacuum.
9. Convection is the slowest process of heat transfer.
10. The position of molecules of the body remain fixed in the case of convection.
11. Heat transfer in a metal rod is by conduction.
12. The basic law of heat conduction is the Fourier's Law.
13. Fourier's law is based on the empirical observation of One-dimensional steady state heat flow through a solid.
14. One dimensional heat flow implies that the temperature is uniform over the surfaces perpendicular to the direction of heat-conduction.
15. Steady state heat transfer implies that the temperature at any point does not change with time.
16. The thermal conductivity 'K' is defined as the heat conducted in unit time, across unit area, through unit thickness, when a temperature gradient of unity is maintained, between the opposite faces.
17. The critical insulation is the thickness of insulation which would give maximum heat flow through it.
18. Heating of rooms by room heaters is an example of forced convection.
19. Heating of circulating water in condenser tubes is an example of free convection.
20. The basic law of heat radiation is Stefan Boltzmann's law.

(C) Multiple Choice Questions :

1. The essential condition for the heat transfer from one body to another is :
   (*a*) both bodies must be in physical contact.
   (*b*) heat content of one body must be more than that of the other.
   (*c*) one of the bodies must have a high value of thermal conductivity.
   (*d*) there must exist a temperature difference between the two bodies.
2. Heat transmission is directly linked with the transport of medium itself *i.e.* there is actual motion of heated particles during;
   (*a*) conduction only
   (*b*) convection only
   (*c*) radiation only
   (*d*) conduction as well as radiation.

3. The material medium between the heat source and receiver is not affected during the process of heat-transmission by;
   (*a*) conduction
   (*b*) convection
   (*c*) radiation
   (*d*) conduction as well as radiation.
4. A satellite in space exchanges heat with the surroundings essentially by;
   (*a*) conduction
   (*b*) convection
   (*c*) radiation
   (*d*) conduction and convection put together.
5. All the three modes of heat transmission are involved in;
   (*a*) melting of ice.
   (*b*) cooling a small metal casting in a quenching bath.
   (*c*) heat flow through the walls of a refrigerator.
   (*d*) automobile engine equiped with a thermo syphon cooling system.
6. Steady State heat flow implies :
   (*a*) negligible flow of heat
   (*b*) no difference of temperature between the bodies
   (*c*) constant heat flow rate *i.e.* heat flow rate independent of time.
   (*d*) inform rate in temperature rise of a body.
7. Most unsteady heat flow occurs;
   (*a*) through the walls of a refrigerator
   (*b*) during anealing of casting
   (*c*) through the walls of a furnace
   (*d*) through lagged (insulated) pipes carrying steam.
8. The fourier's conduction heat equation, $Q = -k.A\frac{dt}{dx}$ pressure,
   (*a*) steady state one dimensional heat flow.
   (*b*) constant value of thermal conductivity.
   (*c*) constant and uniform temperatures at the wall surface.
   (*d*) all of the above.
9. Which of the following is a wrong statement with respect to the Fourier's heat conduction equation ?
   (*a*) Fourier law is valid for the matter regardless of its state.
   (*b*) Fourier law is a vector expression indicating heat flow in the direction of decreasing temperature.
   (*c*) Fourier law can be derived from first principles.
   (*d*) Fourier law helps to define thermal conductivity of the heat conducting medium.
10. Thermal conductivity is defined as the heat flow per unit time,
   (*a*) whom temperature gradient is unity.
   (*b*) when unit temperature difference is maintained across the opposite faces the wall.
   (*c*) through a unit a thickness of the wall.
   (*d*) across unit area when temperature gradient is unity.
11. Molecular transmission of heat is smallest in case of,
   (*a*) gases (*b*) liquids
   (*c*) solids (*d*) alloys.

12. Heat conduction in gases is due to;
(a) motion of electrons
(b) elastic impact of molecules
(c) mixing motion of the difference layers of a gas.
(d) electro magnetic waves.

13. Indicate the metal with highest value of thermal conductivity :
(a) air (b) silver
(c) copper (d) aluminium.

14. Mark the matter with least value of thermal conductivity :
(a) air (b) water
(c) ash (d) window glass.

15. Which of the following forms of water have the highest value of thermal conductivity :
(a) boiling water (b) steam
(c) solid ice (d) melting ice.

16. A composite slab has two layers of different materials with thermal conductivity k_1 and k_2. If each layer has same thickness, the equivalent thermal conductivity of the slab would be;
(a) $k_1 \cdot k_2$ (b) $k_1 + k_2$
(c) $\frac{(k_1 + k_2)}{k_1 . k_2}$ (d) $(2k_1 \cdot k_2)\ (k_1 + k_2)$.

17. Fins are provided on heat transfering surface in order to increase;
(a) heat transfer area (b) heat transfer co-efficient
(c) temperature gradient
(d) mechanical strength of the equipment.

18. For a perfectly black body;
(a) absorptivity $\alpha = 1$, reflectivity $\rho = 0$ and transmissivity, $\tau = 0$
(b) $\rho = 1$ and $\alpha = \tau = 0$
(c) $\tau = 1$ and $\alpha = \rho = 0$
(d) $\alpha + \tau + \rho = 0$

19. The ratio of total emissive power of a body to the total emissive power of a black body at the same temperature is called;
(a) absorptivity (b) transmissivity
(c) reflectivity (d) emissivity.

20. Forced convection in a liquid both is caused by;
(a) density difference brought about by temperature gradients.
(b) molecular energy interaction.
(c) flow of electrons in a random fashion.
(d) intense stiring by an external agency.

ANSWERS

(A) Fill in the Blanks :

1. (i) conduction, (ii) convection, (iii) radiation.
2. KJ/m.k **3.** distance between the two temperature limits.
4. some mechanical means
5. minimise heat transfer through its
6. radiation 7. one
8. the total amount of radiation emitted by a body per unit area and per unit time

9. the total emission from a black body per unit area and per unit time is propertional to the forth power of the absolute temperature of the body.
10. Fourier's
11. conduction
12. free convection, forced convection
13. free
14. forced
15. temperature
16. combination of conduction, convection, radiation
17. absorptivity
18. reflectivity
19. transmissively
20. radiation.

(B) State True or False :

1. True	**2.** False	**3.** True	**4.** False	**5.** False
6. True	**7.** True	**8.** False	**9.** False	**10.** False
11. True	**12.** True	**13.** True	**14.** True	**15.** True
16. True	**17.** True	**18.** False	**19.** False	**20.** True.

(C) Multiple Choice Questions :

1. (*d*)	**2.** (*b*)	**3.** (*c*)	**4.** (*c*)	**5.** (*d*)
6. (*c*)	**7.** (*b*)	**8.** (*d*)	**9.** (*c*)	**10.** (*d*)
11. (*a*)	**12.** (*b*)	**13.** (*b*)	**14.** (*a*)	**15.** (*c*)
16. (*d*)	**17.** (*a*)	**18.** (*a*)	**19.** (*d*)	**20.** (*d*)

(D) Short Answer Type Questions :

Q. 1. What is 'steady state of heat-transfer' ?

Ans. If the rate of heat-transfer is constant, it is called 'steady state of heat-transfer.

Q. 2. What is the 'Unsteady state of heat-transfer' ?

Ans. If the rate of heat-transfer is variable, it is called 'Unsteady state of heat-transfer.

Q. 3. What are the three important modes of heat-transfer ? Name them.

Ans. The three modes of the heat-transfer are as follows :

(*i*) Heat-conduction
(*ii*) Heat-convection, and
(*iii*) Heat-radiation

Q. 4. What is 'Heat-Conduction' ?

Ans. 'Heat-conduction' is the mode of the heat-transfer, due to the elastic impacts of the molecules, in case of gases, due to longitudinal oscillations, in solid non-conductors of electricity and due to the motions of the electrons in metals *i.e.*, conductors of electricity.

Q. 5. What is the expression for 'Fourier's law of heat-conduction ?

Ans. The expression for the Fourier's Law of heat-conduction is;

$$Q = -k.A\left(\frac{dt}{dx}\right)$$

where Q = Rate of heat-transfer
A = Area perpendicular to the direction of heat-transfer
k = Co-efficient of thermal conductivity.
dt = Difference of temperature between the two faces of the body.
dx = Thickness of the body along the heat-flow.
$\frac{dt}{dx}$ = Temperature Gradient.

Q. 6. What is 'thermal conductivity (k) of a substance ?

Ans. The 'Thermal conductivity (k)' of a substance may be defined "The heat conducted in unit time, across the unit area of cross-section through unit thickness of the substance, when a temperature gradient of unity is maintained between the opposite faces.

Q. 7. What is the expression for 'steady-state heat conduction through a composite wall' ?

Ans. Let a composite wall is made of three different materials. The outer surface temperatures of the wall are 't_1' and 't_4'. Let the interface be at temperatures of 't_2' and 't_3'. Let x_1, x_2 and x_3 be the corresponding thickness, and k_1, k_2 and k_3 their thermal conductivities respectively.

The expression for steady state heat conduction through the composite wall is;

$$Q = \frac{(t_1 - t_4)}{\dfrac{x_1}{k_1 . A} + \dfrac{x_2}{k_2 . A} + \dfrac{x_3}{k_3 . A}} = \frac{(t_1 - t_4)}{\sum \dfrac{x}{kA}}.$$

Q. 8. What is the expression for 'Radial heat conduction through the wall of a tube' ?

Ans. Consider a long tube of inside radius 'r_1' outside radius 'r_2' and length 'l'. Also let inside and outside temperatures are assumed to be constant as 't_1' and 't_2' respectively.

Then, the expression for 'Radial heat conduction through the wall of this tube' is :

$$Q = -2xk \frac{(t_2 - t_1)}{\log_e\left(\dfrac{r_2}{r_1}\right)}$$ where 'k' is the thermal conduction.

Q. 9. What is the expression for 'Heat conduction through a thick spherical shell' ?

Ans. Consider a thick spherical shell of inside radius 'r_1' and outside radius 'r_2' with temperatures 't_1' and 't_2' respectively. If 'k' is the term conductivity of the material then.

The expression for heat conduction through this thick spherical shell is;

$$Q = 4\pi k r_1 . r_2 \frac{(t_2 - t_1)}{(r_2 - r_1)}.$$

Q. 10. What is the 'Critical Insulation' ?

Ans. The 'Critical Thickness' is the insulation thickness at which the increase of insulation resistance is equal to the rate of decrease in film resistance, thus giving minimum value of the 'Net Resistance'.

Q. 11. What is 'Heat-convection' ?

Ans. It is the transportation and exchange of heat due to the mixing motion of different parts of the flowing fluid.

Q. 12. What is 'Natural or Free Convection' ?

Ans. When the motion of the fluid is caused by the difference in density or due to temperature, it is called 'Natural or Free Convection'.

Q. 13. What is 'Forced-Convection' ?

Ans. When the motion of the fluid is caused by some mechanical means such as a fan etc. it is called 'forced Convection'.

Q. 14. What is the 'Newton's heat-convection equation' ?

Ans. The Newton's heat convection equation is :

$$Q = h . A . \Delta T \text{ Watt}$$

where Q = Heat-transfered

h = Convective heat-transfer co-efficient in W/m^2 k

A = Area in m^2

Δt = Temperature Gradient in k.

Q. 15. What is the 'Heat-Radiation' ?

Ans. The term 'Heat Radiation' is applied to all kinds of processes which transmit energy by means of electromagnetic waves.

The amount of heat radiation mainly depends upon the nature of the body, its temperature and the kind and extent of its surface.

Q. 16. What is 'Absorptivity ?

Ans. The ratio of the energy absorbed to the total incident energy on a body is called 'Absorptivity'. It is generally denoted by 'α'.

Q. 17. What is 'Reflectivity' ?

Ans. The ratio of the energy reflected to the total incident energy on a body is called 'Reflectivity'. It is generally denoted by 'P'.

Q. 18. What is 'Transmissivity' ?

Ans. The ratio of the energy transmitted through a body to the total incident energy falling on it is called "Transmissivity'. It is generally denoted by 'τ'.

Q. 19. What is a 'Black-Body' ?

Ans. The Black-Body is defined as the body, which absorbs all the radiation incident upon it.

Q. 20. What is the 'Total Emissive Power' ? Also, write down the expression for the 'Stefan Boltzman's Law' for heat emission.

Ans. The total amount of radiation emitted by a body per unit time is defined as the 'Total Emissive Power'.

'The total emission from a black body per unit area and per unit time is proportional to the forth power of the absolute temperature of the body", it is called the 'Stefan Boltzmann's Law'.

Mathematically, it can be expressed as :

$$E_b = \sigma \cdot T_4$$

where 'σ' is a constant called 'Stefan Boltzman's constant and is equal to 5.67×10^{-8} W/m^2 k^4.

REVIEW QUESTIONS

1. What do you mean by 'thermal conduction' 'thermal convection' and 'thermal radiation' ?
2. Explain the phenomenon of heat transfer by conduction. Derive basic equation showing the method of its application in case of composite walls.
3. Derive an expression for heat conduction, through the cylindrical surface composed of two different materials.
4. Derive an expression for flow through a thick sphere, by conduction.
5. (*a*) What do you mean by total emissive power ?
   (*b*) Define convection. What are its different types ?
6. What is a black body ? Explain its concept with reference to emissivily, absorptivity and reflectivity by a black body.
7. Define Stefan Boltzmann's law. Give the value of Stefan Boltzmann's constant. **(Pb., Dec. 1999)**
8. Distinguish between :
   (*i*) Free and forced convection.
   (*ii*) Black and Gray bodies. **(SBTE Haryana, May 1995)**

PROBLEMS FOR PRACTICE

1. If the thermal resistance for a wall is 0.74 W/m^2 k. Calculate the heat loss in watts through the wall having an area of 36 m^2. **[Ans.** 2.664 W]

2. The glass windows of a room have total area of 10 m^2. While its thickness is 6 mm. Find the rate at which the heat escapes from the room per hour by conduction when the temperature difference is 20 °C, Take K for glass as 0.3 W/m k. **[Ans.** 225018 KJ/hr]

3. The inner and outer surface temperature of a 235 mm thick brick wall are 700 °C and 150 °C respectively. Calculate the heat loss per square metre of wall area given that the conductivity is 0.865 W/m k. **[Ans.** 2024.47 W/m]

4. A flat wall of a furnace is composed of two layers of different materials having thickness of 11.5 cm and 60 cm, with thermal conductivities of 0.16 W/m °k and 10.5 W/m °k respectively. If 1000 W/hr of heat passes through for every square meters area. estimate the drop in temperature at the contact layer between the two walls. The temperature inside the furnace is 1000 °C and that at outside layer 150 °C. **[Ans.** 74 °C]

5. A pipe having outside diameter of 15 cm is covered with 5 cm layer of insulations having thermal conductivity of 0.035 W/m °k. The pipe is 40 m long. The inside and outside surface temperature of the insulation are 320 °C and 15 °C respectively.
   (*i*) Neglecting axial heat transfer in the insulation calculate the radial conduction heat transfer rate in KJ/hr.
   (*ii*) What thickness of insulation is needed to reduce the heat loss to 5 KJ/hr. **[Ans.** (*i*) 18.91 KJ/hr, (*ii*) t = 44.25 cm]

6. Temperature difference on two sides of a wall measuring 5 m × 4 m, is 25 °C. If the thermal conductivity of the material of the wall is 0.0004 KW/mC, determine the thickness of wall if the ratio of heat flow is 5000 KJ/h.
   [Ans. 0.144 m]

7. A pipe having 20 cm external diameter is covered by an insulating material 2 cm thick. The thermal conductivity of the insulating material is 0.2 KJ/m/h/°C. If the inner and the outer surface temperatures of the insulating cover are 150 °C and 30 °C respectively determine the rate of heat loss per hour per metre length of pipeline. **[Ans.** 827.1 KJ/hr]

8. The external and internal diameters of a hollow sphere are respectively 400 mm and 300 mm. Temperatures within and outside the sphere are respectively 200 °C and 110 °C. Determine the rate of flow through the wall of the sphere. For the sphere material, take value of k = 0.00212 KW/m °C.
   [Ans. 5374385.4 KJ/hr]

9. The total area of the glass panes used in doors and windows of a bunglow is 25 m^2. The glass is 5 mm thick. Temperature maintained within the room by air conditioning is 15 C and outise atm-air temperature is 35 °C. Find the amount of heat flow into the bunglow through glass panes. Take k = 0.8 W/m-k. **[Ans.** 2888000 KJ/hr]

10. The temperature difference on two sides of a wall measuring 5 m × 4m is 25 °C. If the thermal conductivity of the material of wall is 0.0004 kW/m °C, find the thickness of the wall if the rate of heat flow is 5000 KJ/h.
    [Ans. 1.44 KJ/nm °C, 0.144 m]

□□□

EXPERIMENTS

EXPERIMENTS

EXPERIMENT NO: 1

OBJECT: Determination of temperature by:

1.1. Thermocouple

1.2. Pyrometer

1.3. Infrared thermometer

EQUIPMENT: A furnace, a platinum and platinum rhodium thermocouple, a disappearing filament optical pyrometer, an infrared thermometer and some hot objects.

1.1. Thermocouple:

THEORY:

When two dissimilar metal wires are joined at both ends, an electromotive force (e.m.f.) will appear, if the two ends are at different temperature. The magnitude of the e.m.f. is a function of the kinds of metals used, the length of the wires and the difference in temperature between the hot and cold junctions.

For a given pair of metal wires of fixed length a measurement of the current flow in the circuit will provide an equally satisfactory means of determining the difference between the hot and cold junctions.

It is not necessary for the thermocouple wires to be in contact with each other to form the cold junction. They may be separated by any conductor, provided that there is equality of temperature (thermal equilibrium) at all junctions of the third metal (conductor) and the thermocouple metals. In Fig. 1.1 the instrument acts as a conductor between wires A and B to form the cold junction. When a cold junction is not formed by thermocouple wires, it is common practice to speak of the junction as a reference junction rather than a cold junction.

Fig. 1.2 shows a simple circuit for a thermocouple pyrometer. Instead of connecting the milli-voltmeter directly as in the diagram, it is more accurate if a galvanometer is used. The galvanometer will be balanced for zero current by balancing the voltage of the junction with a voltage across a potentiometer. This balancing is automatically done by indicating and recording instruments available in the market. The accuracy of measurement depends very largely upon the sensitivity of the indicating instruments.

Diagram: (To be drawn on the unruled pages)

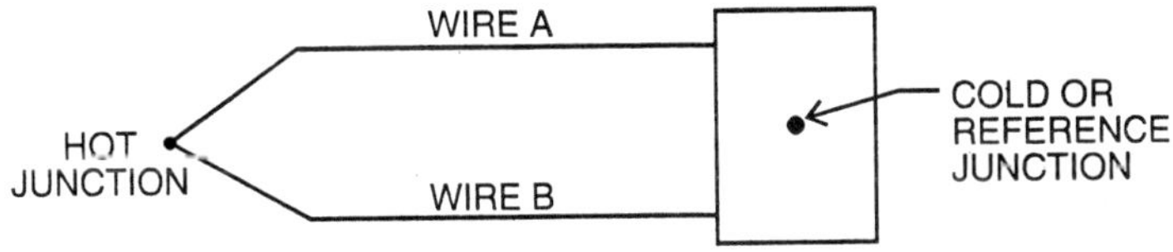

Fig. 1.1.

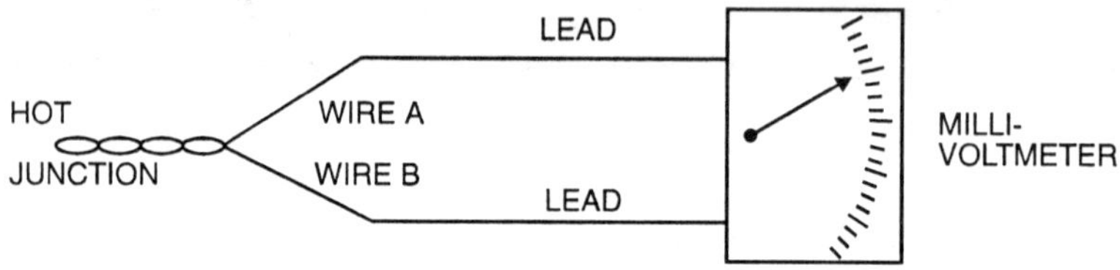

Fig. 1.2.

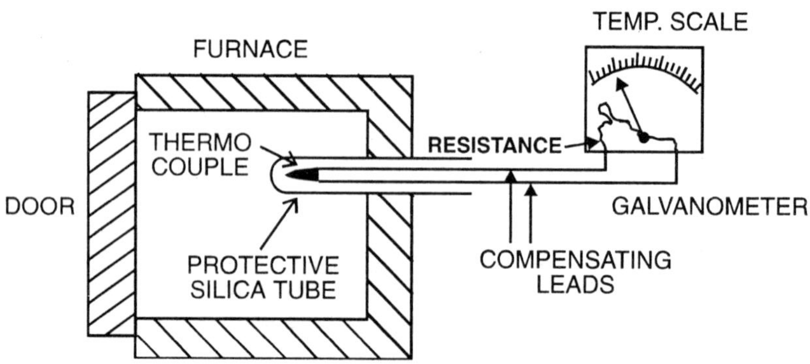

Fig. 1.3.

Use : A thermocouple is useful in determining surface temperatures and temperature of inaccessible locations by indicators located at the convenient points.

PROCEDURE:

- The hot junction (covered by a protective silica tube), made up of platinum and an alloy of platinum and rhodium welded together, is placed in the peep hole in the wall of the hot furnace whose temperature is to be measured (Fig. 1.3)
- The leads from the thermocouple wires are carried to a sensitive galvanometer which constitutes the cold reference junction. The galvanometer pointer deflects proportional to the amount of current flowing in the circuit due to difference of temperature between the two junctions. The galvanometer is so calibrated, that instead of indicating electrical units, it reads in degrees of temp. Thus, we are able to take the reading of the temperature existing inside the furnace directly with the help of the pointer of the galvanometer.

OBSERVATIONS:

(To be noted on unruled pages of practical note book.)

S. No.	Object	Reading of Temperature (in degree°C)
1		
2		
3		

PRECAUTIONS:

(*i*) Calibration of the instrument should be often got checked from some standard physical laboratory or be checked by measuring the temp. of an object of known temperature.

(*ii*) After use the instrument may be safely kept in a box meant for this purpose.

(*iii*) These instruments should never be roughly handled.

1.2. Optical Pyrometer :

THEORY:

In the case of disappearing filament optical pyrometer the brightness of a narrow spectral band of the heated surface is matched with the brightness of the filament of special electric lamp by adjusting the current supplied to the filament. The necessary amperage is a measure of the surface temperature of the heated body. The ammeter used to measure this current is directly calibrated in terms of degrees of temperature, thus making it a direct reading instrument. Figs. 1.4 and 1.5 illustrate the working principle of such a pyrometer.

Use : Optical pyrometers are advantageous when the hot body whose temperature is to be measured is inaccessible to a thermo-couple.

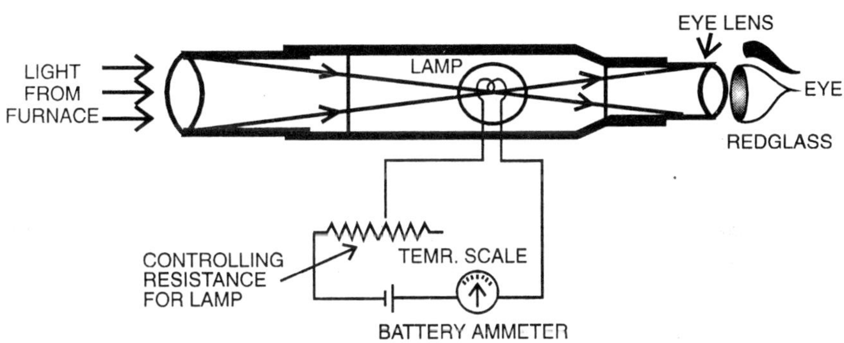

Fig. 1.4.

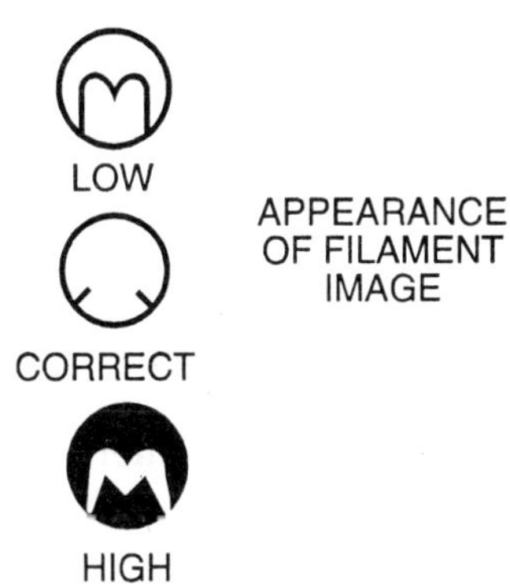

Fig. 1.5.

PROCEDURE :

- The instrument (as shown in Fig. 1.4) is set up in front of the furnace (or the hot object whose temperature is to be measured) and the light from the furnace is viewed through the eye-piece.
- The current through the lamp is varied by a resistance and when a colour match is obtained the lamp filament disappears from sight. The lamp current required to cause this is to be indicated on the ammeter. The ammeter being so calibrated that it reads directly in degrees of temperature. Thus, we are able to take reading directly of temp. of the hot object or furnace from a distance by viewing through this instrument and adjusting the glow of the filament.

OBSERVATIONS:

(To be drawn on unruled page)

(*ii*) Optical Pyrometer:

S. No.	Object	Reading of Temperature (in degree°C)
1		
2		
3		

PRECAUTIONS:

(*i*) Calibration of the instrument should be often got checked from some standard physical laboratory or be checked by measuring the temp. of an object of known temperature.

(*ii*) After use the instrument may be safely kept in a box meant for this purpose.

(*iii*) The instrument should never be roughly handled.

(*iv*) Carefully adjust the resistance of the optical pyrometer electric circuit, so that you obtain proper matching of the light from the object and the filament image as illustrated in Fig. 1.5.

1.3. INFRARED THERMOMETER:

THEORY:

The infrared thermometer measures the radiation emitted by the hot body and by using the Stefan Boltzman's Law of heat radiation the temperature of the hot body is calculated.

Its working principle is similar to that of an optical pyrometer, with a difference that here the heat radiations pass through a filter and got rectified the ultra violet and visible light but allows the infrared rays, having wave length more than the red light which is 800 n.m.

PROCEDURE:

(*i*) The image of the hot body is focused, through the filler with the help of the mirror on the junction in front of it and observed through an eye piece which is fitted in the central opening of the concave mirror. A circular aperture is formed as shown in the Fig. 1.6. Now, when the focusing is correct, the images of the two mirrors will overlap exactly to form a complete circle as shown in Fig. 1.7 (*b*) but if the focussing is incorrect, the two half-images appear to be displaced or mismatched as shown in Fig. 1.7 (*c*).

(*ii*) After achieving the correct focussing, note down the deflection on the galvanometer, θ.

(*iii*) Now, by applying Stefan Boltzman's Law of heat reaction, i.e., $\theta = \sigma (T^4 - T_0^4)$, (where the value of σ lies from 3.8 to 4.2, whereas T and T_0 are the temperatures of the hot body and the temperature of the pyrometer case respectively), the temperature of hot body can be calculated.

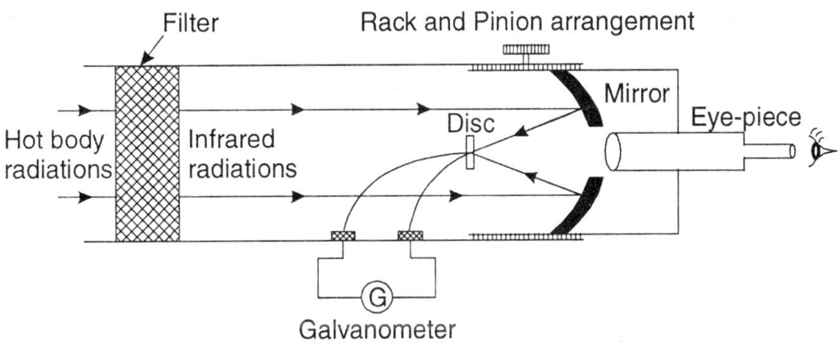

Fig. 1.6

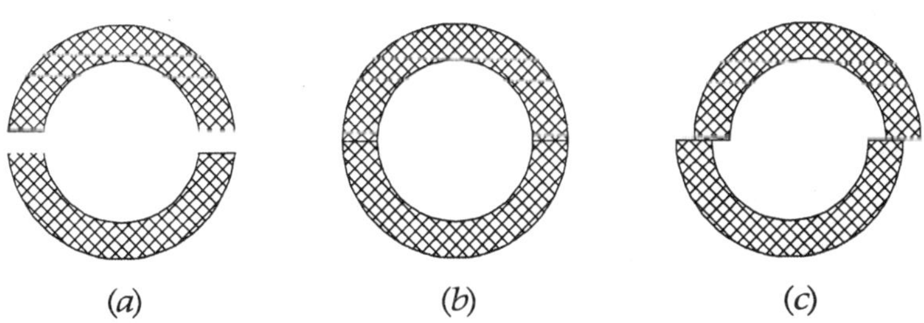

Fig. 1.7.

Precautions :

(*i*) The calibration of the galvanometer should be often checked from a Standard Physical Laboratory.

(*ii*) Handle the instrument very carefully.

(*iii*) Keep the instrument in a box meant for it, after its use, immediately.

EXPERIMENT NO: 2

OBJECT:

Demonstration of mountings and accessories on a boiler.

EQUIPMENT:

A boiler fitted with all its mountings and accessories.

THEORY:

Boiler mountings are the components, generally mounted on the surface of the boiler to accomplish safety during operation and the complete control of the steam generation process.

These are the essential parts of the boiler, without which the boiler operation is not possible.

The following are the various mountings of a boiler.

(*i*) Water level indicator

(*ii*) Pressure Gauge

(*iii*) Safety valve

(*iv*) Junction valve or stop valve.

(*v*) Feed check valve.

(*vi*) Blow off cock

(*vii*) Fusible Plug.

Boiler accessories are those appliances which are installed with a boiler in its neighbouring area to increase the efficiency of the boiler.

These are not the essential parts of the boiler i.e. without installing these the boiler operation can be accomplished although at a lower efficiency.

The following are the various accessories generally installed with a boiler.

(*i*) Feed water pump

(*ii*) Injector

(*iii*) Economiser

(*iv*) Super heater

(*v*) Air preheater

(*vi*) Pressure reducing valve

(*vii*) Steam Drier or separator

(*viii*) Steam Trap

BOILER MOUNTINGS

Boiler mountings are the components generally mounted on the surface of the boiler to accomplish safety during operation and the complete control of steam generation process.

These are the essential parts of the boiler, without which the boiler operation is not possible. The following are the various important mountings of a boiler:

(*i*) Water Level Indicator

(*ii*) Pressure Gauge

(*iii*) Safety Valve
(*iv*) Junction or Stop Valve
(*v*) Feed check Valve
(*vi*) Blow-off-Cock
(*vii*) Fusible Plug

(*i*) WATER LEVEL INDICATOR

It is the device which indicates exact level of water in the boiler. A glass tube is fitted on it, to make the water level inside the boiler visible from the boiler room floor. A common type of glass tube water level indicator is shown in Fig. 2.1.

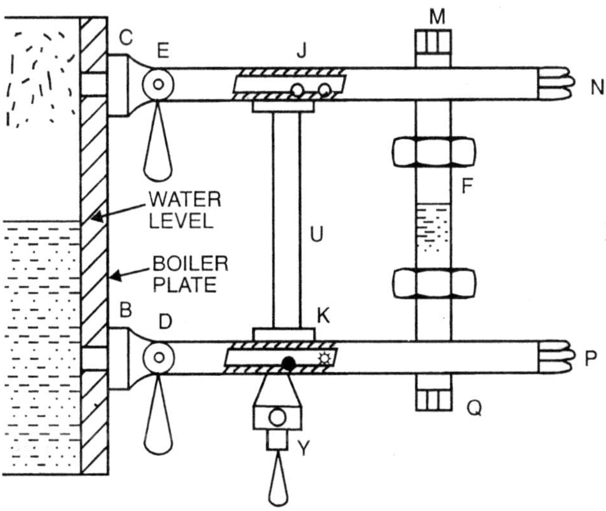

B,C	=	FLANGE
D	=	WATER COCK
E	=	STEAM COCK
M,N,P,Q	=	SCREW CAPS
F	=	GAUGE GLASS
J	=	STEAM OPENING
K	=	WATER OPENING
U	=	METAL TUBE
Y	=	DRAIN COCK

Fig. 2.1.

It is connected to the boiler plate through stuffing boxes in hollow gun metal casting having flanges for bolting to the plate. There are two cocks for controlling the passages of water and steam from the boiler. When these cocks are opened the water stands in the glass tube at the same level as inside the boiler. In case the glass tube gets broken the rush of water from bottom passage and steam from top passage carries the balls in the position shown by dotted circles and shuts off the water and steam. Then the attendant can safely close the cocks and replace the broken glass tube.

(*ii*) PRESSURE GAUGE

This gauge is usually mounted on the front top of the boiler shell. It is mounted on the boiler to show the pressure of steam inside the boiler. Its dial is graduated to read the pressure in different units of measurement. Bourdon's pressure gauge as shown in Fig. 2.2 is commonly used.

The essential element of this gauge is the elliptical spring tube, which is generally made of bronze and is solid drawn. One end of this tube is attached by links to a toothed quadrant which in turn meshes with a small pinion on the central spindle. The other end is connected to steam space. When pressure is applied to the interior of the elliptical tube it tries to straighten up and causes the pointer to move along the scale and indicates the pressure on graduated dial.

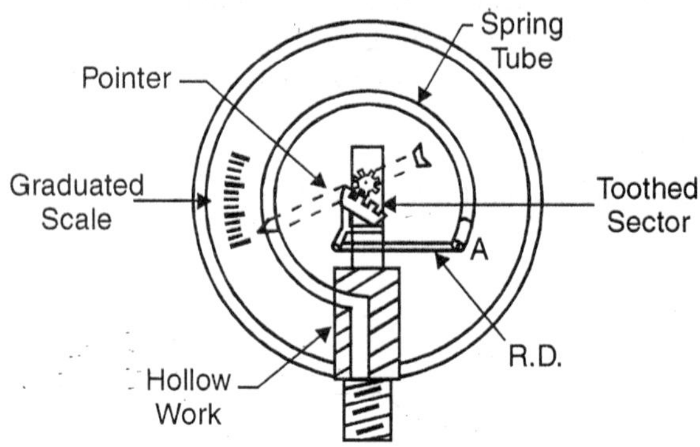

Fig. 2.2.

(*iii*) SAFETY VALVES

All the boilers except the small boilers should have at least two safety valves which will open, when the boiler pressure exceeds the working pressure. Safely valves are classified into two types according to the method of loading the valve i.e.:

(*i*) Weight loaded

(*ii*) Spring loaded.

The weight loaded safety valves are further of following two types:

(*a*) Dead weight safety valves.

(*b*) Lever and weight types of safety valves.

(*a*) Dead Weight Safety Valve

In this type of safety valve, the valve is loaded by the direct application of dead weight as shown in the Fig. 2.3.

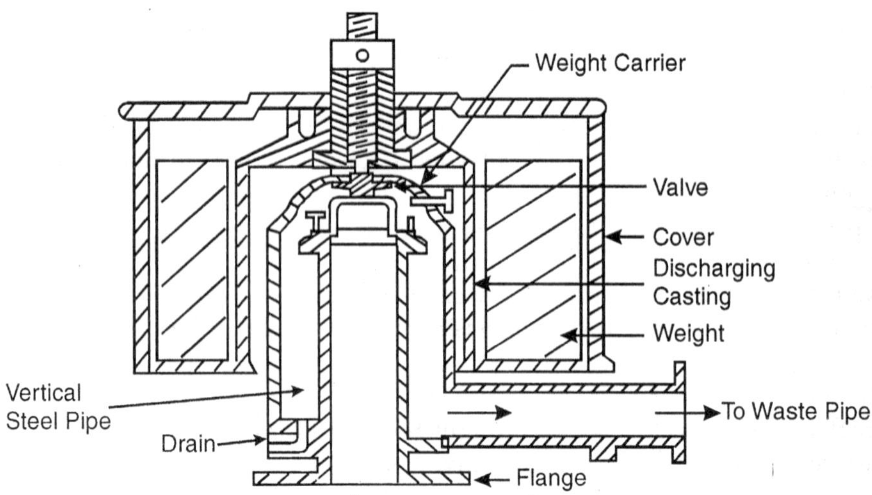

Fig. 2.3. Dead Weight Safety Valve.

In this safety valve the value rests on the valve seat which is fixed on the top of the vertical steel pipe by a ring and screws. The dead weight and the weight of the valve itself balances the pressure on the valve. When the pressure becomes too high, the valve and weight will lift and the surplus steam will escape and it is carried to the waste pipe. Thereafter, the valve again comes down on its seat.

(*b*) Lever Safety Valve

This type of safety valve is suitable for stationary boilers. Lever safety valve is shown in the Fig. 2.4.

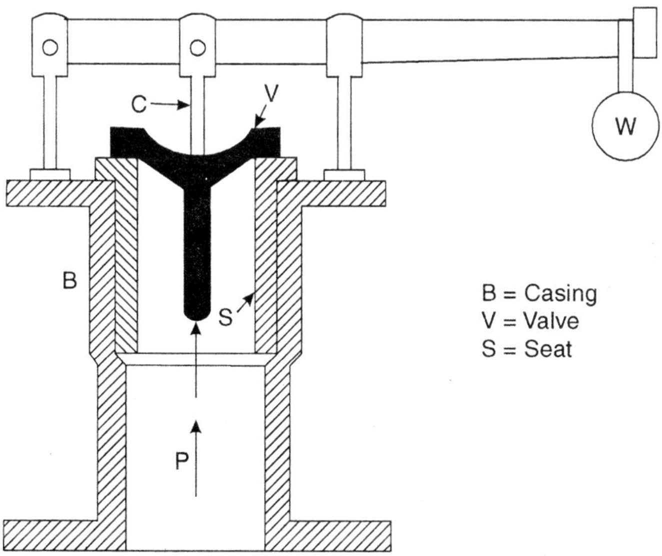

Fig. 2.4. Lever Safety Valve.

The valve is loaded by means of weighted lever which is pivoted about a fulcrum situated close to the valve. The valve load depends upon the lever ratio. Its working and function is the same as discussed in case of dead weight safety valve.

(*c*) Spring Loaded Safety Valve

Spring loaded safety valves are commonly used on locomotive and marine boilers because this type of safety valve is not affected by movement and vibrations. It is shown in the Fig. 2.5.

Spring loaded safety valve consists of two separate valves and seatings. The valves are held down on their seats by the helical spring and lever. The spring is hooked to an arm of the lever midway between the valves. The lower end of the spring is hooked to the shackle which is secured to the valve chest by studs and nuts.

Its disadvantage is that the load on the valves increases as the valve lifts where the steam pressure exceeds the limit. So the pressure required to just lift the valve is less than required to open it fully.

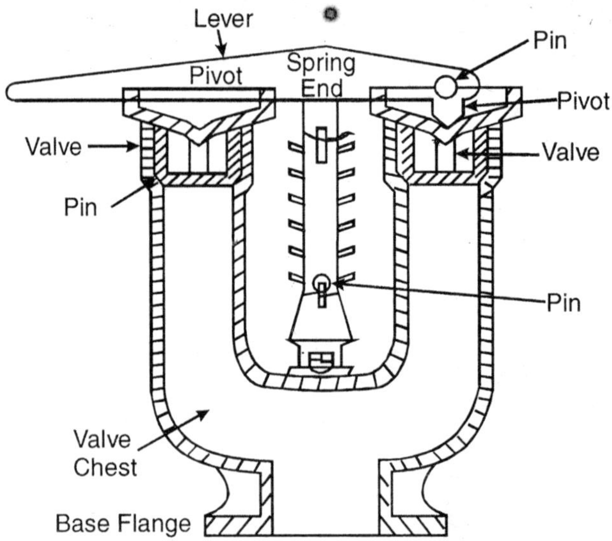

Fig. 2.5. Spring Loaded Safety Valve.

(*d*) Combined High Stream and Low Water Safety Valve

This type of safety valve is also called compound safety valve. It is suitable for stationary boilers.

The principle of working of this safety valve is also the same as lever safety elements are present in this type of safety valve. Constructional view of this type of safety valve is shown in Fig. 2.6.

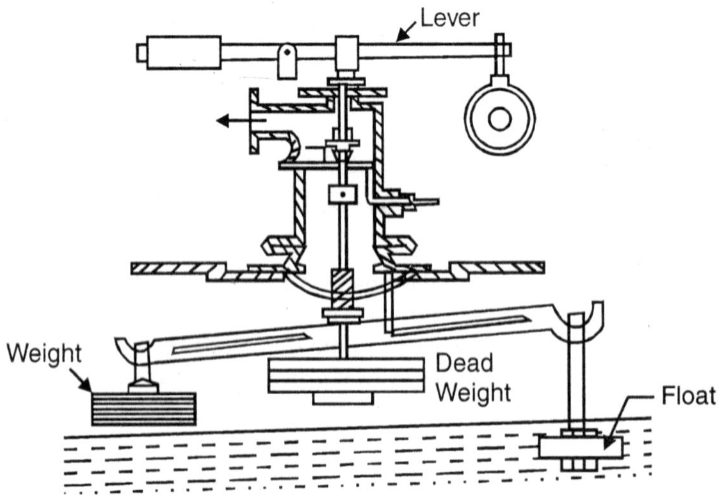

Fig. 2.6. Combined high steam and low water safety valve.

The first safety element consists of high pressure steam valve held down on its seat partly by the dead weights and partly by lever and weight. The second safety element consists of a float device which gives warning of the near approach of dangerously lower water level and then prevents the collapse of the furnace tubes.

(*iv*) Junction or Stop Valve

A valve placed directly on a boiler and connected to the steam pipe which carries steam to the engine or turbine is called junction valve. This valve is necessary for shunting off steam when not required. It is also called as stop valve. A common type of junction valve is shown in Fig. 2.7.

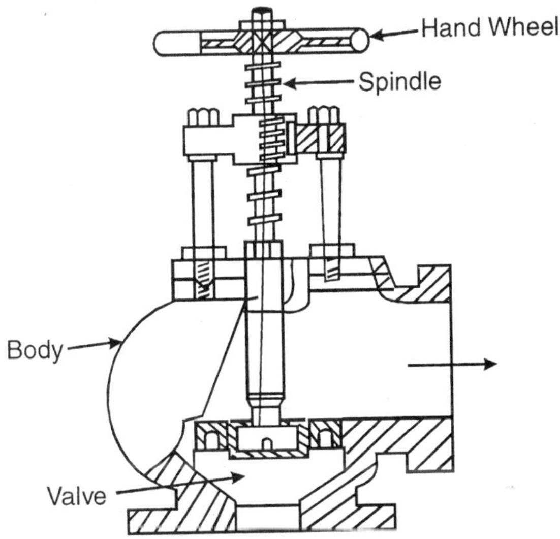

Fig. 2.7. Steam stop or Junction Valve.

The valve spindle passes through a gland and stuffing box fixed in the cover of the valve body. The upper portion of the spindle is threaded. By turning the hand wheel fitted on the spindle, the spindle is raised or lowered and thus the valve disc is raised or lowered to allow the flow of steam or to stop the flow of steam.

(*v*) Feed Check Valve

A valve placed at the boiler end of the feed pipe to control or regulate the flow of water is called the feed check valve. It is a non-return valve which permits flow of water in one direction only and automatically prevents the back flow of water from the boiler when the feed pump is not working. The constructional detail of the feed check valve is shown in Fig. 2.8.

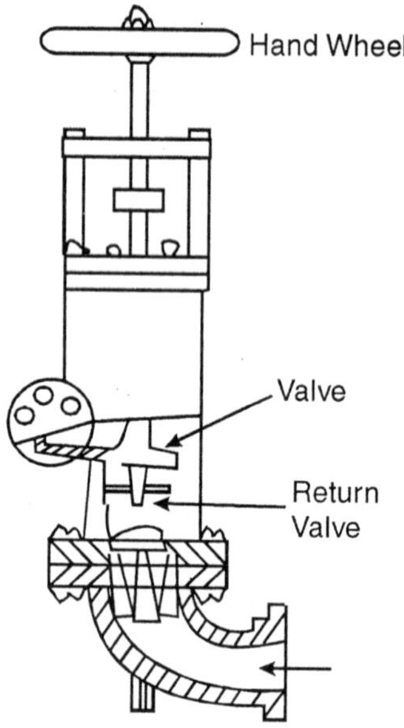

Fig. 2.8. Feed check Valve.

(*vi*) Blow-Off Cock

The constructional detail and self-explanatory working diagram is shown in Fig. 2.9.

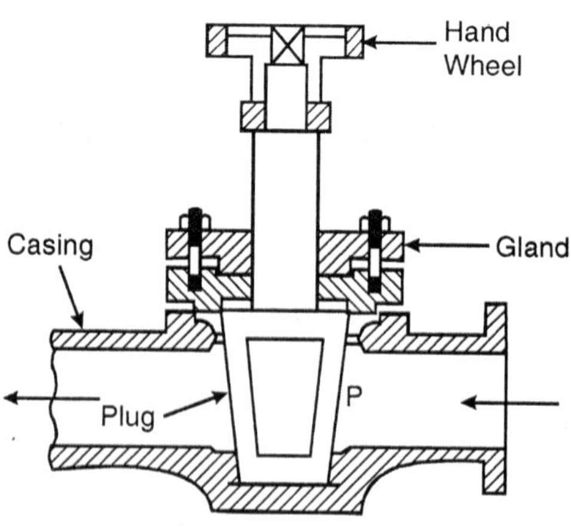

Fig. 2.9. Blow-Off Cock.

The blow-off cock is attached at the lowest joint of every boiler for the purpose of emptying the boiler when necessary and for discharging mud etc. deposited from the feed water at the lowest point.

(*vii*) Fusible Plug

It is the mounting fitted at the crown of the boiler as shown in the Fig. 2.10.

The fusible plug under normal working of a boiler is covered with water which keeps the temperature of the fusible metal below its melting point. But when the water level in the boiler falls too low to uncover the top of the plug the fusible metal quickly melts, the plug drop out and the opening so made allows the steam to rush into the furnace. The steam in this way puts-off the fire or gives warning that the crown of the boiler furnace is in danger of being overheated.

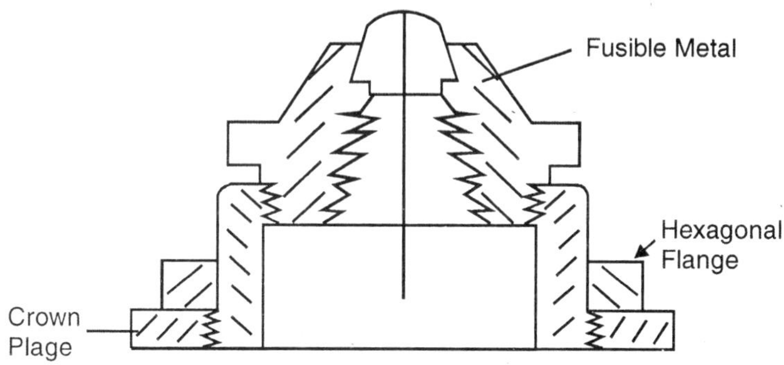

Fig. 2.10. Fusible Plug.

BOILER ACCESSORIES

Boiler accessories are those appliances which are installed with a boiler in its neighbouring area to increase the efficiency of the boiler.

These are not the essential parts of the boiler, so without these the boiler operation can be accomplished although at a lower efficiency. The following are the important boiler accessories:

(*i*) Feed water pump.
(*ii*) Injector
(*iii*) Economiser
(*iv*) Super heater.
(*v*) Air-preheater
(*vi*) Pressure-reducing valve
(*vii*) Steam drier or separator
(*viii*) Steam trap

(*i*) Feed Water Pump

It is used to deliver feed-water to the boiler. It may be either reciprocating or rotary pump. One type of reciprocating feed water pump has been shown in Fig. 2.11.

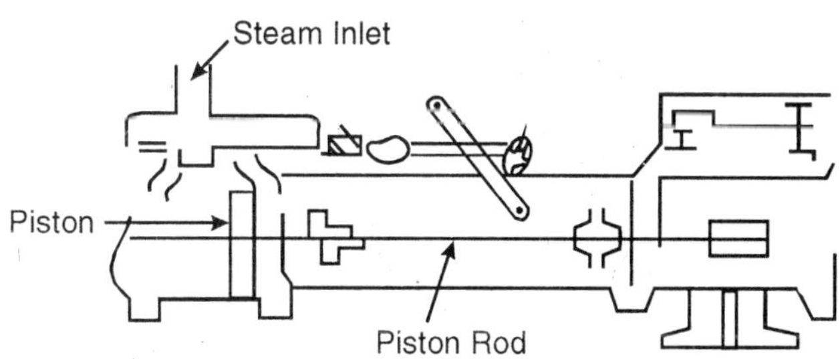

Fig. 2.11. Feed water pump.

In this type of pump there are simple steam engine cylinder and pump cylinder placed side by side. The working of engine cylinder is similar to the working of steam engine. The pump delivers water on each forward and backward stroke, being double acting. There are suction and discharge valves for each side of the pump plunger. The two pumps work alternately, so a continuous flow of water is kept practically.

Rotary feed pumps are of centrifugal type, driven directly by a small steam turbine or by an electric motor. These pumps are also known as turbine pumps.

(*ii*) Injector

It is used in small boilers and locomotive boilers. It can also be fitted as a stand by or as a reserve feed water pump owing to its cheapness and simplicity. It is very simple in working because there are no moving parts or plungers in it. The water is forced into the boiler by the action of steam flowing through a tapered nozzle as shown in the Fig. 2.12.

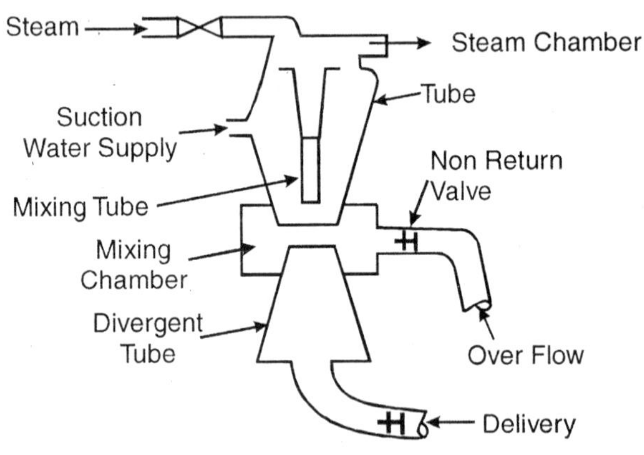

Fig. 2.12. Steam Injector.

The injector is more economical because whole of the heat of the steam utilized in injector is returned to the boiler. The injector, therefore, works as a feed water heater as well as feed water pump. The only disadvantage of this is that, these are not very reliable. Therefore, these are not used in large boilers or marine boilers where large quantity of feed water is required to be pumped. In the working of the injector the steam is admitted through the steam pipe and discharged from the nozzle, and it mixes with the water contained in mixing tube, where condensation of steam takes place. The jet then passes through the mixing chamber and enters the divergent tube which reduces the velocity and increases the pressure of water. Due to this increase of pressure the non-return valve fitted in the delivery pipe is opened and the water enters into the boiler.

(*iii*) Economiser

A considerable heat of the plant is carried away by the flue-gases. So, to utilize this heat, an economiser is placed in the path of the flue gases to pre-heat the feed water. It consists of a number of horizontal or vertical tubes through which the feed water from the pump is passed. The hot gases pass over the external surface of water tubes. The economiser shown in Diagram 2.13 is of vertical type.

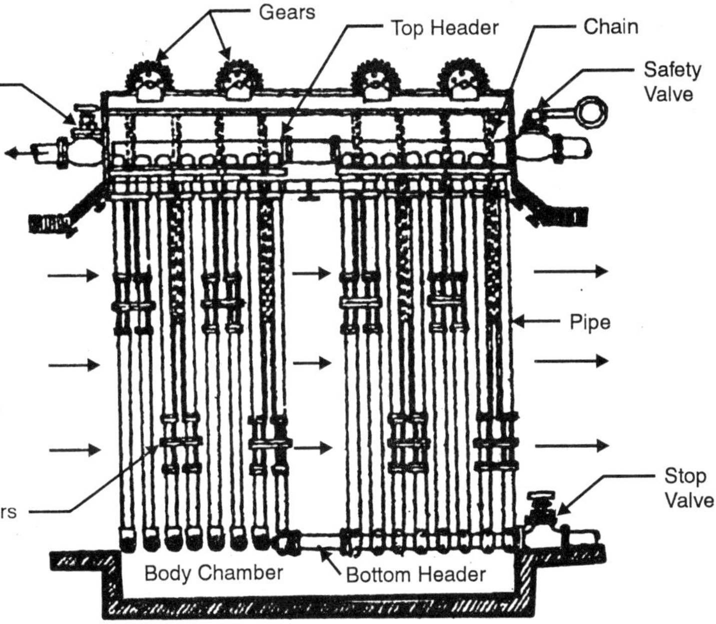

Fig. 2.13. Economiser.

It is fitted with scraper which moves up and down the water tubes by a mechanical drive and in this way the exterior surface of the tube are kept free from root deposits. The advantages of economisers are:

- It increases the heating surface of the boiler and therefore evaporative capacity of the boiler is increased.
- The temperature difference between different parts of the boiler is reduced and, therefore, stresses due to unequal expansion are reduced.
- The heat carried away by the flue gases is taken up by the economiser which results in saving of heat energy. This saving of heat energy results in an increase of the overall efficiency of the plant.

(iv) Super Heater

Super heater used for stationary boiler is shown in Fig. 2.14.

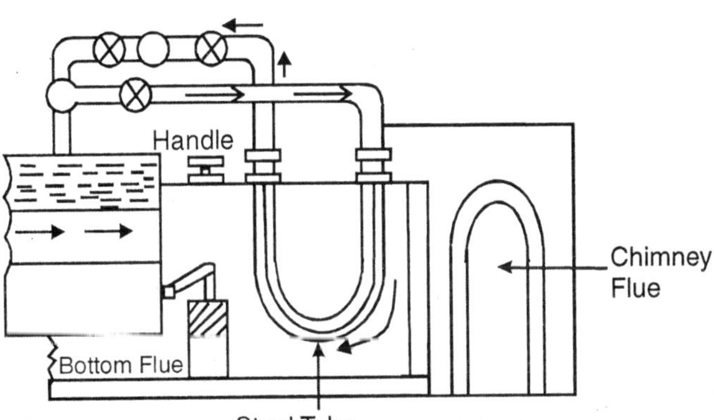

Fig. 2.14. Super Heater.

The use of superheaters enables the wet steam to be completely dried and temperature is raised at constant pressure. The super heater is a nest of steel tubes bent to U-form and expanded into mild steel boxes called headers. These are generally located in the path of furnace gases so that heat is utilized by the superheater from the hot gases.

(*v*) Air-preheater

The air entering the boiler furnace is at a lower temperature as compared to the temperature of exhaust flue gases. Instead of heating this air after it enters the furnace and take part in combustion, it will be desirable to heat this air to some extent. The air is heated by the heat carried away by the flue gases and going as a waste through the chimney.

It is situated between the economiser and the chimney. The plant efficiency is increased by the use of air pre-heater. Air pre-heaters are generally of three types:

1. Tubular Type
2. Plate Type
3. Regenerative Type

A tubular type air pre-heater is shown in the Fig. 2.15.

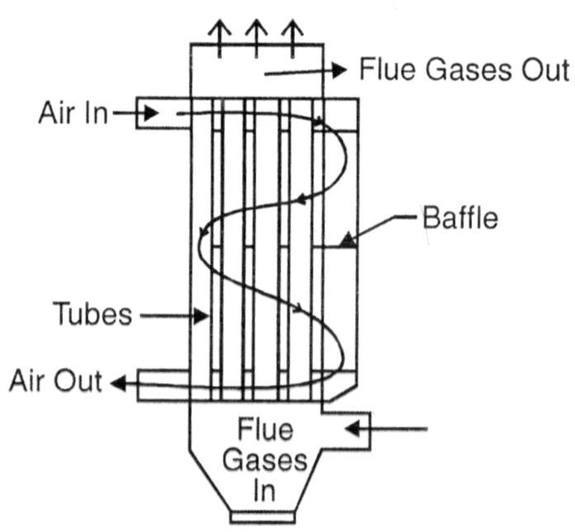

Fig. 2.15. Tubular Air Pre-heater.

Air preheaters are effective and economically justified only in the large steam power plants where high capacity and high pressure are met with. They are not useful for small plants operating at low pressures.

(*vi*) Pressure-Reducing Valve

Sometimes steam is required at a lower pressure than that of the boiler for processing work in an industry or heating purpose. The high pressure steam may be passed through a pressure reducing valve which will maintain a constant reduced pressure on the delivery side.

Fig. 2.16 shows one form of pressure reducing valve in which high pressure steam entering through left hand opening is throttled by the valve 'V' to desired low pressure through the right hand opening.

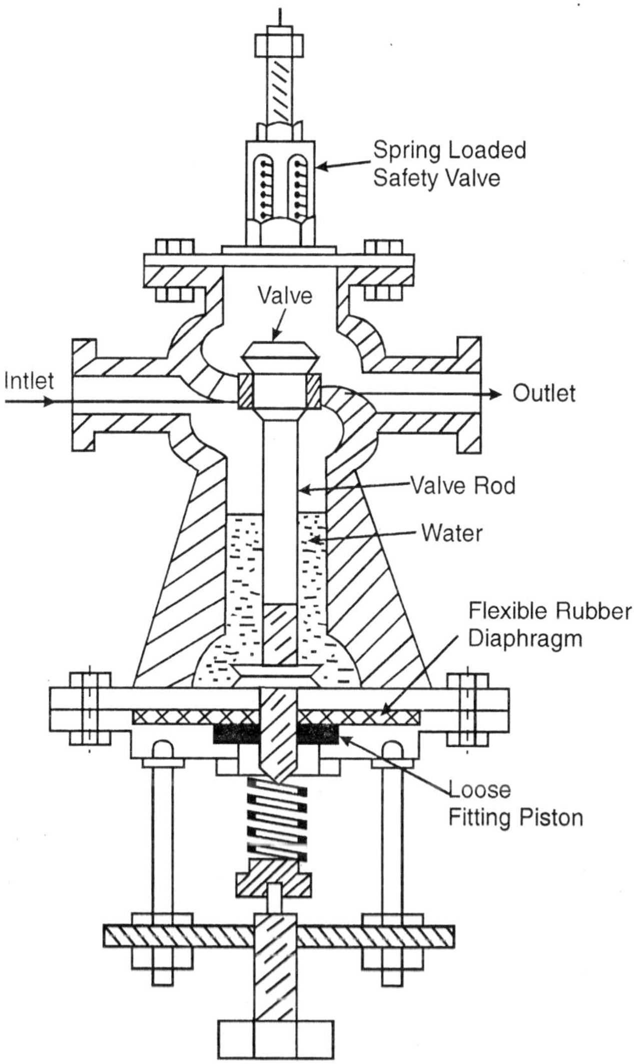

Fig. 2.16. Pressure Reducing Valve.

The valve is connected to a loose fitting position (LFP) through the valve rod (VR). The flexible rubber diaphragm (FRD) prevents leakage past the piston. Since, the diameter of the piston is equal to the diameter of the valve, so upward pressure of steam on the valve is balanced by its downward pressure on the piston. When steam is flowing through the reducing valve, the pressure of reduced pressure steam on the top of the valve balanced by the force of the spring. The force exerted by the spring may be adjusted to the desired value with the help of adjusting screw.

Any rise or fall of steam pressure at inlet will be followed by a momentary rise or fall of pressure on the top of the valve, which being forced upward by the constant force of the spring, will fall or rise till the steam pressure above the valve becomes the pressure to which the spring is adjusted.

(*vii*) Steam Drier or Separator

The function of a drier or steam separator is to separate water particles from steam before it is supplied to a steam engine or turbine. It is provided as close as possible to the engine in the main steam pipe line leading from the boiler to the engine.

A simple form of steam separator is shown in Fig. 2.17.

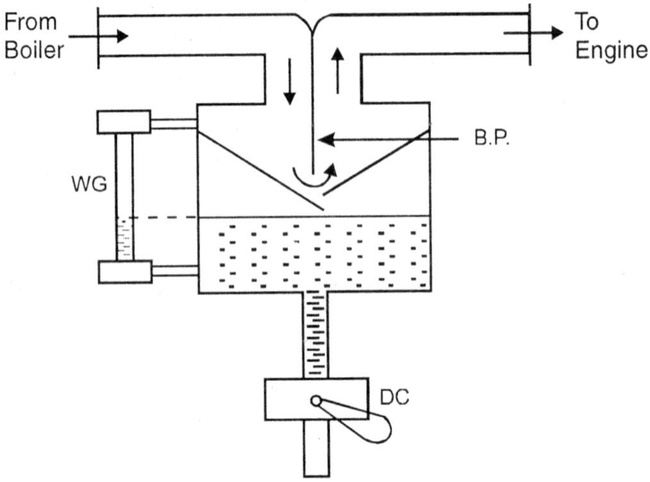

Fig. 2.17. Steam Drier or Separator.

It consists of a cylindrical vessel (CV) fitted with baffle plates (BP). The steam striking the baffle plate is suddenly deflected so that its direction of flow is changed. Consequently the water particles, due to their greater weight and interia after striking with baffle plates fall to the bottom of the separator. The water gauge indicates the level of water collected at the bottom of separator. The drain cock (DC) fitted at the bottom of the separator helps to drain away the water separator from incoming steam. The steam free from water particles goes to the engine through the outlet pipe as shown in the Fig. 2.17.

(*viii*) Steam Trap

Steam traps are devices used to automatically drain away or return to the boiler, the water collected in the steam pipe or steam jacket as a result of partial, condensation of steam, without allowing the steam to escape.

Figure 2.18 shows a common type of steam trap known as bucket steam trap. It consists of a casing (C) which contains sufficient water, to float the bucket (B) which carries a vertical spindle (VS). The valve (V) fitted to the top of the spindle fits in a pipe (P). The guide tube (GT) guides the vertical spindle (VS).

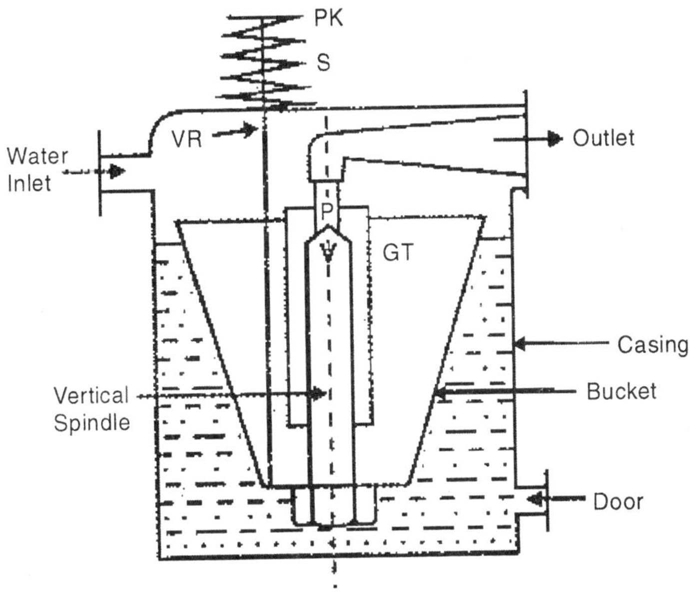

Fig. 2.18. Steam Trap.

The steam trap is so located that the water formed due to condensation of steam in the steam pipe, flows into it by gravity. When the water rises high enough in the casing (C), it overflows into the bucket-(B) which ultimately sinks down and the valve (V) opens. The water is then forced up through the guide tube (GT).

The door (D) provided at the lower end of the trap is used to remove periodically the sediments or dust deposited at the bottom of the casing. The trap can be tested at any time by blowing-off by pressing the knob by hand.

Precautions :

(*i*) As far as possible study the mountings and accessories in as mounted conditions, and do not disturb their settings.

(*ii*) If due to some reason the setting of any part gets disturbed again set it with the guidance of boiler incharge and properly check it again before boiler operation.

(*iii*) Mishandling should be avoided.

(*iv*) If facing any problem take advice from the boiler incharge.

EXPERIMENT NO: 3

OBJECT:

Study of boilers (through industrial visits)

Visit to sukhjit starch & chemical mills ltd. phagwara

The Sukhjit Starch & Chemical Mills Ltd. is situated on G.T. Road Phagwara near the bus stand of city. It is producing Starch and Glucose since last five decades.

The mill is utilising 'Cochran' and 'Fludised' Bed Combustion Boilers for steam production required for the process of starch production.

(*i*) COCHRAN BOILER

The cochran boiler is an improvement over the simple vertical boiler in which the heating surface is greatly increased by providing a large number of parallel tubes. Fig. 3.1 shows the brief outline of such a boiler. The combustion of coal takes

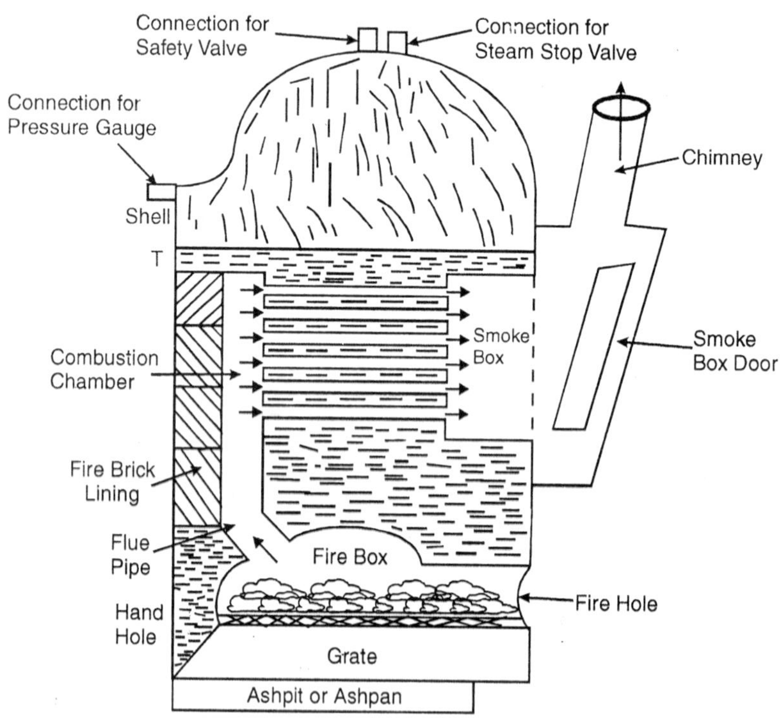

Fig. 3.1 Cochran Boiler.

place on the grate (G) inside the fire box (FB) which is hemispherical in shape so that the unburnt fuel, if any, is deflected back to the grate. In order to ensure complete combustion the flue gases from the fire box are not allowed to pass directly into the fire tubes (T) but via the combustion chamber (CC) which is provided with a fire brick lining (FBL) to prevent the shell (S) from being damaged due to overheating by the flue gases. The latter then pass through a large number of horizontal fire tubes and in doing so convert water into steam which collects over the water space at the boiler from where it can be taken out through the connection for steam stop valve (CSSV). Finally, the flue gases escape to the atmosphere via the smoke box (B) and the chimney (C). The smoke box door (SBD) enables the cleaning and inspection of the smoke box and the fire tubes. The crown of the shell (S) is also hemispherical in shape. Hence, for a given weight of the plate material of the shell it has the maximum space and strength. At the top of the crown of the shell there is a man hole (M) through which a man can enter the boiler and clean it.

The coal can be introduced on the grate through the fire hole (FH). The boiler may also be oil fired in which case an oil burner will be fitted at the fire hole, there will be no grate and the connections of safety valve and pressure gauge are also shown in the diagram. This boiler is provided with usual mountings and is a self-contained unit. It is a made in sizes up to about 2.743 m diameter and 5.791 m height. Its maximum evaporative capacity from and at 100°C is 4735 kg. per hour.

The ratio of heating surface to grate area varies from 10 to 25 in this case. The steam may be generated upto a pressure of 10 kgf/cm^2. It also requires a smaller floor area and is the most popular form amongst the vertical boilers. The design is quite compact possessing a great external and internal accessibility.

(*ii*) FLUIDISED BED COMBUSTION BOILER

The constructional features of this type of boiler are shown in Fig. 3.2. It is a water tube boiler. The speciality of this type of boiler is that in this boiler cheaper and easily available fuels like rice husk, saw dust or ground nut shell etc. can be used inspite of coal or coke that are being used in almost all type of conventional boilers.

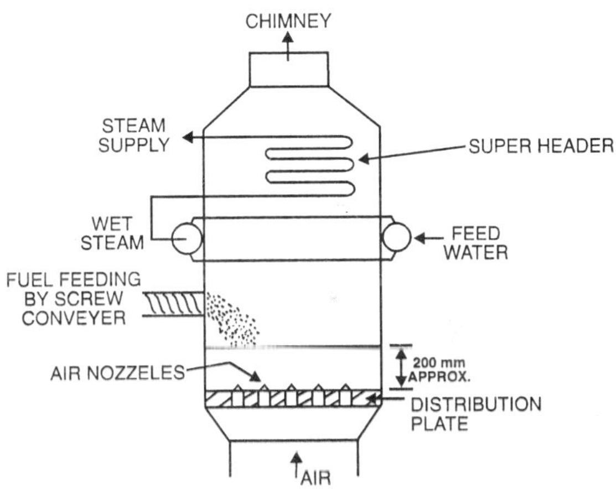

Fig. 3.2.

The fluidised bed consists of a rectangular box, bottom of which is the distributor plate. Air nozzles are arranged in a specific order on this distributor plate. This plate also has provisions for the fuel nozzles through which the fuel is admitted pneumatically from below the fluidised bed in such a way, so as to cover the length and breadth of the furnace resulting in optimum combustion.

In some cases the fuel may be admitted over the bed with the help of screw conveyer as shown also inspite of at the bottom of bed. The bottom of the distributor plate consists of air box, through which high pressure air from forced-draft fan is admitted into the fluidised bed via the air nozzles.

The temperature of the fluidised bed which is normally 800 to 860°C is controlled by immersing the water bed coils, which also helps in steam generation. The flue gas emerging from the bed then passes through the convective zone of the boiler, the heat recovery system, the dust collection system and finally the induced draft fan before finally passing into the atmosphere through the chimney.

EXPERIMENT NO: 4

OBJECT:

Study of air compressors.

Equipment:

A single stage reciprocating compressor, double stage reciprocating compressor, multi-stage reciprocating air-compressor, fan, rotary displacement blower, turbo blower, vane-type blower, centrifugal compressor, axial flow compressor.

Study:

Description of Reciprocating Air Compressor

Single Stage Reciprocating Air Compressor

(*i*) **Single Stage, Single Acting Reciprocating Air Compressor.** A single stage single acting reciprocating compressor has been shown in Fig. 4.1.

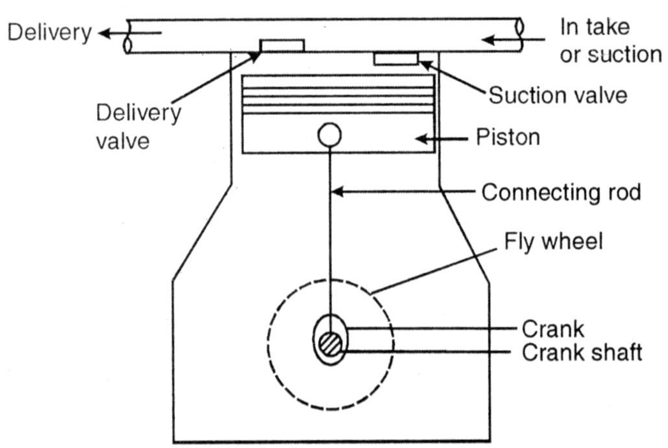

Fig. 4.1. Single stage acting reciprocating air compressor.

It is a positive displacement compressor from which the delivery is intermittent. It essentially consists of a piston and cylinder. The piston reciprocates inside the cylinder. It is driven by crankshaft through a connecting rod. There are two valves mounted in the head of the cylinder. These valves are the suction valve and the delivery valve. Both the valves, usually operate as a result of pressure difference across them.

(*ii*) **Single Stage Double Acting Reciprocating Air Compressor.** When the atmospheric air is compressed to the desired pressure in a single operation and the compression is carried out by the two sides of the piston, then the compressor is known as 'single stage, double acting reciprocating air compressor'. A single stage double acting reciprocating compressor has been shown in Fig. 4.2.

In this compressor, air is compressed by both the faces of the piston as it reciprocates inside the cylinder. The crank end side of the cylinder is provided with a stuffing box through which the piston rod reciprocates. The function of stuffing box is to make the crank end side of the cylinder air tight. The piston rod is connected to the connecting rod through a cross-head.

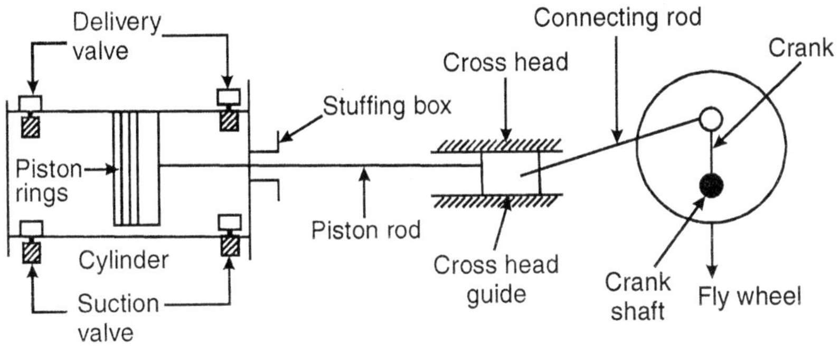

Fig. 4.2. Single stage double acting reciprocating air compressor.

Multi Stage Reciprocating Air Compressor

When the required delivery pressure of compressed air is achieved by the use of two or more than two stages/cylinders, then the compressor is said to be the 'multi-stage compressor'.

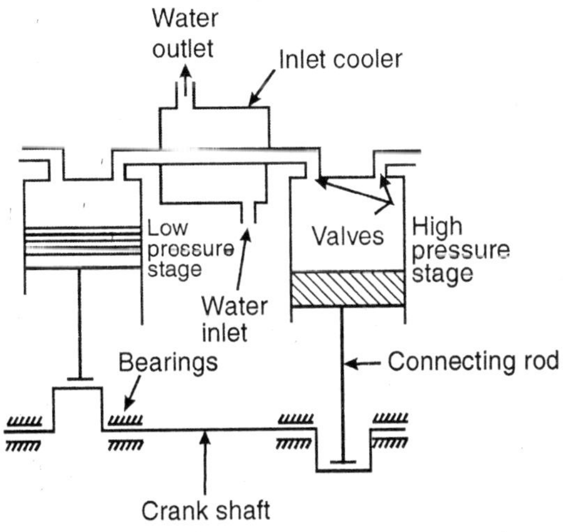

Fig. 4.3. Two stage reciprocating air compressor.

Multi-stage compressor is employed whenever it is not possible to achieve the required very high pressure of air in a single stage compressor.

A two stage reciprocating compressor with intercooler has been shown in Fig. 4.3.

In this the first cylinder is known as the low pressure cylinder and it is the first stage of compression. The second cylinder receives the compressed air through the intercooler and is known as high pressure cylinder. It is the second stage of compression. The dimensions of the H.P. cylinder is smaller than that of L.P.

Cylinder, because the volume of air is decreased as it is compressed. The intercooler saves amount of work required during the second stage of compression. The L.P. and H.P. cylinders are coupled and are driven by the same prime-mover.

Rotary Air Compressors

The rotary air compressors run at very high speeds. These are used for supplying large quantity of air at relatively low pressures. Due to their high speed these can be directly coupled with steam turbine or gas turbine shafts. Pressures upto nearly 11 bar can be obtained with a multi stage turbo compressor. Rotary compressors may be subdivided into

(*i*) Fans

(*ii*) Rotary displacement blowers

(*iii*) Turbo blowers

(*iv*) Turbo compressors.

(*i*) **Fans.** The fan consists of a single rotating ring of curved vanes rotating at high speed and producing a flow of air at a comparatively low pressure. The fan produces a gauge pressure of less than 0.07 bar. It is generally used for air circulation, production of air cooling currents low pressure blast air and ventilation.

(*ii*) **Rotary displacement blowers.** In this type of compressor the air is compressed in the space formed by two sets of mutually engaging cam surfaces. As the two surfaces engage, the space between them is reduced in volume, so that the air between them is compressed. One type of rotary displacement blower (the root compressor) is shown in Fig. 4.4.

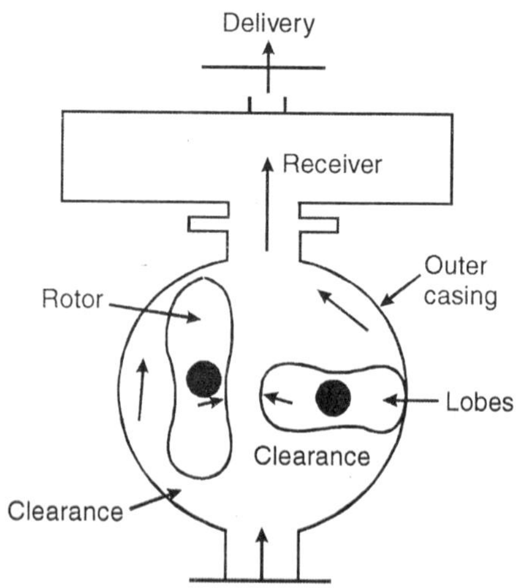

Fig. 4.4. Rotary displacement blower (Root Compressor).

The pressure Ratio upto 3.6 : 1 can be obtained in this type of blower. Higher pressure may be obtained by placing two or more blowers in series, having intercooler between each stage. The blower can be coupled directly to a turbine shaft and it occupies a very small space compared with other compressors for the same output. It is used for super charging I.C. engines, for producing scavenging air for two stroke diesel engine and as a compressor in gas turbine plants.

(*iii*) **Turbo blowers.** This is a low pressure type of rotary compressor and it consists of one or more rotating rings of curved blades. The complete blade ring is known as an impeller as shown in Fig. 4.5.

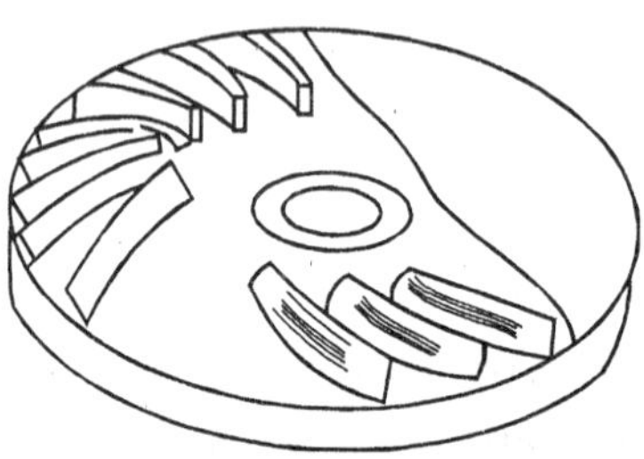

Fig. 4.5. Impeller of turbo blower.

Several impellers may be keyed on to a common shaft. Each impeller is surrounded by a diffuser which converts, the high velocity of the air leaving the impeller into pressure energy. A vane in the outer casing then guides the air stream into the next impeller. Each impeller and its diffuser combined is known as a stage and each stage of blower may increase the pressure of the air by a ratio of 1.6 : 1. The turbo blowers produce air pressure upto 2.5 bar. These are used for ventilators for producing blast air in work shops and for engine super chargers.

(*iv*) **Turbo compressors.** The working principle of turbo compressor is the same as that of turbo blower. But the turbo compressor usually has more number of stages. Delivery air pressure upto 12 bar is obtained and as many as 20 impellers in series may be used. Turbo compressors are designed to run at one speed only. They further can be classified as follows :

(*a*) **Vane Blower.** It consists of rotor located eccentrically in a cylindrical outer casing, a set of vanes mounted on the rotor and rotating eccentrically in the outer casing, inlet and delivery ports as shown in Fig. 4.6.

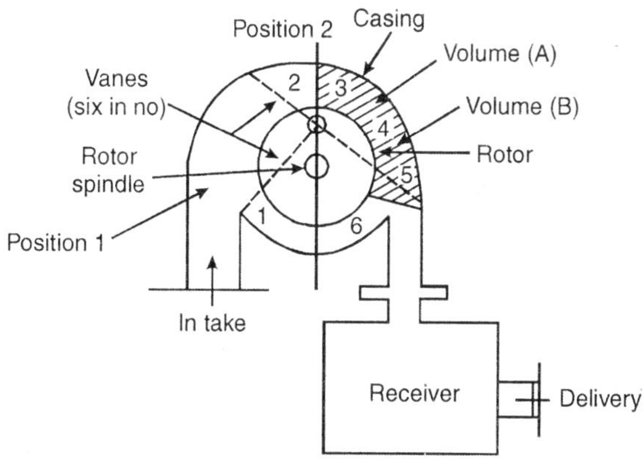

Fig. 4.6. Vane blower.

Each vane will start collecting air after it leaves the portion marked by position 1 in the figure and no more air will be sucked after vane moves to position 2.

Then compression of trapped air could commence due to decreasing space volume, between the rotor and the casing. When the vane travels from position 3 to 4, the air occupying volume A is compressed, into a smaller volume B. This type of blower is used for air capacities upto 150 m^3/min and pressure ratio upto 8 : 5 : 1.

(*b*) **Centrifugal Compressor.** A centrifugal compressor is shown in Fig. 4.7.

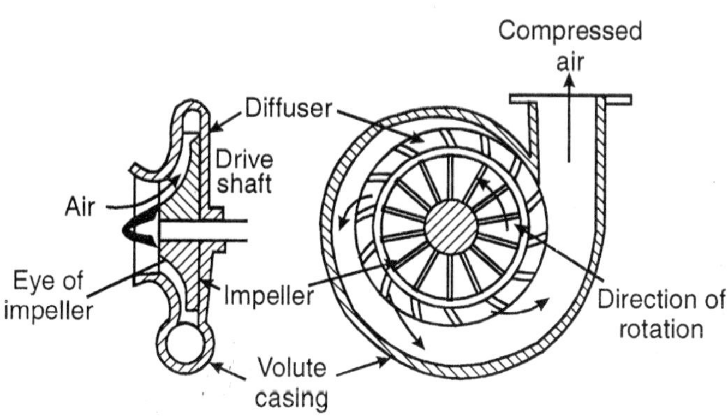

Fig. 4.7. Centrifugal compressor.

This type of compressor mainly consists of the following elements

(*i*) **Series of Inlet Buckets.** These are attached to the shaft and rotate with it.

(*ii*) **Rotating Impeller.** The impeller is a disc fitted with radial vanes. The air supplied to these vanes by the inlet buckets.

(*iii*) **Volute Casing.** The casing surrounds the rotating impeller.

(*iv*) **Diffuser.** The diffuser is housed in radial portion of the casing.

The air enters the eye of the impeller with a low velocity and at atmospheric pressure. Depending upon the centrifugal action of the impeller, the air moves readily outwards and during its movement, it is guided by the impeller vanes. The impeller traverses energy of the drive to the air causing a rise both in static pressure and temperature and an increase in velocity. The air now enters the diverging passage called diffuser, where it is efficiently slowed down. The kinetic energy of the air is converted into pressure energy with the result that there is further rise in static pressure. In case of multi stage centrifugal compressor a number of single stage centrifugal compressors are arranged on one shaft. Where the delivery of one unit leads to the intake of the next for further compression. Delivery from the last unit is made use of for doing the required work.

(*c*) **Axial Flow Compressors.** If the air flowing through the compressor is forced to flow in a direction parallel to its axis, then this compressor is called as an 'axial flow compressor'.

It consists of several moving blade rings attached to a common rotating shaft which are separately by rings of fixed blades attached to the casing as shown in Fig. 4.8.

The circular area is usually reduced from inlet to outlet of the compressor. This is done to keep the flow of velocity constant throughout the compressor length. In the diverging passage, there is a rise in pressure due to diffusion. The fixed blades serve two purposes;

(*i*) They convert a part of the kinetic energy of the air into pressure energy and this conversion is achieved by diffusion process carried out in the diverging blade passage.

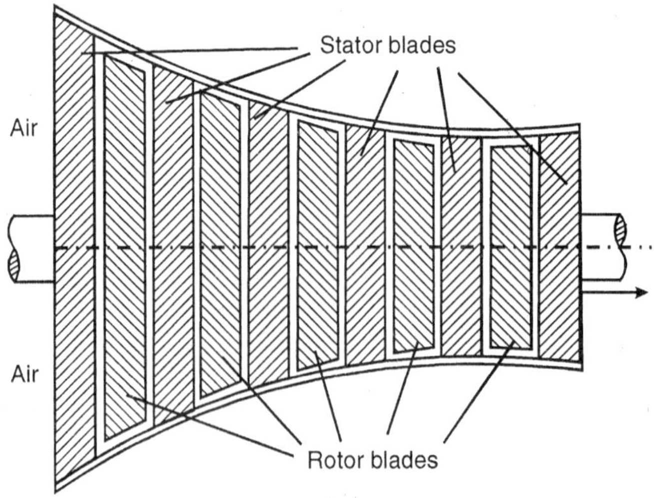

Fig. 4.8. Axial flow compressor.

(*ii*) Guide and redirect the air flow so that entry to the next stage is without shock. Axial flow compressors are used to produce pressure upto 5.5 bar and delivering about 1,00,000 kg of air per hour.

EXPERIMENT NO: 5

OBJECT:

Demonstration of heat-transfer through conduction, convection and Radiation.

Heat transfer. It is defined as the process by which heat is transferred from a body at higher temperature to the body at lower temperature.

There are three modes of heat transfer from one point to another when temperature differential exists :

(*i*) Conduction

(*ii*) Convection

(*iii*) Radiation.

5.1. OBJECT : Heat-transfer through conduction.

Equipment:

A cup of hot water, a cold spoon.

Theory : The process of heat transfer, within a material medium, in which heat energy is transferred from molecule to molecule without the transference of molecules of the medium is called conduction.

When one end of a rod is kept at higher temperature, then heat is transferred from the end at higher temperature to the end at lower temperature by conduction. This transfer of heat continues till the temperature of the two ends are equal.

The conduction process can be well explained on the basis of molecular theory. The molecules at the hot end of the rod have more kinetic energy due to higher temperature while the molecules at the cold end have lesser kinetic energy on account of lower temperature. Molecules at hot end begin to vibrate with larger amplitude and hand over kinetic energy to the neighbouring molecules. This speeds up the neighbouring molecules which further transfer their kinetic energy to the next molecules. In this way, energy is transferred from molecule to molecule in the rod from hot end to cold end.

As this process continues, the molecules at hot end go on transferring energy to the neighbouring molecules which further transfer energy to the next molecules and so on. In this way, heat is transferred through the rod by conduction.

Heat is transmitted through solids by conduction.

Procedure : When we put a cold spoon into a cup of hot water, its lower end becomes hot as shown in Fig. 5.1 (*a*). Then the upper portion also gets warmed, even though it is not touching water. Clearly, heat is transferred from water to the bowl of the spoon and up the handle. Water molecules have more kinetic energy due to higher temperature whereas the molecules of cold spoon have lesser kinetic energy. When the spoon is put in the cup, some of the water molecules hit the spoon molecules and speed them up. The result is that water molecules are slowed down while spoon molecules are speeded up.

Spoon molecules gain kinetic energy and hand over this energy to the neighbouring molecules in the handle and cause them to go faster as shown in Diagram 5.1 (*b*). This transfer of heat from molecule in the spoon is called conduction. The water molecules impart energy only, but they themselves do not get transferred to the spoon.

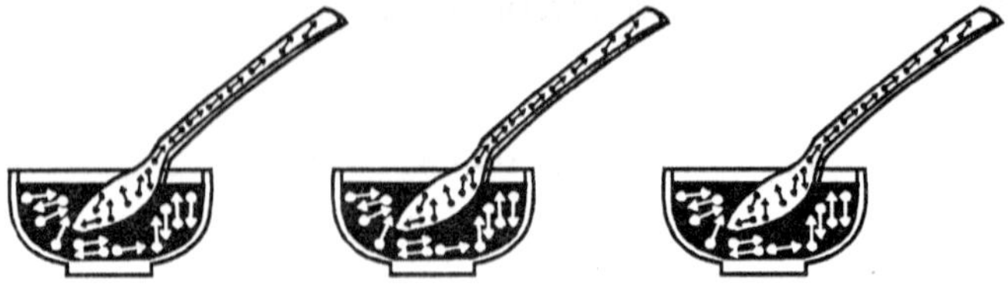

(a) (b) (c)

Fig. 5.1.

This transfer of heat continues till water and spoon molecules have acquired the same kinetic energy, i.e., temperature as shown in Fig. 5.1 (c).

Thus in conduction, heat is transferred from molecule to molecule within the material, without any transferrence of molecules themselves. Also this transfer of heat continues till the two bodies are at the same temperature.

5.2. OBJECT : Heat transfer through convection.

Equipments:

A box with a glass front and fitted with two chimneys, a burning candle, a smoking paper.

Theory : The process of heat transfer in which heat is transferred from one point to another by the actual motion of the heated particles is known as convection.

Liquids and gases transfer heat by convection. When a liquid or gas is heated, its molecules are speeded up and density decreases, i.e., molecules are separated quite apart. These molecules possessing high kinetic energy rise upward and thus carry the heat with them.

Procedure : Take a box with glass front and fitted with two chimneys as shown in Fig. 5.2. Place a burning candle under one of the chimneys and hold a smoking paper above the other. It will be seen that smoke enters the box through this chimney and goes out through the chimney above the burning candle. The air above the burning candle is heated so that it expands and rises up and carries away heat. Cool air comes from the other chimney to take its place. This cool air is again heated, expands and rises above. In this way, heat of candle flame is transferred by convection currents.

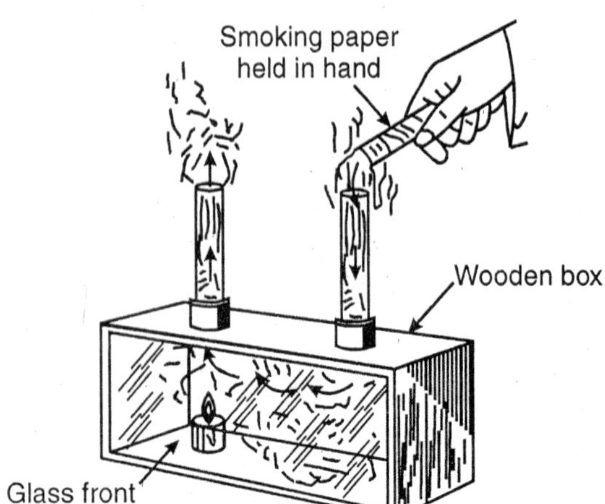

Fig. 5.2.

5.3. OBJECT : Heat transfer through radiation.

Equipments:

A wire filament, glass bulb, thermometer.

Theory :

The process of heat transfer in which heat is transferred from one place to another in a straight line with the velocity of light without heating the intervening medium is known as radiation such as;

(*i*) The sum transfers heat to earth by radiation. The space between the earth's atmosphere and the sun is a vacuum containing practically no molecules. The heat energy comes to the earth in the form of waves.

(*ii*) An open fire place gives heat to a room mainly by radiation.

(*iii*) All hot bodies emit heat radiations.

Procedure : Heat radiations require no material medium for their propagation.

Enclose a wire filament in a glass bulb which can be exhausted as shown in Fig. 5.3. Insert the thermometer inside the bulb and evacuate the bulb. Now pass current through the filament. Thermometer will indicate a rise in temperature there by showing that heat has travelled from filament to thermometer through vacuum. This shows that heat radiations require no material medium for their propagation.

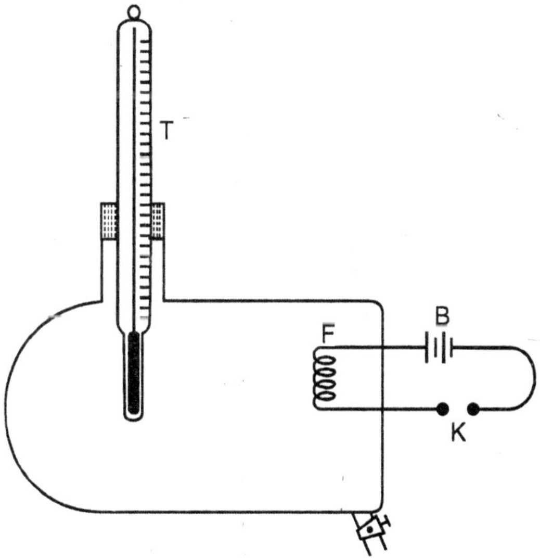

Fig. 5.3.

□□□

TABLES

Table 1. Properties of Saturated Steam with Temperature as Parameter

Temp. °C	Pressure bar	Specific Volume m^3/kg		Specific Enthalpy kJ/kg			Specific Entropy kJ/kg k		
		Sat. Liquid v_f	Sat. Vapour v_g	Sat. Liquid h_f	Evap. h_{fg}	Sat. Vapour h_g	Sat. Liquid s_f	Evap s_{fg}	Sat. Vapour s_g
0	006108	.00100	206.278	-0004	25016	2501 3	0001	9.1566	9.1565
01	006113	.00100	206.136	001	2501.6	25016	0000	9.1562	9.1562
1	006567	.00100	192.577	4.16	24990	25032	0.0152	9.1147	9.1299
2	007056	.00100	179.889	837	2496.7	2505.0	0.0305	90710	9.1015
3	.007577	.00100	168.132	12.57	24943	25069	0.0457	90316	9.0773
4	.008131	.00100	157.232	16.78	2491.9	2508.7	0.0610	89904	9.0514
5	.008721	.00100	147.120	20.98	2489.6	2510.6	0.0761	89496	9.0257
6	.009349	.00100	137.734	25.20	2487.2	2512.4	0.0912	89090	90001
7	.010016	.00100	129.017	29.39	2484.8	2514.2	0.1062	8.8688	8.9751
8	.010724	.00100	120.917	33.60	2482.5	2516.1	01212	8.8289	8.9501
10	.012276	.00100	106.379	42.01	2477.7	2519.8	01510	8.7498	8.9008
11	.013123	.00100	99.857	46.20	2475.4	2521.6	0.1658	8.7107	8.8765
12	014022	.00100	93.784	50.41	2473.0	25234	0.1806	86718	88524
13	.014974	.00100	88124	54.60	2470.7	2525.3	01953	8.6332	8.8285
14	.015983	00100	82.848	58.80	24683	2527.1	0.2099	8.5949	8.8048
15	.017051	.00100	77.926	62.99	2465.9	25289	02245	85569	87814
16	018181	.00100	73,333	67.19	2463.6	25308	02390	85191	87582
17	.019376	.00100	69.044	71.38	2461.2	25326	0.2535	84516	87351
18	.020640	.00100	65.038	75.58	2458.8	2534.4	02679	8.4443	37123
19	.021975	.00100	61.293	79.77	2456.5	2536.2	0.2823	8.4073	86897
20	.02339	.00100	57.791	83.96	2454.1	2538.1	0.2966	83706	8.6672
21	.02487	.00100	54.514	88.14	2451.8	2539.9	0.3109	8.3341	8.6450
22	.02645	.00100	51.447	92.33	2449.4	2541.7	03251	82979	8.6229
23	.02810	.00100	48.574	96.52	2447.5	2543.5	03393	8.2618	8.6011
24	.02985	.00100	. 45.883	100.70	2444.7	25454	0.3534	8.2261	8.5794
25	.03169	.00100	43.360	104.89	2442.3	2547.2	03674	8 1905	8J5580
26	.03361	.00100	40.994	109.07	2439.9	2549.0	03814	8.1552	85167
27	.03567	.00100	31.774	113.25	24376	2550.8	0.3954	8.1202	8.5156
28	.03712	.00100	36.690	117.43	2435.2	2552.6	0.4093	8.0854	8.4946
29	.04001	.00100	34733	12161	2432.8	2554.5	0.4231	8.0508	8.4739
30	.04246	.00100	12.894	125.79	2430.5	2556.3	0.4369	8.0164	84533
31	.04496	.00100	31.165	129.97	2428.1	2558.1	0.4507	7.9822	8.4129
32	.04759	.00101	29540	134 15	24257	2559.9	04644	79483	8.4127
33	.05034	.00101	28011	138.33	24234	2561.7	0.4781	7 9146	8.1927
34	.05124	.00101	26571	142.50	2421.0	2563.5	0.4917	7.8811	8.3728
35	.05628	0010 1	25216	146.68	2418.6	2565.3	05053	7.8478	8.3531
36	.05947	.00101	23.940	150.86	24162	2567.1	0.5188	78147	8.3336
37	.06281	.00101	22.737	155.03	2113.9	28689	0.5323	7.7819	8.3142
38	.06612	.01101	21.602	159.21	2411.5	2570.7	0.5458	7.7492	8.2759
39	.06999	.00101	20.533	163.39	2409.1	2572.5	0.5592	7.7167	8.2759

Temp. °C	Pressure bar	Specific Volume m^3/kg		Specific Enthalpy kJ/kg			Specific Entropy kJ/kg k		
		Sat. Liquid v_f	Sat. Vapour v_g	Sat. Liquid h_f	Evap. h_{fg}	Sat. Vapour h_g	Sat. Liquid s_f	Evap s_{fg}	Sat. Vapour s_g
40	.07384	.00101	19.523	167.57	2406.7	2574.3	0.5725	7.6845	8.2570
41	.07786	.00101	18.570	171.74	2404.3	2576.1	0.5858	7.6524	8.2383
42	.08208	.00101	17.671	175.91	2401.9	2577.9	0.4951	7.6206	8.2197
43	.18649	.00101	16.821	180.10	2399.5	2579.6	0.6123	7.5889	8.2012
44	.09111	.00101	16.018	184.27	2397.2	2581.4	0.6255	7.5574	8.1829
45	.09593	.01010	15.260	188.45	2394.8	2583.2	0.6390	7.5261	8.1650
46	.10101	.001010	14.540	192.61	2392.4	2585.0	0.6523	7.4954	8.1470
47	.10622	.001011	13.861	196.82	2390.0	2586.8	0.6650	7.4642	8.1292
48	.11181	.001011	13.223	201.00	2387.6	2588.6	0.6782	7.4334	8.1110
49	.11753	.001012	12.512	205.12	2385.2	2590.3	0.69I4	7.4013	8.0941
50	.12354	.001012	12.034	209.31	2382.7	2592.1	0.7040	7.3720	8.0761
51	0.1296	.001013	11.499	213.4	2380.5	2593.9	0.716	7.344	8.060
52	0.1361	.001013	10.890	217.6	2378.1	2595.7	0.729	7.314	8.043
53	0.1429	.001014	10.488	221.8	2375.7	2597.5	0.742	7.284	8.026
54	0.1500	.001014	10.022	226.0	2373.2	599.2	0.755	7.284	8.009
55	0.1574	.001015	9.5789	230.2	2370.8	2601.8	0.768	7.225	7.993
56	0.1651	.001015	9.1587	234.4	2368.4	2602.7	0.780	7.196	7.976
57	0.1731	.001016	8.7598	238.5	2365.9	2604.5	0.793	7.166	7.959
58	0.1815	.001016	8.3808	242.7	2363.5	2606.2	0.806	7.137	7.943
59	1.1902	.001017	8.0208	246.9	2361.1	2608.0	0.818	7. 109	7.927
60	0. 1992	.001017	7.6785	251.1	23586	2609.7	0.831	7.080	7.911
61	0.2086	.001018	73532	255.3	2356.2	2611.4	0.844	7.051	7.895
62	0.2184	.001018	7.0437	259.5	2353.7	2613.2	0.856	7.023	7.879
63	0.2286	.001019	6.7493	263.6	2351.3	2614.9	0.868	6.995	7.163
64	0.2391	0.001019	6.4690	267.8	2348.8	2616.6	0.881	6.967	7.848
65	0.2501	0.001020	6.2023	272.0	2346.3	2618.4	0.893	6.939	7.832
66	0.2615	0.001020	5.9482	276.2	2343.9	2620.1	0.906	6.911	7.817
67	0.2733	0.001021	5.7062	280.4	2341.4	2621.8	0.918	6.884	7.802
68	0.2856	0.001022	5.4756	284.6	2338.9	2623.5	0.930	6.856	7.786
69	0.2984	0.001022	5.2558	188.8	2336.4	2625.2	0.943	6.829	7.771
70	0.3116	0.001023	5.0463	293.0	2334.0	2626.9	0.955	6.802	7.757
71	0.3254	0.001024	4.8464	297.2	2331.5	2628.6	0.967	6.775	7.742
72	0.3396	0.001024	4.6557	301.4	2329.0	2630.3	0.979	6.748	7.727
73	0.3543	0.001025	4.4737	305.5	2326.5	2632.0	0.991	6.721	7.712
74	0.3696	0.001025	4.3000	309.7	2324.0	2633.7	1.003	6.695	7.698
75	0.3855	0.001026	4.1341	313.9	2321.5	2635.4	1.015	6.668	7.683
76	0.4019	0.001027	3.9757	318.1	2318.9	2637.1	1.027	6.642	7.669
77	0.4189	0.001027	3.8243	322.3	2316.4	2638.7	1.039	6.616	7.655
78	0.4365	0.001028	3.6796	326.5	2313.9	2640.4	1.051	6.590	7.641
79	0.4547	0.001029	3.5413	330.7	2311.4	2642.1	1.063	6.564	7.627
80	0.4736	0.001029	3.4091	334.9	2308.8	2643.8	1.075	6.538	7.613
81	0.4931	0.001030	3.2826	339.1	2306.3	264S.4	1.087	6.512	7.599
82	0.5133	0.001031	3.1616	343.3	2303.8	2647.1	1.099	6.487	7.586
83	0.5342	0.001031	3.0458	347.5	2301.2	264S.7	1.111	6.461	7.572

Temp. °C	Pressure bar	Specific Volume m^3/kg		Specific Enthalpy kJ/kg			Specific Entropy kJ/kg k		
		Sat. Liquid v_f	Sat. Vapour v_g	Sat. Liquid h_f	Evap. h_{fg}	Sat. Vapour h_g	Sat. Liquid s_f	Evap s_{fg}	Sat. Vapour s_g
84	0.5557	0.001032	2.9350	351.7	2298.6	2650.4	1.123	6.436	7.559
85	0.5780	0.001033	2.8288	355.9	2296.1	2652.0	1.134	6.411	7.545
86	0.6011	0.001033	2.7272	360.1	2293.5	2653.6	1.146	6.386	7.532
87	0.6249	0.001034	2.629S	364.3	2290.9	2655.3	1.158	6.361	7.519
88	0.6495	0.001035	2.5365	368.5	2288.4	2656.9	1.169	6.337	7.506
89	0.6749	0.001035	2.4470	372.7	2285.8	2658.5	1.181	6.312	7.493
90	0.7011	0.001036	2.3613	376.9	2283.2	2660.1	1.193	6.217	7.480
91	0.7282	0.001037	2.2791	381.1	2280.6	2661.7	1.204	6.263	7.467
92	0.7561	0.001038	2.2002	385.4	2278.0	2663.4	1.216	6.239	7.454
93	0.7849	0.001038	2.1245	389.6	2275.4	2665.0	1.227	6.215	7.442
94	0.8146	0.001039	2.0519	393.8	2272.8	2666.6	1.239	6.190	7.429
95	0.8453	0.001040	1.9822	398.0	2227.2	2668.1	1.250	6.167	7.417
96	0.8769	0.001041	1.9153	402.2	2267.5	2669.7	1.261	6.143	7.404
97	0.9094	0.001041	1.8510	406.4	2264.9	2671.3	1.273	6.119	7.392
98	0.9430	0.001042	1.7893	410.6	2262.2	2672.9	1.284	6.095	7.380
99	0.9776	0.001043	1.7300	414.8	2259.6	2674.4	1.296	6.072	7.361
100	1.01325	0.001044	1.6730	419.1	2256.9	2676.0	1.307	6.048	7.355
102	1.088	0.001045	1.5665	427.5	2251.6	2679.1	1.329	6.002	7.331
104	1.167	0.001047	1.4662	435.9	2216.3	2682.2	1.352	5.956	7.308
106	1.250	0.001048	1.3742	444.4	2240.9	2685.3	1.374	5.910	7.284
108	1.339	0.001050	1.2889	452.9	2235.4	2688.3	1.396	5.865	7.261
110	1.433	0.001052	1.2099	461.3	2230.0	2691.3	1.419	5.820	7.239
112	1.532	0.001054	1.1366	469.8	2224.5	2694.3	1.440	5.776	7.216
114	1.636	0.001055	1.0685	478.3	2219.0	2697.2	1.462	5.732	7.194
116	1.747	0.001057	1.0052	486.7	2213.4	2700.2	1.484	5.688	7.172
118	1.863	0.001059	0.94634	495.2	2207.9	2703.1	1.506	5.45	7.151
120	1.985	0.001061	0.89152	503.7	2202.2	2706.0	1.528	5.602	7.129
122	2.115	0.001063	0.84045	512.2	2196.6	2708.8	1.549	5.559	7.101
124	2.250	0.001064	0.79283	520.7	2190.9	2711.6	1.571	5.517	7.087
126	2.393	0.001066	0.74840	529.2	2185.2	2714.4	1.592	5.475	7.067
128	2.544	0.001068	0.70691	537.8	2179.4	2717.2	1.613	5.433	7.046
130	2.701	0.001070	0.66814	546.3	2173.6	2719.9	1.634	5.392	7.026
132	2.867	0.001072	0.63188	554.8	2167.8	2722.6	1.655	5.351	7.006
134	3.041	0.001074	0.59795	563.4	2161.9	2725.3	1.676	5.310	6.986
136	3.223	0.001076	0.56618	572.0	2155.9	2727.9	1.697	5.270	6.967
138	3.414	0.001078	0.53641	580.5	2150.0	2730.5	1.718	5.229	6.447
140	3.614	0.001080	0.50849	589.1	2144.0	2733.1	1.739	5.189	6.928
142	3.823	0.001082	0.48230	597.7	2137.9	2735.6	1.760	5.150	6.910
144	4.012	0.001084	0.45771	606.3	2131.8	2738.1	1.780	5.111	6.891
146	4.271	0.001086	0.43460	614.9	2125.7	2740.6	1.801	5.071	6.172
148	4.510	0.001089	0.41288	623.5	2119.5	2743.0	1.821	5.033	6.854
150	4.760	0.001091	0.39245	632.1	2113.2	2745.4	1.842	4.994	6.836
155	5.433	0.001096	0.34644	635.8	2097.4	2751.2	1.892	4.899	6.791
160	6.181	0.001102	0.30676	675.5	2081.3	2756.7	1.943	4.805	6.748

Temp. °C	Pressure bar	Specific Volume m^3/kg		Specific Enthalpy kJ/kg			Specific Entropy kJ/kg k		
		Sat. Liquid v_f	Sat. Vapour v_g	Sat. Liquid h_f	Evap. h_{fg}	Sat. Vapour h_g	Sat. Liquid s_f	Evap s_{fg}	Sat. Vapour s_g
165	7.008	0.001108	0.27240	697.3	2064.8	2762.0	1.992	4.713	6.705
170	7.920	0.001114	0.24255	719.1	2047.9	2767.1	2.042	4.621	6.663
175	8.924	0.001121	0.21654	741.1	2030.7	2771.8	2.091	4.531	6.622
180	10.027	0.001128	0.19380	763.1	2013.2	2776.3	2.139	4.443	6.582
185	11.233	0.001135	0.17386	785.3	1995.2	2780.4	2.188	4.355	6.542
190	12.551	0.001142	0.15632	807.5	1976.7	2784.3	2.236	4.268	6.504
195	13987	0.001149	0.14084	829.9	1957.9	2787.8	2.283	4.182	6.465
200	15.549	0.001156	0.12716	852.4	1938.6	2790.9	2.331	4.097	6.428
205	17.243	0.001164	0.11503	875.0	1918.8	2793.8	2.378	4.013	6.391
210	19.077	0.001172	0.10424	897.4	1898.5	2796.2	2.425	3.929	6.354
215	21.060	0.001181	0.094625	920.6	1877.6	2798.3	2.471	3.846	6.318
220	23.198	0.001190	0.086038	943.7	1856.2	2799.9	2.518	3.764	6.282
225	25.501	0.001199	0.078349	966.9	1834.3	2801.2	2.564	3.682	6.246
230	27.976	0.001209	0.071450	990.3	1811.7	2802.0	2.610	3.601	6.211
235	30.632	0.001219	0.065245	1013.8	1788.5	2802.3	2.656	3.519	6.176
240	33.478	0.001229	0.059654	1037.6	1764.6	28022	2.702	3.439	6.141
245	36.523	0.001240	0.054606	1061.6	1740.0	2801.6	2.748	3.358	6.106
250	39.776	0.001251	0.050037	1085.8	1714.7	2800.4	2.794	3.277	6.071
255	43.246	0.001263	0.045896	1110.2	1688.5	2798.7	2.839	3.197	6.036
260	46.943	0.001276	0.042134	1134.9	1661.5	2796.4	2.885	3.116	6.001
265	50.877	0.001289	0.038710	1159.9	1633.5	2793.5	2.931	3.035	5.966
270	55.058	0.001303	0.035588	1185.2	1604.6	2789.9	2.976	2.954	5.930
275	59.496	0.001317	0.032736	1210.9	1574.7	2785.5	3.022	2.873	5.895
280	64.202	0.001332	0.030126	1236.8	1543.6	2780.4	3.068	2.790	5.859
285	69.186	0.001349	0.027733	1263.2	1511.3	2774.5	3.115	2.707	5.822
290	74.461	0.001366	0.025535	1290.0	1477.6	2767.6	3.161	2.624	5.785
295	80.037	0.001384	0.023513	1317.3	1442.6	2759.8	3.208	2.539	5.747
300	85.927	0.001404	0.021649	1345.1	1406.0	2751.0	3.255	2.453	5.708
305	92.144	0.001425	0.019927	1373.4	1367.7	2741.1	3.303	2.366	5.669
310	98.700	0.001448	0.018334	1432.4	1327.6	2730.0	3.351	2.277	5.628
315	105.61	0.001473	0.016856	1432.1	1285.5	2717.6	3.400	2.186	5.586
320	112.89	0.001500	0.015480	1462.6	1241.1	2703.7	3.450	2.092	5.542
325	120.56	0.001529	0.014195	1494.0	1194.0	2688.0	3.501	1.996	5.497
330	128.63	0.001562	0.012989	1526.5	1143.6	2670.2	3.551	1.896	5.449
335	137.12	0.001598	0.011854	1560.3	1089.5	2649.7	3.606	1.792	5.391
340	146.05	0.001639	0.010780	1595.5	1030.7	2626.2	3.662	1.681	5.343
345	155.45	0001686	0.0097631	1632.5	966.4	2598.9	3.719	1.564	5.283
350	165.35	0.001741	0.0087991	1671.9	895.7	2567.7	3.780	1.438	5.218
355	175.77	0.001809	0.0078592	1716.6	813.8	2530.4	3.849	1.295	5.144
360	186.75	0.001896	0.0069398	1764.2	721.3	2485.4	3.921	1.139	5.060
365	198.33	0.002016	0.0060116	1818.0	610.0	2428.0	4.002	0.956	4.961
370	210.54	0.002214	0.0049728	1890.2	452.6	2342.8	4.111	0.704	4.814
374.15	221.20	0.003170	0.003170	2107.4	0.0	2107.4	4.443	0.000	4.443

Table 2. Saturated Properties of Steam with Pressure as Parameter

Press. bar	Temp. °C	Specific Volume m^3/kg		Specific Enthalpy kJ/kg			Specific Entropy kJ/kg k		
		Sat. Liquid v_f	Sat. Vapour v_g	Sat. Liquid h_f	Evap. h_{fg}	Sat. Vapour h_g	Sat. Liquid s_f	Evap s_{fg}	Sat. Vapour s_g
006113	0.01	.00100	206.136	0.01	2501.3	2501.4	0000	9.1562	9.1562
.007	1.89	.00100	181.255	7.91	2496.9	2504.8	0288	9.0775	9.1064
.008	3.77	.00100	159.675	15.81	2492.5	2508.1	0575	8.9999	9.0573
.009	5.45	.00100	142.789	22.89	2488.5	2511.4	0.829	8.9112	9.0142
.010	6.98	.00100	129.208	29.30	2484.9	2514.2	1059	8.8697	8.9756
.011	8.37	.00100	118.042	35.17	2481.6	2516.8	I268	8.8140	8.9408
.012	9.66	.00100	108.696	40.58	2478.6	2519.1	1460	8.7611	8.9091
.013	10.86	.00100	100.755	45.60	2475.7	2521.1	1637	8.7162	8.8799
.014	11.98	.00100	93.922	50.31	2473.1	2521.4	1802	8.6727	8.8529
.015	13.03	.00100	87.980	54.71	2470.6	2525.1	I957	8.6122	8.8279
.016	14.02	.00100	82.763	58.87	2468.3	2527.1	2102	8.5941	8.8044
.017	14.95	.00100	78.146	62.80	2466.0	2528.8	2238	8.5586	8.7825
.018	15.84	.00100	74.030	66.54	2463.9	2530.5	2368	8.5250	8.7618
.019	16.69	.00100	70.337	70.10	2461.9	2512.0	2491	8.493 1	8.7422
.020	17.51	.001001	67.004	73.48	2460.0	2511.5	2607	8.4629	8.7237
0.21	61.15	0.001018	73068	2559	2355.8	2611 7	0X45	7(|47	7XV7
0.22	62.16	0.001018	69946	260 1	2153.3	2613 5	OX5X	7 OIX	7X76
0.23	63.14	0.001019	67089	264.2	2150.9	26152	0 X70	6991	7X(,I
0.24	64.08	0.001019	64462	'2682	23486	2616 X	OXX2	6 9<>4	7 X46
0.25	64.99	0.001020	62040	2720	21464	26183	0 XVI	6 919	7X12
0.26	65.87	0.001020	5 9799	2757	214*2	26199	1) 904	6915	7 XIV
0.27	66.72	0.001021	5.7720	2792	214! 1	2621 3	0915	6X91	7X06
0.28	67.55	0.001021	5 57X4	2X2 7	2.14) 0	2622 7	0925	6 K68	7791
0.29	68.15	0.001022	5 3979	2X6 12	11 U |	2624 I	II 915	6K47	7 7X1
0.30	69.11	0.001022	5.2290	2X9 3	2.1K. 1	2(>25 4	\ 0 944	6825	7 769
0.31	69.88	0.001023	50707	292 5	23342	2626.7	0 953	6805	7 758
0.32	70.62	0.001023	4.9220	2956	2332 4	2628.0	0 962	67X5	7747
0.33	71.33	0.001024	4.7820	2986	21306	2629 2	0 97 1	6766	7737
0.34	72.03	0.001024	46501	301 5	232X9	26304	0 980	6747	7727
0.35	72.71	0001025	4.5255	304}	232 '.2	2631 5	0 9X8	6729	7717
0.36	73.38	0.001025	4.4076	307.1	2325.5	2632.7	0.996	6.711	7.707
0.37	74.02	0.001025	4.2958	309.8	2323.9	2633.7	1.004	6.694	7.6W
0.38	74.66	0.001026	4.1898	312.5	2322.3	2634.8	1.011	6.677	7.6N
0.39	75.28	0.001026	4.0891	3151	23208	2635.9	1.019	6.661	7.610
0.40	75.89	0.001027	3.9932	317.7	2319.2	2636.9	1.026	6.645	7671
0.41	76.48	0.001027	3.9019	320.2	2317.7	2637.9	1.033	6.629	7.662
0.42	77.06	0001027	3.8148	322.6	2316.3	2638.9	1.040	6.614	7.654
0.43	77.64	0001228	3.7317	325.0	23148	2639.8	1.047	6.599	7.646
0.44	78.19	0.001028	36522	327.3	2313.4	2640.7	1.054	6.585	7.631
0.45	78.74	0.001028	35761	329.6	2312.0	2641.7	·*060^	6.570	7.631

Press. bar	Temp. °C	Specific Volume m^3/kg		Specific Enthalpy kJ/kg			Specific Entropy kJ/kg k		
		Sat. Liquid v_f	Sat. Vapour v_g	Sat. Liquid h_f	Evap. h_{fg}	Sat. Vapour h_g	Sat. Liquid s_f	Evap s_{fg}	Sat. Vapour s_g
0.46	79.28	0.001029	3.5032	331.9	2310.7	2642.6	1067	6JSY	7.623
0.47	79.81	0.001029	3.4334	334.1	2309.3	2643.4	1.073	6.543	???
0.48	80.33	0.001029	3.3663	336.3	2308.0	2644.3	1.079	6.529	760*
0.49	80.84	0.001030	3.3020	338.5	2306.7	26452	085	6.516	7.602
0.50	81.35	0001030	3.2401	340.6	2305.4	2646.0	.091	6.504	7.5»5
0.52	82.33	0001031	3.1233	344.7	2302.9	2647.6	.103	6.478	7.5SI
0.54	83.28	0001031	3.0148	3487	2300.5	2649.2	.114	6.45V	7569
0.56	84.19	0.001032	2.9139	352.5	2298.2	2650.7	.125	6.431	7.556
0.58	85.09	0.001033	2.8197	356.3	2295.9	2652.1	.135	6.409	7.544
0.60	85.95	0001033	2.7317	359.9	2293.6	26536	.146	6.387	7.513
0.62	86.80	0001034	2.6491	363.5	2291.5	26549	.155	6.366	7.522
0.64	87.62	0001034	25715	366.9	2289.4	2656.1	.165	6.346	7.511
0.66	88.42	0001035	2.4985	3703	2287.3	2657.6	.174	6.3*0	7.500
0.68	89.20	0001036	24297	373.6	2285.3	2658.8	.183	6.307	7490
0.70	89.96	0001036	2.3647	376.8	2283.3	26601	.192	6.288	74M
0.72	90.70	0001037	23031	379.9	2281.4	2661.3	.201	6.270	7.471
0.74	91.43	0001037	2.2448	383.0	22795	2662.4	.209	6.253	7.462
0.76	92.14	0001038	2 1895	385.9	2277.6	2663.6	.217	6235	7.452
0.78	92.83	0001038	2 1369	388.9	2275.8	2664.7	225	6.219	7.444
0.80	41.51	0001039	20869	391.7	2274.0	2665.8	.233	6.202	7.435
0.82	94.18	0.001039	2.0392	394.5	2272.3	2666.8	.241	6.186	7.427
0.84	94.83	0001040	19938	397.3	2270.6	2667.9	.248	6.171	7.419
0.86	95.47	0001040	19504	4000	2268.9	2668.9	.255	6.155	7.411
0.88	96.10	0001041	19089	402.6	2267.3	2669.9	.263	6140	7.403
0.90	96.7I	0001041	18691	4052	22656	2670.9	270	6126	7.395
0.92	97.32	0.001042	18311	407.8	2264.0	2671.8	1.277	6.111	7.388
0.94	97.91	0001042	1.7946	410.3	2262.5	2672.7	1.283	6.098	7381
0.96	98.50	0.001043	1.7596	4127	2260.9	2673.7	1290	6084	7.374
0.98	99.07	0.001043	1.7260	415.1	2259.4	2674.6	1.296	6.070	7.367
1.00	99.63	0.001043	1.6937	41*5	2257.9	2675.4	1.303	6.057	7360
1.01325	100.0	0.001044	1.6730	419.1	2256.9	2676.0	1.307	6.048	7.355
1.05	101.0	0.001045	1.6181	423.3	2254.3	2677.6	1.318	6.025	7343
1.10	102.3	0.001046	.5492	428.8	2250.8	2679.6	1333	5.995	7.328
1.15	103.6	0.001047	.4861	434.2	2247.4	2681.6	1.347	5.966	7313
1.20	104.8	0.001048	.4281	439.4	2244.1	2683.4	1.361	5.937	7.298
1.25	106.0	0.001049	.3746	444.4	2240.9	2685.2	1374	5911	7285
1.30	1071	0.001050	3250	449.2	2237.8	26870	1.387	5.885	7.272
1.35	108.2	0001050	.2791	453.9	22348	26887	1.399	5860	7.259
1.40	1093	0.001051	2363	458.4	2231.9	2690.3	1.411	5836	7.247
1.45	110.4	0.001052	.1963	462.8	2229.0	2691.8	1423	5.812	7.235

Press. bar	Temp. °C	Specific Volume m^3/kg		Specific Enthalpy kJ/kg			Specific Entropy kJ/kg k		
		Sat. Liquid v_f	Sat. Vapour v_g	Sat. Liquid h_f	Evap. h_{fg}	Sat. Vapour h_g	Sat. Liquid s_f	Evap s_{fg}	Sat. Vapour s_g
1.50	1114	0.001053	1.1590	4671	2226.2	26934	1.434	5.790	7223
1.60	1133	0001055	1.0911	4754	22209	2696.2	1.455	5.747	7.202
1.70	1152	0001056	1.0309	483.2	2215.7	2699.0	1475	5.706	7181
1.80	116.9	0.001058	0.97718	490.7	2210.8	27015	1.494	5.668	7.162
1 90	118.6	0001059	092895	4979	2206.1	2704.0	1.513	5.631	7144
2.00	120.2	0001061	0.88540	504.7	22016	2706.3	1 530	5.597	7.127
2 10	1218	0.001062	0.84586	511.3	2 197 2	2708.5	1.547	5564	7111
2.20	123.3	0001064	0.80980	517.6	2193.0	27106	1.563	5532	7095
2.30	124.7	0.001065	0.77677	523.7	2188.9	27126	1.578	5502	7.080
2.40	1261	0001066	0.74641	529.6	2184.9	2714.5	1593	5473	7066
2.50	1274	0001068	0.71840	535.4	21810	27164	1607	5.445	7.052
2.60	128.7	0.001069	0.69247	540.9	2177.3	27182	1621	5418	7039
2.70	1300	0.001070	0.65840	546.2	21736	27199	1.643	5392	7026
2.80	131.2	0.001071	0.64600	551.5	21701	27215	1.647	5367	7.014
2.90	132.4	0001072	062509	556.5	21666	2723 1	1660	5343	7.002
3.0	1335	0001074	0.60553	561.4	2163.2	27247	1.672	5319	6.991
3.1	134.7	0001075	058718	566.2	2159.9	2726.1	1.683	5.297	6.980
3.2	1358	0.001076	0.56995	570.9	2156.7	2727.6	1.695	5.274	6.969
3.3	13<i.S	0.001077	055373	5755	2153.5	2729.0	1.706	5253	6.959
3.4	137.9	0.(V”»78	053843	5799	2150.4	27303	1.717	5232	6.949
3.5	1389	0.001079	0.52397	584.3	2147.3	2731.6	.727	5212	64)9
3.6	1399	0001080	051029	588.5	2144.4	2732.9	.738	5.192	6».X)
3.7	1408	0.001081	0.49733	592. 7	2141.4	2734.1	.748	5.173	6.921
3.8	141.8	0.001082	0.45802	5968	2138.6	2735.3	.758	5.154	. 6.9 II
3.9	1427	0.001083	0.47733	600.8	2135.7	2736.5	.767	5136	690)
4.0	1436	0001084	0.46220	604.7	21329	2737.6	.776	5118	6X94
4.1	1445	0.001085	045159	6085	2130.2	2738.7	.786	5.100	6816
4.2	145.4	0.001086	0.44147	6123	2127.5	2739.8	.795	5.083	6.S7I
4.3	1463	0.001087	0.43181	616.0	21249	2740.9	.803	5067	6870
4.4	147.1	0.001088	0.42257	619.6	2122.3	27419	.812	5050	6162
4.5	1479	0.001089	0.41373	623.2	2119.7	27429	1.820	5.034	6X55
4.6	1487	0001090	0.40526	626.7	2117.2	27439	1829	5019	6X47
4.7	1495	0.001090	0.39714	630.1	21147	2744 X	1.837	5.003	6X40
4.8	1503	0.001091	0.38934	633.5	2112.2	2745.7	1.845	4.988	6.83)
4.9	151. 1	0001092	038186	636.8	2109.8	2746.6	1.853	4973	6X26
5.0	151 9	0001093	037466	640.1	2107.4	27475	1.860	4959	6X19
5.2	1533	0.001095	036106	6465	2102.7	2749.3	1.875	4.931	6806
5.4	1548	0.001096	034844	652.8	2098.1	2750.9	1.890	4903	6.79)
5.6	156.2	0.001098	0.33669	658.8	2093.7	2752.5	1.904	4877	6.781
5.8	157.5	0001099	0.32572	6647	2089.3	2754.0	1.918	4.851	6769

Press. bar	Temp. °C	Specific Volume m^3/kg		Specific Enthalpy kJ/kg			Specific Entropy kJ/kg k		
		Sat. Liquid v_f	Sat. Vapour v_g	Sat. Liquid h_f	Evap. h_{fg}	Sat. Vapour h_g	Sat. Liquid s_f	Evap s_{fg}	Sat. Vapour s_g
6.0	1588	0.001101	031546	670.4	2085.0	2755.5	1.931	4827	675X
6.2	1601	0001102	030584	676.0	2080.8	2756.9	1.944	*4.803	6746
6.4	1614	0001104	0.29680	681.5	2076.7	27582	1.956	4779	67)6
6.6	1626	0001105	0.28829	686.8	2072.7	2759.5	1968	4.757	6725
6.8	163.8	0.001107	0.28026	692.0	2068.8	2760:8	1.980	4735	6715
7.0	1650	0001108	027268	697 1	2064.9	2762.0	1.992	4713	6.705
7.2	1661	0.001110	0.26550	702.0	2061.1	2763.2	2.003	4693	6696
7.4	167.2	0001111	025870	706.9	2057.4	2764.3	2.014	4.672	6686
7.6	1683	0.001112	0.25224	7117	2053.7	2765.4	2025	4652	6677
7.8	169.4	0.001114	0.24610	716.3	2050.1	2766.4	2035	4633	6.668
8.0	170.4	0001115	0.24026	720.9	2046.S	2767.5	2.046	4614	6660
8.2	1714	0.001116	0.23469	725.4	2043.0	2768.5	2.056	4.595	6651
8.4	172.4	0.001 IK	0.22938	729.9	2039.6	276*4	2.066	4.577	664)
8.6	173.4	0.001119	0.22431	7342	2036.2	2770.4	2.075	4.560	66)5
8.8	174.4	0.001120	0.21946	738.J	2032.8	2771.3	2.0(5	4542	6627
9.0	175.4	0001121	021482	742.6	2029.5	27721	2.094	4525	6.619
9.2	176.3	0001123	021037	7468	2026.2	27730	2.103	4508	6612
9.4	177.2	0001124	020610	7508	2023.0	2773.8	2.112	4492	6604
9.6	178.1	0001125	020201	754.8	2019.8	2774.6	2.121	4476	6597
9.8	179.0	0001126	0 19808	758.7	2016.7	2775.4	2.130	4.460	6590
10.0	179.9	0001127	0.19430	762.6	2013.6	2776.2	2138	4 445	6583
10.5	182.0	0001130	0 18548	772.0	2006.0	27780	2.159	4407	6566
11.0	184 1	0001133	017739	781.1	1998.6	2779.7	2.179	4.371	6550
11.5	186.0	0001136	0.17002	789.9	1991.4	2781.3	2.198	4.336	6534
12.0	1880	0001139	0.16321	798.4	1984.3	2782.7	2.216	4303	6519
12.5	1898	0001141	0.15696	806.7	1977.5	2784.2	2234	4.271	6505
13.0	1916	0001144	0.15114	814.7	1970.7	27854	2251	4.240	6491
13.5	1933	0001146	0.14S76	822.5	1964.2	27867	2267	4.211	6478
14.0	1950	0.001149	0.14073	830.1	1957.7	27878	2284	4 181	6465
14.5	1967	0001151	013606	837.5	1951.4	27889	2.299	4 153	6453
15.0	1983	0001154	0.13167	844.6	1945.3	27899	2314	4.126	6441
15.5	1998	0001156	0.12756	851.7	1939.2	2790.8	2.329	4 100	6429
16.0	2014	0001159	0.12370	858:5	1933.2	27917	2.344	4.074	. 6418
16.5	202.9	0001161	0.12006	865.2	1927.3	2792.6	2.358	4.049	6.407
17.0	204.3	0001163	0.11664	871.8	1921.6	2793.4	2.371	4.025	6.396
17.5	2057	0001166	0.11340	8782	1915.9	2794.1	2.384	4.001	6385
18.0	207.1	0.001168	011033	8845	1910.3	2794.8	2398	3978	6.375
18.5	208.5	0.001170	0.10742	890.7	1904.8	2795.5	2.410	3.955	6.365
19.0	2098	0001172	0.10467	896.8	1899.3	2796.1	2.423	3.933	6356
19.5	211 1	0.001174	0.10204	902.7	1894.0	2796.7	2.435	3911	6.346

Press. bar	Temp. °C	Specific Volume m^3/kg		Specific Enthalpy kJ/kg			Specific Entropy kJ/kg k		
		Sat. Liquid v_f	Sat. Vapour v_g	Sat. Liquid h_f	Evap. h_{fg}	Sat. Vapour h_g	Sat. Liquid s_f	Evap s_{fg}	Sat. Vapour s_g
20.0	212.4	0.001177	0.099549	908.6	1888.7	2797.2	2.447	3.890	6.337
20.5	213.6	0001179	0.097171	914.3	1883.4	2797.7	2.459	3.869	6.328
21.0	2148	0001181	0094902	919.9	1878.3	2798.2	2.470	3.849	6.319
21.5	2160	0001183	0.092735	925.5	1873.2	2798.6	2.481	3.829	6.310
22.0	217.2	0 001185	0.090663	930.9	1868.1	27991	2.492	3.809	6.301
22.5	218.4	0.001187	0.088680	936.3	1863.1	2799.4	2.503	3.790	6.293
23.0	219.6	0.001189	0.086780	941.6	1858.2	2799.8	2.514	3.771	6.285
23.5	220.7	0.001191	0.084958	946.8	1853.3	2800.1	2.524	3.753	6.277
24.0	2218	0.001 193	0.083209	951.9	1848.5	2800.4	2.534	3.735	6.269
24.5	2229	0.001195	0.081530	957.0	1843.7	2800.7	2.544	3.717	6.261
25.0	2239	0.001197	0.079915	961.9	1839.0	2800.9	2554	3.699	6.254
26.0	2260	0001201	0076865	971.7	1829.7	2801.4	2.574	3.665	6.2)9
27.0	228 1	0001205	0.074033	9812	1820.5	2801.7	2.592	3632	6.224
28.0	2300	0001209	0071396	990.5	1811.5	2802.0	2.611	3600	6211
29.0	2320	0.001213	0068935	9995	1802.7	2802.2	2.628	3569	6197
30.0	2338	0001216	0.066632	1008.3	1794.0	2802.3	2646	3.538	6. 114
31.0	2357	000 1220	0.064473	10170	1785.4	2802.3	2662	3.509	6171
32.0	2374	0.001224	0062443	1025.4	1776.9	2802.3	2.679	3480	6119
33.0	2392	0.001227	0060533	10337	1768.6	2802.3	2694:	3452	6.146
34.0	2409	0001231	0058731	1041.8	1760.3	2802.1	2.710	3424	6.134
35.0	2425	0.001235	0.057028	1049.7	1752.2	2802.0	2.725	3398	6.123
36.0	2442	0001238	0055417	1057.5	17442	2801.7	2.740	3.371	6.111
37.0	2458	0001242	0.053889	1065.2	1736.2	2801.4	2.755	3346	6.100
38.0	247.3	0.001245	0.052439	1072.7	1728.4	2801.1	2.769	3.321	6 WO
39.0	2488	0001249	0.051061	1080.1	1720.6	2800.8	2783	3296	6.07*
40.0	2503	0001252	0049749	1087.4	17129	2800.3	2.797	3.272	6.06*
41.0	251 8	0.001256	0.048499	1094.6	1705.3	2799 9	2.810	3248	60*
42.0	2532	0.001259	0.047306	1101.6	1697.8	2799.4	2823	3.225	6.0*1
43.0	254.7	0.001262	0.046167	1108.5	1690.3	2798.9	2.836	3202	6,031
44.0	2561	0.001266	0.045078	1115.4	1682.9	2798.3	2849	3.180	6.021
45.0	2574	0.001269	0.044035	1122.1	16756	2797.7	2.861	3158	60»
46.0	258.8	0.001273	0.043036	1128.8	1668.3	27970	2.874	3.136	6.010
47.0	260.1	0.001276	0.042078	1135.3	1661.0	2796.4	2.886	3115	6.080
48.0	261.4	0.001279	0.04 II 58	1141.8	1653.9	2795.7	2897	3.094	5.9M
49.0	262.7	0.001283	0.040275	1148.2	16468	2794.9	2.909	3.073	5,9*1
50.0	263.9	0.001286	0.039425	1154.5	1639.7	2794.2	2.921	3.053	!.»»
52.0	2664	0001293	0.037820	1166.9	1625.7	2792.6	2.943	3.013	5.9M,
54.0	268.8	0.001299	0.036330	1179.0	1611.8	27908	2.965	2.974	!»»
56.0	271.1	0.001306	0.034942	1190.8	1598.2	2789.0	2.986	2.936	51M
58.0	273.4	0.001312	0.033646	1202.4	15846	2787.0	3.007	2.899	5.M
60.0	275.6	0.001319	0.032433	1213.7	1571.3	2785.0	3.027	2.863	5.1*1
62.0	277.7	0.001325	0.031295	1224.9	1558.0	2782.9	3.047	2.828	517!
64.0	279.8	0.001332	0.030225	1235.8	1544.8	2780.6	3.066	2.794	5.M
66.0	281.9	0.001338	0.029218	1246.5	1531.8	2778.3	3.085	2.760	514!
68.0	283.9	0.001345	0.028267	1257.1	1518.8	2775.9	3.104	2.727	5.1)1

Press. bar	Temp. °C	Specific Volume m^3/kg		Specific Enthalpy kJ/kg			Specific Entropy kJ/kg k		
		Sat. Liquid v_f	Sat. Vapour v_g	Sat. Liquid h_f	Evap. h_{fg}	Sat. Vapour h_g	Sat. Liquid s_f	Evap s_{fg}	Sat. Vapour s_g
70.0	2858	0001351	0027368	12675	15060	2773.4	3 122	2694	5X16
72.0	287.7	0001358	0026517	1277.7	14932	2770.9	3 140	2662	5802
74.0	2896	0001365	0025711	12878	14805	2768.2	3157	2631	5788
76.0	291 4	0001371	0024944	12977	14678	27655	3 174	2600	5.774
78.0	29V 2	0001378	0.024215	13075	14553	27627	3191	2569	5760
80.0	2950	0001384	0.023521	13172	1442.7	27599	3.208	2539	5747
82.0	2967	0001391	0.022860	13267	14303	2757.0	3224	2510	5734
84.0	298.4	0001398	0022228	13361	14178	2754.0	3.240	2481	5721
86.0	300 1	0001404	0021624	13454	1405.5	2750.9	3256	2452	5.708
88.0	101 7	0001.411	0021046	13547	1393.!	2747.8	3.271	2.423	5695
90	.101 .1	0001418	0.020493	1363.8	13808	2744.6	3287	2395	5682
92	104 •)	0001425	0.019962	13728	13b85	2741.3	3302	2.367	5669
94	1065	0001432	0.019453	1381.7	13563	27380	3317	2.340	5657
96	1080	0001439	0.018964	13906	1344.1	27347	3332	2313	5644
98	1095	0.001446	0.018493	13994	1331.9	2731 2	3.346	2286	5632
100	311.0	0001453	0018041	1408 1	13197	27277	1.361	.^39	5620
105	314.6	0001470	0.016981	1429.5	1289.2	2718.7	33%	2.194	5590
110	3180	0001489	0016007	1450.6	12588	2709.3	3430	2 129	5560
115	321.4	0001508	0.015114	1471.3	12282	2699.5	.V464	2066	5530
120	3246	OM01527	0014285	1491.7	11975	2698.2	3.497	2003	5500
125	3278	0001547	0013518	1511 9	11665	2678.4	3.530	1 941	5471
130	3308	0.001567	0.012800	1531.9	11351	26670	3561	1880	5441
135	3338	0001588	0.012130	1551.8	11033	2655.1	3593	1818	5411
140	3366	0001611	0011498.	1571.5	10709	2642.4	3624	1.756	5380
145	3394	0.001634	0010905	1591.3	10379	2629.2	3655	1694	5349
150	34? 1	0001658	0010343	16109	10042	2615.1	3686	1632	5318
I55	344.8	0001683	0009813	1630.7	9698	26004	3.716	570	5286
160	347.3	0.001710	0009310	16504	934.5	25849	3.747	506	5253
165	3497	0001739	0008833	1670.4	898.5	25689	3778	442	5220
170	3523	0.001770	0.008372	1691.6	860.0	25516	3811	375	5 186
175	354.6	0.001803	0.007927	1713.3	820.0	2533.3	4.844	.306	5150
180	3570	0.001840	0.007497	1734.8	779.0	2513.9	3.877	236	5113
185	3592	0001881	0.007082	1756.5	736.5	24930	3.910	164	5074
190	3614	0.001926	0.006676	1778.7	6918	2470.5	3.943	1.090	5.03J
195	363.6	0.001978	0.006276	1801.9	643.9	2445.8	3.978	1.011	4989
200	.1657	0002037	0005875	1826.6	591.6	2418.2	4.015	0.926	4941
205	3678	0002110	0005462	18542	532.0	2386.2	4056	0.830	4816
210	3698	0002202	0.005023	1886.3	4612	2347.5	4 105	0717	4822
215	3718	0.002342	0.004509	1929.4	365.2	2294.6	4 170	0.566	4736
220	3737	0002668	0.003735	2010.3	186.3	2196.6	4.293	0 288	4581
221.2	374.15	0003170	0.003170	2107.4	000.0	2107.4	4.443	0.000	4.443

Table 3. Specific Volume of Superheated Steam

Absolute Pressure in bar (p)	Saturation Temperature in °C (ts)	Specific Volume (v) in m/kg at Various Temperatures in °C										
		100	150	200	250	300	350	400	500	600	700	800
0.02	17.5	86.08	976.3	109.2	120.7	132.2	143.8	55.3	178.4	201.5	224.6	247.6
0.04	29.0	43.03	48.81	54.58	60.35	66.12	71.89	77.66	89.20	100.7	112.3	123.8
0.06	36.2	28.68	32.53	36.38	40.23	44.08	47.93	51.77	59.47	67.16	74.84	82.54
0.08	41.5	21.50	24.40	27.28	30.17	33.06	35.94	38.03	44.60	50.37	56.14	61.91
0.10	45.8	17.20	19.51	21.83	24.14	26.45	28.75	31.06	35.68	40.30	44.91	49.53
0.15	54.0	11.51	13.06	14.61	16.16	17.71	19.25	20.80	23.89	26.98	30.07	33.16
0.20	60.1	8.585	9.748	10.19	120.7	13.22	14.37	15.53	17.48	20.15	22.45	24.76
0.25	65.0	6.874	7.808	8:737	9.665	10.59	11.52	12.44	14.29	16.14	17.99	19.84
0.30	69.1	5.714	6.493	7.268	8.040	8.811	9.581	10.35	11.89	13.43	14.70	16.51
0.35	72.7	4.898	5.568	6.233	6.896	7.557	8.128	8.879	1020	11.52	12.84	14.16
0.40	75.9	4.279	4.866	5.448	6.028	6.607	7.185	7.763	8.918	10.07	11.23	12.38
0.45	78.7	3.803	4.325	4.844	5.360	5.875	6.389	6.903	7.930	8.957	9.984	10.99
0.50	81.3	3.418	3.889	4.356	4.821	5.284	5.747	6.209	7.134	8.057	8.891	9.904
0.60	86.0	2.844	3.238	3.628	4.016	4.402	4.788	5.174	5.944	6.714	7.484	8.254
0.70	90.0	2.434	2.773	3.108	3.441	3.772	4.103	4.434	5.095	5.755	6.415	7.074
0.80	93.5	2.126	2.425	2.718	3.010	3.300	3.590	3.879	4.457	5.035	5.613	6.190
0.90	96.7	1.887	2.153	2.415	2.674	2.933	3.190	3.448	3.962	4.475	4.989	5.502
1.00	99.6	1.696	1.936	2.172	2.406	2.639	2.871	3.103	3.565	4.028	4.490	4.952
1.50	111.4	-	1.285	1.444	1.601	1.757	1.912	2.067	2.376	2.685	2.993	3.301
2.00	120.2	-	0.9595	1.080	1.199	1.316	1.433	1.549	1.781	2.013	2.244	2.475
2.5	127.4	-	0.7641	0.8620	0.9574	1.052	1.145	1.239	1.424	1.610	1.795	1.980
3.0	133.5		0.6337	0.7164	0.7964	0.8753	0.9535	1.031	1.187	1.341	1.496	1.650
3.5	138.9	-	0.5406	0.6123	0.6814	0.7493	0.8166	0.8835	1.017	1.149	1.282	1.414
4.0	143.6	-	0.47.7	0.5343	0.5952	0.6549	0.7139	0.7725	0.8892	1.005	1.121	1.237
4.5	147.9	-	0.4165	0.4738	0.5284	0.5817	0.6343	0.6865	0.7905	0.8939	0.9971	1.100
5.0	151.8	-	-	0.4250	0.4744	0.5226	0.5701	0.6172	0.7108	0.8040	0.8969	0.9896
6.0	158.8	-	-	0.3520	0.3939	0.4344	0.4742	0.5136	0.5918	0.6696	0.7471	0.8245
7.0	165.0	-	-	0.2999	0.3364	0.3714	0.4057	0.4396	0.5069	0.5737	0.6402	0.7066
8.0	170.4	-	-	0.2608	0.2932	03241	0.3543	0.3842	0.4432	0.5017	0.5600	0.6181
9.0	175.4	-	-	0.2303	0.2596	0.2874	0.3144	0.3410	0.3936	0.4458	0.4976	0.5493
10.0	179.9	-	-	0.2059	0.2328	0.2580	0.2824	0.3065	0.3540	0.4010	0.4477	0.4943
11.0	184.1	-	-	0.1859	0.2108	0.2339	0.2563	0.2782	0.3215	0.3644	0.4069	0.4492
12.0	188.0	-	-	0.1692	0.1924	0.2139	0.2345	0.2547	0.2945	0.3338	0.3729	0.4112
13.0	191.6	-	-	0.1551	0.1769	0.1969	0.2161	0.2348	0.2716	0.3080	0.3441	0.3800
14.0	125.0	-	-	0.1429	0.1636	0.1823	0.2002	0.2177	0.2520	0.2859	0.3194	0.3528
15.0	198.3	-	-	0.1324	0.1520	0.1697	0.1865	0.2029	0.2350	0.2667	0.2980	0.3292
16.0	201.4	-	-	-	0.1419	0.1587	0.1745	0.1900	0.2202	0.2499	0.2793	0.3086
17.0	204.3	-	-	-	0.1329	0.1489	0.1640	0.1786	0.2070	0.2351	0.2628	0.2904
18.0	207.0	-	-	-	0.1250	0.1402	0.1546	0.1684	0.1954	0.2219	0.2481	0.2742
19.0	209.8	-	-	-	0.1179	0.1325	0.1461	0.1593	0.1849	0.2101	0.2350	0.2597
20.0	212.4	-	-	-	0.1115	0.1255	0.1386	0.1511	0.1756	0.1995	0.2232	0.2467
22.0	217.2	-	-	-	0.1004	0.1134	0.1255	0.1370	0.1593	0.1812	0.2028	0.2242
24.0	221.8	-	-	-	0.09108	0.1034	0.1146	0.1252	0.1458	0.1659	0.1858	0.2054
26.0	226.0	-	-	-	0.08321	0.09483	0.1053	0.1153	0.1344	0.1530	0.1714	0.1895
28.0	230.0	-	-	-	0.07644	0.08751	0.09740	0.1067	0.1246	0.1419	0.1590	0.1759

Absolute Pressure in bar (p)	*Saturation Temperature in °C (ts)*	*Specific Volume (v) in m/kg at Various Temperatures in °C*										
		100	*150*	*200*	*250*	*300*	*350*	*400*	*500*	*600*	*700*	*800*
30.0	233.8				0.07055	0.08116	0.09053	0.09931	0.1161	0.1323	0.1483	0.1641
32.0	237.4				0.06538	0.07559	0.08451	0.09283	0.1087	0.1239	0.1390	0.1538
34.0	240.9				0.06080	0.07068	0.07920	0.08711	0.1021	0.1165	0.1307	0.1447
36.0	244.2				0.05670	0.06630	0.07448	0 08202	0.09626	0.1100	0.1234	0.1366
38.0	247.3				0.05302	0.06237	0.07025	0.07747	0.09104	0.1041	0.1168	0.1294
40.0	250.3					0.05883	0.06645	0.07338	0.08634	0.09876	0.1109	0.1229
42.0	253.2					0.05563	0.06300	0.06967	0.08209	0.09397	0.1056	0.1170
44.0	256.0					0.05270	0.05986	0.06630	0.07823	0.08961	0.1007	0.1116
46.0	258.8					0.05003	0.05699	0.06322	0.07470	0.08562	0.09626	0.1967
48.0	261.4					0.04757	0.05436	0.06039	0.07147	0.08197	0.09219	0.1022
50.0	263.9					0.04530	0.05194	0.05779	0.06849	0.07862	0.08845	0.09809
55.0	269.9					0.04343	0.04666	0.05213	0.06202	0.07131	0.08031	0.08912
60.0	275.6					0.03615	0.04222	0.04738	0.05659	0.06518	0.07348	0.08159
65.0	280.8					0.03258	0.03848	0.04338	0.05203	0.06003	0.06774	0.07526
70.0	285.8					0.02946	0.03523	0.03992	0.04809	0.05559	0.06279	0.06980
75.0	290.5					0.02672	0.03243	0.03694	0.04469	0.05176	0.05852	0.06509
80.0	295.0					0.02426	0.02995	0.03431	0.04170	0.04839	0.05477	0.06096
85.0	299.2					0.02191	0.02776	0.03200	0.03908	0.04544	0.05148	0.05732
90.0	303.3						0.02579	0.02993	0.03674	0.04280	0.04853	0.05408
95.0	307.2						0.02403	0.02808	0.03465	0.04045	0.04591	0.05119
100.0	311.0						0.02242	0.02641	0.03276	0.03832	0.04355	0.04858
110.0	318.0	-	-	-	-	-	0.01961	0.02351	0.02950	0.03466	0.03947	0.04408
120.0	324.6	-	-	-	-	-	0.01721	0.02108	0.02679	0.03160	0.03607	0.04033
130.0	330.8	-	-	-	-	-	0.01510	0.01902	0.02449	0.02902	0.03319	0.03716
140.0	336.6	-	-	-	-	-	0.01321	0.01723	0.02251	0.02680	0.03072	0.03444
150.0	342.1	-	-	-	-	-	0.01146	0.01566	0.02080	0.02488	0.02859	0.03209
160.0	347.3	-	-	-	-	-	0.00976	0.01428	0.01929	0.02320	0.02672	0.03003
170.0	352.3	-	-	-	-	-	-	0.01303	0.01797	0.02172	0.02507	0.02821
180.0	357.0	-	-	-	-	-	-	0.01191	0.01679	0.02040	0.02360	0.0659
190.0	361.4	-	-	-	-	-	-	0.01089	0.01573	0.01922	0.02229	0.02515
200.0	365.7	-	-	-	-	-	-	0.00995	0.01477	0.01816	0.02111	0.02385
210.0	369.8	-	-	-	-	-	-	0.00907	0.01391	0.01720	0.02004	0.02267
220.0	373.7	-	-	-	-	-	-	0.00825	0.01312	0.01633	0.01907	0.02160
221.2	374.15	-	-	-	-	-	-	0.00816	0.01303	0.01622	0.01895	0.02135

Table 4. Enthalpy of Superhead Steam.

Absolute Pressure in bar (p)	Saturation Temperature in °C (ts)	Enthalpy (h) in kJ/kg at Various Temperatures in °C										
		100	150	200	250	300	350	400	500	600	700	800
0.02	17.5	2688.5	2783.7	2880.0	2977.7	3076.8	3177.5	3279.7	3489.2	3705.6	3928.8	4158.7
0.04	29.0	2688.3	2783.5	2879.9	2977.6	3076.8	3177.4	3279.7	3489.2	3705.6	3928.8	4158.7
0.06	36.2	2688.0	2783.4	2879.8	2977.6	3076.7	3177.4	3279.6	3489.2	3705.6	3928.8	4158.7
0.08	41.5	2688.8	2783.2	2879.7	2977.5	3076.7	3177.3	3279.6	3489.1	3705.5	3928.8	4158.7
0.10	45.8	2688.5	2783.1	2879.6	2977.4	3076.6	3177.3	3279.6	3489.1	3705.5	3928.8	4158.7
0.15	54.0	2686.9	2782.4	2879.5	2977.3	3076.5	3177.2	3279.5	3489.1	3705.5	3928.7	4158.7
0.20	60.1	2686.3	2782.3	2879.2	2977.1	3076.4	3177.1	3279.4	3489.0	3705.4	3928.7	4158.7
0.25	65.0	2685.7	2782.0	2879.0	2977.0	3076.3	3177.0	3279.3	3489.0	3705.4	3928.7	4158.6
0.30	69.1	2685.1	2781.6	2878.7	2976.8	3076.1	3176.9	3279.3	3488.9	3705.4	3928.7	4158.6
0.35	72.7	2684.5	2781.2	2878.5	2976.7	3076.0	3176.8	3279.2	3488.9	3705.3	3928.7	4158.6
0.40	75.9	2683.8	2780.9	2878.2	2976.5	3075.9	3176.8	3279.1	3488.8	3705.3	3928.6	4158.6
0.45	78.7	2683.2	2780.5	2878.0	2976.3	3075.8	3176.7	3279.1	3488.8	3705.2	3928.6	4158.5
0.50	81.3	2682.6	2780.1	2877.7	2976.1	3075.7	3176.6	3279.0	3488.7	3705.2	3928.6	4158.5
0.60	86.0	2681.3	2779.4	2877.3	2975.8	3075.4	3176.4	3278.8	3488.6	3705.1	3928.5	4158.5
0.70	90.0	2680.0	2778.6	2876.8	2975.5	3075.2	3176.2	3278.7	3488.5	3705.0	3928.4	4158.4
0.80	93.5	2678.8	2777.8	2876.3	2975.2	3075.0	3176.0	3278.5	3488.4	3705.0	3928.4	4158.4
0.90	96.7	2677.5	2777.1	2875.8	2974.8	3074.7	3175.8	3278.4	3488.3	3704.9	3928.3	4158.3
1.00	99.6	2676.2	2776.3	2875.4	2974.5	3074.5	3175.6	3278.2	3488.1	3704.8	3928.2	4158.3
1.50	111.4	-	2772.5	2872.9	2972.9	3073.3	3174.7	3277.5	3487.6	3704.4	3927.9	4158.0
2.00	120.2	-	2768.5	2870.5	2971.2	3072.1	3173.8	3276.7	3487.0	3704.0	3927.6	4157.8
2.5	127.4	-	2764.5	2868.0	2969.6	3070.9	3172.8	3275.9	3486.5	3703.6	3927.3	4157.6
3.0	133.5	-	2760.4	2865.5	2967.9	3069.7	3171.9	3275.2	3486.0	3703.2	3927.0	4157.3
3.5	138.9	-	2756.3	2863.0	2966.2	3068.4	3170.9	3274.4	3485.4	3702.7	3926.7	4157.1
4.0	143.6	-	2752.0	2860.4	2964.5	3067.2	3170.0	3273.6	3484.9	3702.3	3926.4	4156.9
4.5	147.9	-	2746.7	2857.8	2962.8	3066.0	3169.1	3272.9	3484.3	3701.9	3926.1	4156.7
5.0	151.8	-	-	2855.1	2961.1	3064.8	3168.1	3272.1	3483.8	3701.5	3925.8	4156.4
6.0	158.8	-	-	2849.7	2957.6	3062.3	3166.2	3270.6	3482.7	3700.7	3925.1	4155.5
7.0	165.0	-	-	2844.2	2954.0	3059.8	3164.3	3269.0	3481.6	3699.9	3924.5	4155.9
8.0	170.4	-	-	2838.6	2950.4	3057.3	3162.4	3267.5	3480.5	3699.1	3923.9	4155.0
9.0	175.4	-	-	2832.7	2946.8	3054.7	3160.5	3266.0	3479.4	3698.2	3923.3	4154.5
10.0	179.9	-	-	2826.8	2943.0	3052.1	3158.5	3264.4	3478.3	3697.4	3922.7	4154.1
11.0	184.1	-	-	2820.7	2939.3	3049.6	3156.6	3262.9	3477.2	3696.6	3922.0	4153.6
12.0	188.0	-	-	2814.4	2935.4	3046.6	3154.6	3261.3	3476.1	3695.8	3921.4	4153.1
13.0	191.6	-	-	2808.0	2931.5	3044.7	3152.7	3259.7	3475.0	3695.0	3920.8	4152.7
14.0	125.0	-	-	2801.4	2927.6	3041.7	3150.7	3258.2	3473.9	3694.1	3920.2	4152.2
15.0	198.3	-	-	2794.7	2923.5	3038.9	3148.7	3256.6	3472.8	3693.3	3919.6	4151.7
16.0	201.4	-	-	-	2919.4	3036.2	3146.7	3255.0	3471.7	3692.5	3918.9	4151.3
17.0	204.3	-	-	-	2915.3	3033.5	3144.7	3253.5	3470.6	3691.7	3918.3	4150.8
18.0	207.0	-	-	-	2911.0	3030.7	3142.7	3251.9	3469.5	3690.9	3917.7	4150.3
19.0	209.8	-	-	-	2906.7	3027.9	3140.7	3250.3	3468.4	3690.0	3917.1	4149.8
20.0	212.4	-	-	-	2902.4	3025.0	3138.6	3248.7	3467.3	3689.2	3916.5	4149.4
22.0	217.2	-	-	-	2893.4	3019.3	3134.5	3245.5	3465.1	3687.6	3915.2	4148.4
24.0	221.8	-	-	-	2884.2	3013.4	3130.4	3242.3	3662.9	3685.9	3914.0	4147.5
26.0	226.0	-	-	-	2874.7	3007.4	3126.0	3239.0	3460.6	3684.3	3912.7	4146.6
28.0	230.0	-	-	-	2864.9	3001.3	3121.9	3235.8	3458.4	3682.6	3911.5	4145.6

Absolute Pressure in bar (p)	*Saturation Temperature in °C (ts)*	*Enthalpy (h) in kJ/kg at Various Temperatures in °C*										
		100	*150*	*200*	*250*	*300*	*350*	*400*	*500*	*600*	*700*	*800*
30.0	233.8	-	-	-	2854.8	2995.1	3117.5	3232.5	3456.2	3681.0	3910.3	4149.4
32.0	237.4	-	-	-	2893.4	2988.7	3113.2	3229.2	3454.0	3679.3	3909.0	4143.8
34.0	240.9	-	-	-	2884.2	2982.2	3108.2	3225.9	3451.7	3677.7	3907.8	4142.8
36.0	244.2	-	-	-	2874.7	2975.6	3104.2	3222.5	3449.5	3676.1	3906.6	4141.9
38.0	247.3	-	-	-	2864.9	2968.9	3099.7	3219.1	3447.2	3674.4	3905.3	4141.0
40.0	250.3	-	-	-	-	2962.0	3095.1	3215.7	3445.0	3672.8	3904.1	4140.0
42.0	253.2	-	-	-	-	2955.0	3090.4	3212.3	3442.7	3671.1	3902.8	4139.1
44.0	256.0	-	-	-	-	2947.8	3085.7	3208.8	3440.5	3669.5	3901.6	4138.2
46.0	258.8	-	-	-	-	2940.5	3080.9	3205.3	3438.2	3667.8	3900.3	4137.2
48.0	261.4	-	-	-	-	2933.1	3076.1	3201.8	3435.9	3666.2	3899.1	4136.3
50.0	263.9	-	-	-	-	2925.5	3071.2	3198.3	3433.7	3664.5	3897.9	4135.3
55.0	269.9	-	-	-	-	2905.8	3058.7	3189.3	3427.9	3660.4	3894.8	4133.0
60.0	275.6	-	-	-	-	2885.0	3045.8	3180.1	3422.2	3656.2	3891.7	4130.7
65.0	280.8	-	-	-	-	2863.0	3032.4	3770.8	3416.4	3652.1	3888.3	4128.8
70.0	285.8	-	-	-	-	2839.0	3018.7	3161.2	3410.6	3647.9	3885.4	4126.0
75.0	290.5	-	-	-	-	2814.1	3004.5	3151.6	3401.7	3643.7	3882.4	4123.7
80.0	295.0	-	-	-	-	2786.6	2989.9	3141.6	3398.8	3639.5	3879.2	4121.3
85.0	299.2	-	-	-	-	2757.1	2974.7	3131.5	3392.8	3635.4	3876.1	4119.0
90.0	303.3	-	-	-	-	-	2959.0	3121.2	3386.8	3631.1	3873.0	4116.7
95.0	307.2	-	-	-	-	-	2942.7	3110.7	3380.7	3687.0	3869.9	4114.4
100.0	311.0	-	-	-	-	-	2925.8	3099.9	3374.6	3622.7	3866.8	4112.0
110.0	318.0	-	-	-	-	-	2889.6	3077.8	3362.2	3614.2	3860.5	4107.3
120.0	324.6	-	-	-	-	-	2849.7	3054.8	3349.6	3605.7	3854.3	4102.7
130.0	330.8	-	-	-	-	-	2805.0	3030.7	3336.8	3597.1	3848.0	4098.0
140.0	336.6	-	-	-	-	-	2754.2	3005.6	3323.8	3588.5	3841.7	4093.3
150.0	342.1	-	-	-	-	-	2694.8	2979.1	3310.6	3579.8	3835.4	4088.6
160.0	347.3	-	-	-	-	-	2620.8	2951.3	3297.1	3571.0	3829.1	4084.0
170.0	352.3	-	-	-	-	-	-	2921.7	3283.5	3563.2	3822.8	4079.3
180.0	357.0	-	-	-	-	-	-	2890.3	3269.6	3553.4	3816.5	4074.6
190.0	361.4	-	-	-	-	-	-	2856.7	3255.4	3544.5	3610.2	4070.0
200.0	365.7	-	-	-	-	-	-	2820.5	3241.1	3535.5	3803.8	4065.3
210.0	369.8	-	-	-	-	-	-	2781.3	3226.5	3526.5	3797.5	4060.5
220.0	373.7	-	-	-	-	-	-	2738.8	3211.7	3517.4	3791.1	4055.9
221.2	374.15	-	-	-	-	-	-	2734.5	3210.7	3516.4	3789.1	4054.7

Table 5. Entropy of Superhead Steam.

Absolute Pressure in bar (p)	*Saturation Temperature in °C (ts)*	*Enthalpy (h) in kJ/kg at Various Temperatures in °C*										
		100	*150*	*200*	*250*	*300*	*350*	*400*	*500*	*600*	*700*	*800*
0.02	17.5	9.193	9.433	9.648	9844	10.023	10.191	10.131	10.641	10.904	11.146	11.371
0.04	29.0	8.893	9.113	9.328	9.524	9.705	9.874	10.31	10.321	10.585	10.827	11.051
0.06	36.2	8.685	9.925	9.141	9.337	9.518	9.686	9.844	10.134	10.397	10.639	10.864
0.08	41.5	8.552	8.792	9.008	9.204	9.385	9.554	9.711	10.001	10.265	10.507	10.731
0.10	45.8	8.449	6.689	8.905	9.101	9.282	9.450	9.608	9.898	10.162	10.404	10.628
0.15	54.0	8.261	8.502	9.718	8.915	9.096	9.264	9.422	9.712	9.975	10.217	10.442
0.20	60.1	8.126	8.368	8.584	8.781	8.962	9.130	9.288	9.578	93842	10.084	10.309
0.25	65.0	8.022	8.264	8.481	8.678	8.859	9.028	9.186	9.476	9.739	9.981	10.206
0.30	69.1	7.936	8.179	8.396	8.593	8.774	8.943	9.101	9.391	9.654	9.897	10.121
0.35	72.7	7.864	8.107	8.323	8.522	8.703	8.892	9.030	9.320	9.583	9.826	10.050
0.40	75.9	7.801	8.045	8.263	8.460	8.641	9.810	8.968	9.258	9.522	9.765	9.989
0.45	78.7	7.745	7.990	8.208	8.405	8.587	8.755	8.914	9.204	9.467	9.709	9.936
0.50	81.3	7.695	7.941	8.159	8.356	8.538	8.707	8.865	9.155	9.419	9.661	9.886
0.60	86.0	7.609	7.855	8.074	8.272	8.454	8.622	8.781	9.071	9.334	9.576	9.801
0.70	90.0	7.535	7.783	8.002	8.200	8.382	8.551	8.709	9.000	9.363	9.505	9.730
0.80	93.5	7.470	7.720	7.940	8.138	8.320	8.489	8.648	8.938	9.214	9.444	9.669
0.90	96.7	7.413	7.664	7.884	8.083	8.266	8.435	8.593	8.884	9.147	9.389	9.614
1.00	99.6	7.362	7.614	7.836	8.034	8.217	8.386	8.544	8.836	9.098	9.341	9.565
1.50	111.4	-	7.419	7.644	7.845	8.028	8.198	8.356	8.647	8.911	9.153	9.378
2.00	120.2	-	7.279	7.507	7.710	7.894	8.64	8.223	8.514	8.778	9.020	9.245
2.5	127.4	-	7.169	7.400	7.604	7.789	7.960	8.119	8.410	8.674	8.917	9.142
3.0	133.5	-	7.077	7.312	7.518	7.703	7.874	8.034	8.326	8.590	8.833	9.058
3.5	138.9	-	6.998	7.237	7.444	7.631	7.802	7.962	8.254	8.518	8.761	8.986
4.0	143.6	-	6.929	7.171	7.380	7.567	7.740	7.899	8.192	8.456	8.699	8.925
4.5	147.9	-	6.866	7.112	7.323	7.512	7.684	7.844	8.137	8.402	8.645	8.870
5.0	151.8	-	-	7.059	7.272	7.461	7.634	7.795	8.088	8.353	8.596	8.821
6.0	158.8		-	6.638	6.876	8.089	7.257	7.421	7.718	7.985	8.229	8.455
7.0	165.0	-	-	6.587	6.831	7.034	7.214	7.379	7.677	7.944	8.188	8.415
8.0	170.4	-	-	6.539	5.788	6.994	7.175	7.340	7.639	7.906	8.151	8.378
9.0	175.4	-	-	6.494	6.748	9.956	7.139	7.305	7.603	7.871	8.116	8.343
10.0	179.9	-	-	6.692	6.926	7.125	7.303	7.467	7.763	8.029	8.273	8.500
11.0	184.1	-	-	6.638	6.876	7.078	7.257	7.421	7.718	7.985	8.229	8.455
12.0	188.0	-	-	6.587	6.831	7.034	7.214	7.379	7.677	7.944	8.188	8.415
13.0	191.6	-	-	6.539	6.788	6.994	7.175	7.340	7639	7.906	8.151	8.378
14.0	125.0	-	-	6.474	6.748	6.956	7.139	7.305	6.603	7.871	8.116	8.343
15.0	198.3	-	-	6.451	6.710	6.921	7.104	7.271	7.570	7.839	8.084	8.311
16.0	201.4	-	-	-	6.674	6.887	7.072	7.239	7.540	7.808	8.054	8.281
17.0	204.3	-	-	-	6.640	6.856	7.042	7.210	7.511	7.779	8.025	8.252
18.0	207.0	-	-	-	6.607	6.826	7.013	7.182	7.48.	7.552	7.997	8.226
19.0	209.8	-	-	-	6.576	6.797	7.986	7.155	7.457	7 727	7.973	8.200
20.0	212.4	-	-	-	6.545	6.770	6.960	7.130	7.432	7.702	7.949	8.176
22.0	217.2	-	-	-	6.488	6.718	6.911	7.082	7.386	7.657	7.904	8.132
24.0	221.8	-	-	-	6.434	6.670	6.866	7.308	7.344	7.615	7.862	8.091
26.0	226.0	-	-	-	6.382	6.625	6.824	6.998	7.305	7.577	7.825	8.053
28.0	230.0	-	-	-	6.333	6.582	6.784	6.960	7.269	7.541	7.789	8.018

Absolute Pressure in bar (p)	Saturation Temperature in °C (ts)	*Enthalpy (h) in kJ/kg at Various Temperatures in °C*										
		100	150	200	250	300	350	400	500	600	700	800
30.0	233.8	-	-	-	6.286	6.542	6.747	6.925	7.235	7.508	7.756	7.986
32.0	237.4	-	-	-	6.240	6.504	6.712	6.891	7.203	7.477	7.726	7.955
34.0	240.9	-	-	-	6.195	6.467	6.679	6.860	7.172	7.447	7.697	7.927
36.0	244.2	-	-	-	6.151	6.432	6.647	6.829	7.144	7.420	7.669	7.900
38.0	247.3	-	-	-	6.109	6.397	6.616	6.801	7.117	7.393	7.643	7.874
40.0	250.3	-	-	-	-	6.364	6.587	6.773	7.091	7.368	7.619	7.850
42.0	253.2	-	-	-	-	6.332	6.559	6.747	7.066	7.344	7.595	7.826
44.0	256.0	-	-	-	-	6.301	6.532	6.722	7.043	7.321	7.573	7.804
46.0	258.8	-	-	-	-	6.270	6.505	6.697	7.020	7.299	7.551	7.783
48.0	261.4	-	-	-	-	6.240	6.479	6.674	6.998	7.278	7.531	7.763
50.0	263.9	-	-	-	-	6.211	6.455	6.651	6.977	7.258	7.511	7.743
55.0	269.9	-	-	-	-	6.139	6.395	6.597	6.928	7.210	7.464	7.697
60.0	275.6	-	-	-	-	6.069	6.339	6.546	6.882	7.166	7.422	7.635
65.0	280.8	-	-	-	-	6.001	6.285	6.499	6.839	7.126	7.382	7.617
70.0	285.8	-	-	-	-	5.933	6.233	6.454	6.799	7.088	7.346	7.581
75.0	290.5	-	-	-	-	5.864	6.184	6.411	6.762	7.053	7.311	7.547
80.0	295.0	-	-	-	-	5.794	6.135	6.369	6.726	7.019	7.279	7.516
85.0	299.2	-	-	-	-	5.744	6.088	6.330	6.692	6.987	7.249	7.486
90.0	303.3	-	-	-	-	-	6.041	6.292	6.660	6.957	7.220	7458
95.0	307.2	-	-	-	-	-	5.995	6.254	6.629	6.929	7.129	7.430
100.0	311.0	-	-	-	-	-	5.949	6.218	6.599	6.901	7.166	7.406
110.0	318.0	-	-	-	-	-	5.857	6.148	6.543	6.850	7.117	7.358
120.0	324.6	-	-	-	-	-	5.764	6.081	6.491	6.802	7.072	7.315
130.0	330.8	-	-	-	-	-	5.666	6.016	6.441	6.758	7.030	7.274
140.0	336.6	-	-	-	-	-	5.562	5.951	6.394	6.716	6.991	7.237
150.0	342.1	-	-	-	-	-	5.487	5.888	6.349	6.676	6.954	7.201
160.0	347.3	-	-	-	-	-	5.311	5.824	6.305	6.639	6.919	7.168
170.0	352.3	-	-	-	-	-	-	5.760	6.264	6.603	6.886	7.137
180.0	357.0	-	-	-	-	-	-	5.695	6.223	6.569	6.854	7.107
190.0	361.4	-	-	-	-	-	-	5.628	6.184	6.536	6.824	7.078
200.0	365.7	-	-	-	-	-	-	5.559	6.146	6.504	6.795	7.051
210.0	369.8	-	-	-	-	-	-	5.486	6.108	6.474	6.768	7.025
220.0	373.7	-	-	-	-	-	-	5.410	6.072	6.444	6.741	7.000
221.2	374.15	-	-	-	-	-	-	5.399	6.068	6.441	6.738	6.994

Table 6. Specific Volume of Superheated Steam.

Absolute Pressure in bar	*Specific Volume (v) in m^3/kg at Various Temperatures in °C*										
	400	425	450	475	500	550	600	650	700	750	800
250	0.00601	0.00788	0.00917	0.01021	0.01113	0.01272	0.01413	0.01542	0.01663	0.01779	0.01891
260	0.00528	0.00730	0.00862	0.00966	0.01057	0.01213	0.01351	0.01476	0.01594	0.01707	0.01815
270	C 004 54	0.00675	0.00810	0,00914	0.01004	0.01159	0.01293	0.01416	0.01530	0.01640	0.01744
280	0.00382	0.00625	0.00761	0.00866	0.00956	0.01 108	0.01240	0.01359	0.01471	0.01577	0.01679
290	0.00324	0.00575	0.00716	0.00821	0.00910	0.01061	0.01 190	0.01307	0.01416	0.01520	0.01619
300	0.00283	0.00528	0.00674	O.f10779	0.00868	0.01017	0.01 144	0.01258	0.01365	0.01465	0.01562
320	000239	000442	0.005%	0.00703	0.00791	0.00937	0.01060	0.01170	0.01272	001368	0.01459
340	0.00218	0.00371	0.00527	0.00636	0.00724	0.00866	0.00986	0.01092	0.01190	0.01281	0.01369
360	0.00206	0.00318	0.00466	0.00576	0.00663	0.00804	0.00920	0.01023	0.01117	0.01205	0.01288
380	0.00197	0.00281	0.00412	0.00522	0.00610	0.00748	0.00861	0.00961	0301051	001136	0.01217
400	0.00191	0,00256	0.00367	0.00475	0.00562y	0.00698	0.00809	0.00905	0.00993	0.01075	0.01152
450	0.00(80	0.00220	0.00291	0.00380	0.00462	0.00593	0.00698	0.00789	0.00870	0.00945	0.01016
500	0.00173	0.00201	0.00249	0.00316	0.00388	0.00511	0.00611	0.006%	0.00722	0.00842	0.00908
550	0.00167	0.00190	0.00224	0.00275	0.00334	0.00447	0.00541	0.00621	0.00693	0.00758	0.00820
600	0.00163	0.00181	0.00208	0.00248	0.00295	0.00395	0.00484	0.00560	0.00627	0.00689	0.00746
650	0.00160	0.00175	0.00197	0.00228	0.00267	0.00354	0.00436	0.00509	0.00572	0.00630	0.00685
700	0.00157	0.00170	0.00189	0.00214	0.00247	0.00322	0.00397	0.00465	0.00526	0.00581	0.00632
800	0.00152	0.00163	0.00177	0.001%	0.00219	0.00276	0.00338	0.00397	0.00452	0.00502	0.00548
900	0.00148	0.00157	0.00169	0.00184	0.00201	0.00246	0.00297	0.00348	0.003%	0.00442	0.00484
1000	0.00145	0.00152	0.00163	0.00175	0.00189	0.00225	0.00267	0.00311	0.00354	0.00395	0.00434

Table 7. Enthalpy of Superheated Steam.

Absolute Pressure in bar	*Enthalpy (h) in KJ/kg at Various Temperatures in °C*										
	400	425	450	475	500	550	600	650	700	750	800
250	2582.0	2808.3	2954.3	3067.6	3165.9	3337.0	3489.9	3633.4	3771.9	3907.7	4041.9
260	2511.7	2773.8	2930.2	3048.7	3150.2	3325.3	3480.6	3625.7	3765.5	3902.2	4037.2
270	2427.5	2737.3	2905.3	3029.3	3134.2	3313.5	3471.3	3618.1	3759.0	3896.8	4032.5
280	2330.7	2698.8	2879.6	3009.5	3118.1	3301.5	3461.9	36104	3752.6	3891.3	4027J
290	2237.0	2658.0	2853.0	2989.3	3101.6	3289.5	3452.5	3602.7	3746.2	3885.8	4023.2
300	2161.8	2613.4	2825.6	29686	3085.0	3277.4	3443.0	3595.0	3739.7	3880.3	4018.5
320	2065.5	2517.1	2768.2	2926.1	305 /.O	3252:9	3423.9	3579.5	3726'.8	3869.4	4009.1
340	2011.8	2418.8	2707.7	2882.0	3016.0	3228.0	3404.7	3563.9	3713.8	3858.4	3999.8
360	1977.5	2332.6	2644.5	2836.4	2980.3	3202.8	3385.4	3548.3	3700.8	3847.4	3990.4
380	1953.0	2263.2	2579.1	2789.6	2943.8	3177.3	3365.9	3532.7	3687.8	3836.4	3981.1
400	1934.1	2209.3	2515.6	2742.1	2906.8	3151.6	3346.4	3517.0	3674.8	3825.5	3971.7
450	1900.6	2120.5	2384.2	2624.2	2813.5	3086.5	3297.4	3477.8	3642.4	3798.1	3948.4
500	1877.7	2068.4	2293.2	2521.7	2723.0	3021.1	3248.3	3438.9	3610.2	3770.9	3925.3
550	1860.7	2034.0	2231.4	2441.9	2639.9	2957.1	3199.6	3400.4	3578.3	3744.0	3902J
600	1847.3	2008.9	2187.1	2379.9	2570.6	2896.2	3151.6	3362.4	3547.0	3717.4	3879.6
650	1836.6	1989.9	2154.9	2332.5	2514.0	2840.2	3105.0	3325.3	3516.3	3691.4	3857J
700	1827.8	1974.7	2129.9	2295.4	2467.1	2790.7	3060.4	3289.0	3486.3	3665.8	3835J
800	1814.2	1952.1	2094.1	2242.5	2397.4	2708.0	2980.3	3220.3	3428.7	3616.7	3792.1
900	1804.6	1936.0	2069.3	2206.9	2349.9	2643.0	2913.5	3158.2	3374.6	3570.1	37524
1000	1797.6	1924.4	2051.2	2181.3	2316.1	2593.8	2857.5	3105.3	3324.4	3526.0	3714J

Table 8. Entropy of Superheated Steam.

Absolute Pressure in bar	*Entropyes in KJ/kg at Various Temperatures in °C*										
	400	*425*	*450*	*475*	*500*	*550*	*600*	*650*	*700*	*750*	*800*
250	5.146	5.476	5.682	5.836	5.966	6.180	6.360	6.520	6.666	6.803	6.931
260	5.033	5.416	5.636	5.797	5.931	6.151	6.334	6.496	6.643	6.780	6.909
270	4.900	5.353	5.590	5.759	5.897	6.122	6.308	6.472	6.620	6.759	.6.888
280	4.750	5.304	5.544	5.721	5.864	6.095	6.283	6.448	6.598	6.737	6.868
290	4.606	5.222	5.497	5.682	5.830	6.066	6.258	6.426	6.577	6.717	6848
300	4.490	5.150	5.450	5.644	5.797	6.039	6.234	6.403	6.556	6.697	6.829
320	4.339	4.998	5.353	5.567	5.732	5.985	6.187	6.360	6.516	6.659	6.792
340	4.253	4.846	5.245	5.490	5.667	5.933	6.142	6.319	6.477	6.622	6.757
360	4.195	4.712	5.152	5.413	5.603	5.882	6.098	6.279	6.440	6.587	6.723
380	4.153	4.604	5.050	5.316	5.539	5.832	6.055	6.241	6.404	6.553	6.691
400	4 119	4.515	4.951	5.259	5.476	5.784	6.014	6.204	6.370	6.521	6.661
450	4.055	4.376	4.747	5.073	5.323	5.665	5.914	6.115	6.289	6.445	6.589
500	4.008	4.286	4.603	4.913	5.178	5.553	5.821	6.033	6.214	6.375	6.522
550	3.970	4.223	4.501	4.787	5.047	5.446	5.732	5.956	6.144	6.310	6.461
600	3.938	4.174	4.425	4.686	4.937	5.346	5.648	5.883	6.078	6.248	6.403
650	3.910	4.134	4.366	4.607	4.846	5.256	5.568	5.814	6.015	6.191	6.349
700	3.886	4.100	4.318	4.543	4.769	5.175	5.493	5.748	5.956	6.136	6.298
800	3 843	4.043	4.243	4.443	4.649	5.038	5.360	5.627	5.587	6.035	6.203
900	3.843	3.998	4.185	4.372	4.560	4.928	5.247	5.520	5.748	5.944	6.118
1000	3.774	3.849	4.137	4.314	4.491	4.839	5.151	5.427	5658	5.860	6.040

Index

D

E

J

K

L

M

N

O

P

U

V

W

Z

❑❑❑

Win Prizes

ATTENTION READERS

Win Prizes

We request for your frank assessment regarding various aspects of the book as given below :

THERMODYNAMICS-I

By : B.S. UBHI

Please Fill up the Sheet in Neat and Clean Writing

(i) What topics of your syllabus, important for your examination point of view, are not covered in this book ?

....

....

....

....

(ii) What are the chapters wherein the treatment of the subject matter is not systematic or organised or updated ?

....

....

....

....

(iii) Have you come across misprints/mistakes/factual inaccuracies in the book ? Please specify the chapters and the page numbers.

....

....

....

....

(iv) What other books on the same subject have you found/heard better than the present book ? Please specify in terms of price and quality.

....

....

....

....

(v) Further suggestions and comments for the improvement of this book.

....

....

....

....

(PS : If need be please attach a separate sheet to write your views)

OTHER DETAILS

(i) Who recommended you this book ? (Please tick (ü) in the box near the option relevant to you.

☐ Teacher ☐ Friends ☐ Bookseller

....

....

....

....

(ii) Name of the recommending teacher, his designation and address:

....

....

....

....

(iii) Name and address of the bookseller you purchased the book from :

....

....

....

....

(iv) Name and address of your institution (Please mention the University or Board as the case may be)

....

....

....

....

(v) Your name and complete address :

....

....

....

....

(vi) Course you are enrolled in :

....

....

....

....

The best assessment will be awarded each month : The award will be in the form of our publications as decided by the Editorial Board.

Please mail the filled up coupon at your earliest to :

S.K. KATARIA & SONS®
4885/109, Prakash Mahal, Dr. Subhash Bhargav Lane,
Opposite Delhi Medical Association, Daryaganj, New Delhi-110002 (India)
+91-11-43551243 Telefax: 011-23243489
Email: katariabooks@yahoo.com; katariabook@gmail.com
Website : www.skkatariaandsons.com,